Springer Complexity

Springer Complexity is an interdisciplinary program publishing the best research and academic-level teaching on both fundamental and applied aspects of complex systems—cutting across all traditional disciplines of the natural and life sciences, engineering, economics, medicine, neuroscience, social and computer science.

Complex Systems are systems that comprise many interacting parts with the ability to generate a new quality of macroscopic collective behavior the manifestations of which are the spontaneous formation of distinctive temporal, spatial or functional structures. Models of such systems can be successfully mapped onto quite diverse "real-life" situations like the climate, the coherent emission of light from lasers, chemical reaction-diffusion systems, biological cellular networks, the dynamics of stock markets and of the internet, earthquake statistics and prediction, freeway traffic, the human brain, or the formation of opinions in social systems, to name just some of the popular applications.

Although their scope and methodologies overlap somewhat, one can distinguish the following main concepts and tools: self-organization, nonlinear dynamics, synergetics, turbulence, dynamical systems, catastrophes, instabilities, stochastic processes, chaos, graphs and networks, cellular automata, adaptive systems, genetic algorithms and computational intelligence.

The two major book publication platforms of the Springer Complexity program are the monograph series "Understanding Complex Systems" focusing on the various applications of complexity, and the "Springer Series in Synergetics", which is devoted to the quantitative theoretical and methodological foundations. In addition to the books in these two core series, the program also incorporates individual titles ranging from textbooks to major reference works.

Editorial and Programme Advisory Board

Péter Érdi
Center for Complex Systems Studies, Kalamazoo College, USA,
and Hungarian Academy of Sciences, Budapest, Hungary

Karl J. Friston
Institute of Cognitive Neuroscience, University College London, London, UK

Hermann Haken
Center of Synergetics, University of Stuttgart, Stuttgart, Germany

Janusz Kacprzyk
System Research, Polish Academy of Sciences, Warsaw, Poland

Scott Kelso
Center for Complex Systems and Brain Sciences,
Florida Atlantic University, Boca Raton, USA

Jürgen Kurths
Nonlinear Dynamics Group, University of Potsdam,
Potsdam, Germany

Linda E. Reichl
Center for Complex Quantum Systems, University of Texas, Austin, USA

Peter Schuster
Theoretical Chemistry and Structural Biology, University of Vienna,
Vienna, Austria

Frank Schweitzer
Systems Design, ETH Zurich, Zurich, Switzerland

Didier Sornette
Entrepreneurial Risk, ETH Zurich, Zurich, Switzerland

Springer Series in Synergetics

Founding Editor: H. Haken

The Springer Series in Synergetics was founded by Herman Haken in 1977. Since then, the series has evolved into a substantial reference library for the quantitative, theoretical and methodological foundations of the science of complex systems.

Through many enduring classic texts, such as Haken's Synergetics and Information and Self-Organization, Gardiner's Handbook of Stochastic Methods, Risken's The Fokker Planck-Equation or Haake's Quantum Signatures of Chaos, the series has made, and continues to make, important contributions to shaping the foundations of the field.

The series publishes monographs and graduate-level textbooks of broad and general interest, with a pronounced emphasis on the physico-mathematical approach.

Alexander Balanov, Natalia Janson,
Dmitry Postnov, Olga Sosnovtseva

Synchronization

From Simple to Complex

With 237 Figures

Dr. Alexander Balanov
Loughborough University
Department of Physics
Loughborough
Leicestershire LE11 3TU
United Kindom
a.balanov@lboro.ac.uk

Dr. Natalia Janson
Loughborough University
School of Mathematics
Ashby Road
Loughborough
Leicestershire LE11 3TU
United Kingdom
n.b.janson@lboro.ac.uk

Prof. Dmitry Postnov
Saratov State University
Department of Physics
Astrakhanskaya 83
Saratov, Russia 410026
postnov@chaos.ssu.runnet.ru

Dr. Olga Sosnovtseva
Technical University of Denmark
Department of Physics
Building 309
2800 Lyngby, Denmark
olga@fysik.dtu.dk

ISBN 978-3-540-72127-7   e-ISBN 978-3-540-72128-4

DOI 10.1007/978-3-540-72128-4

Library of Congress Control Number: 2008929606

© 2009 Springer-Verlag Berlin Heidelberg

This work is subject to copyright. All rights are reserved, whether the whole or part of the material is concerned, specifically the rights of translation, reprinting, reuse of illustrations, recitation, broadcasting, reproduction on microfilm or in any other way, and storage in data banks. Duplication of this publication or parts thereof is permitted only under the provisions of the German Copyright Law of September 9, 1965, in its current version, and permission for use must always be obtained from Springer. Violations are liable to prosecution under the German Copyright Law.

The use of general descriptive names, registered names, trademarks, etc. in this publication does not imply, even in the absence of a specific statement, that such names are exempt from the relevant protective laws and regulations and therefore free for general use.

Typesetting and Production: VTEX, Vilnius
Cover design: WMX Design GmbH, Heidelberg, Germany

Printed on acid-free paper

9 8 7 6 5 4 3 2 1

springer.com

To our parents

Preface

This book is written by scientists who live in different countries (United Kingdom, Denmark, Russia), but who have graduated from, and were established as researchers at the same place: The Laboratory of Nonlinear Dynamics, Department of Physics, Saratov State University, Russia. Being apart for many years, we have united in one team again to write this book. Why?

We aim to summarize both classical results that are crucial for the understanding of the concept of synchronization, and an up-to-date account of the accompanying fascinating phenomena. The main theme that runs throughout the book is that interaction between complex systems is governed by the same universal principles. We strive to explain the material in a way that the newcomers to the field would hopefully appreciate, namely,

- From simple calculations to advanced theoretical approaches
- From simple dynamics to complex behavior
- From mathematical and physical to general perspectives

Assuming only the basic knowledge of mathematics, our book takes the reader to the frontiers of what is currently known about this research area.

The classical approach to synchronization we have learned by heart during our regular and inevitably hot discussions, and most of the results on the new synchronization phenomena we obtained together. It is therefore difficult to separate scientific contribution and to compare the efforts made by each co-author, so we decided to arrange the list of authors in alphabetic order to emphasize an equal investment of their time, ideas and enthusiasm.

This book would not have been possible without the help of many people. First of all, we are deeply indebted to our teacher Prof. V.S. Anishchenko who has introduced us to Nonlinear World and who patiently taught us to properly speak the language of science. We are grateful to our teachers and colleagues Prof. V.V. Astakhov and T.E. Vadivasova for their active support and many invaluable discussions during the years. We extend our thanks to Prof. E. Mosekilde, Prof. P. McClintock, Prof. S.K. Han, who are our closest collaborators in the field of synchronization, and to Prof. N.-H. Holstein-Rathlou, Prof. D. Marsh, and Prof. H. Braun, with whom we have been enjoying collaborations in the field of modeling of biological systems. Our special thanks are due to Prof. E. Schöll who has encouraged us to write this book. We acknowledge fruitful discussions with our colleagues A. Pikovsky, M. Rosenblum, M. Zaks, J. Kurths, L. Schimansky-Geier, A. Neiman, A. Nikitin, and A. Silchenko on various aspects of synchronization. We gratefully acknowledge the help of S. Malova with references and of P. Sherbakov with experiments. We would also like to warmly thank Victoria Sosnovtseva for making funny illustrations especially for this book. Finally, we would like to express our sincere gratitude to our families for their constant support and inspiration.

Over the years our studies were supported by the Russian Foundation for Basic Research (Russia), the U.S. Civilian Research and Development Foundation (USA), Engineering and Physical Sciences Research Council (UK), Medical Research Council (UK), The Leverhulme Trust (UK), Forskningsrådet for Natur og Univers (Denmark), and the European Union through the Network of Excellence BioSim.

Loughborough, *Alexander Balanov*
Saratov, *Natalia Janson*
Lyngby, *Dmitry Postnov*
May 2008 *Olga Sosnovtseva*

Contents

Preface vii

1   **Introduction** 1

Part I General Mechanisms of Synchronization

---

2   **General Remarks** 9
    2.1   What Are We Going to Talk About? 9
    2.2   Topics to Consider 10
    2.3   Self-Sustained Oscillations: A Key Concept in Synchronization
          Theory 12
          2.3.1   Features of Self-Oscillations 12
          2.3.2   Features of Self-Oscillating Systems 13
          2.3.3   Modern Revisions of the Definition of a Self-Sustained
                 System 16
          2.3.4   Self-Sustained Oscillations and Attractors 17
          2.3.5   Synchronization as a Control Tool 17
    2.4   Duality of the Description of Synchronization 17
    2.5   Oscillations Helping Each Other Out 18
    2.6   Terms of Bifurcations Theory 19

3   **1 : 1 Forced Synchronization of Periodic Oscillations** 21
    3.1   Phase of Quasiharmonic Oscillations 24
    3.2   Derivation of Truncated Equations for Phase Difference and
          Amplitude 26
    3.3   Amplitude of Unperturbed Oscillations at Small Non-linearity 31
    3.4   Analysis of Truncated Equations for Weak Forcing 32
    3.5   Derivation of Truncated Equations in Descartes Coordinates 34
    3.6   Analysis of Truncated Equations in Descartes Coordinates 37
    3.7   Synchronization Region from the Truncated Equations:
          Non-bifurcational Approach 45
    3.8   Fourier Power Spectra at Strong Forcing 50
    3.9   Phase Locking and Suppression: Numerical Simulation 56

x Contents

	3.9.1	Phase Locking	56
	3.9.2	Suppression of Natural Dynamics	60
3.10	Phase Locking and Suppression: Experiment		62
	3.10.1	Amplitudes from Oscilloscopes	64
3.11	Beat Frequency: Theory, Simulations and Experiment		67
	3.11.1	Theory	67
	3.11.2	Numerical Simulation	71
	3.11.3	Experiment	72

4   1 : 1 Mutual Synchronization of Periodic Oscillations 75

4.1   Truncated Equations for Weakly Non-linear Oscillators 77
4.2   Periodic Oscillators with Dissipative Coupling 80
   4.2.1   Symmetric Solutions 81
   4.2.2   Asymmetric Solutions 83
   4.2.3   Oscillation Death 84
4.3   Dissipative Coupling: Numerical Simulation 84
   4.3.1   Locking 85
   4.3.2   Bifurcations 86
   4.3.3   Suppression 87
4.4   Reactive Coupling 89
   4.4.1   Locking 90
   4.4.2   Suppression 92
   4.4.3   Bifurcations 92
   4.4.4   Phase Multistability 94
4.5   Reactive Coupling and the Saddle Torus 95
   4.5.1   Hypothesized Structure of the Phase Space 96
4.6   Generality of Bifurcational Transitions at Reactive Coupling 97
4.7   Experiment 99
   4.7.1   Phase Locking 100
   4.7.2   Suppression 101
4.8   Comparison of Synchronization Transitions in Forced and in
      Mutually Coupled Oscillators 103

5   Homoclinic Mechanism of Synchronization of Periodic Oscillations .. 105

5.1   Global Bifurcation 108
   5.1.1   Features of a Homoclinic Bifurcation of a Cycle 110
5.2   Homoclinics Inside Synchronization Tongue? 111
5.3   How Homoclinics Leads to Synchronization 114
5.4   Synchronization in a Bacteria–Viruses Model 117
5.5   Summary 120

6 $n : m$ Synchronization of Periodic Oscillations 121

6.1   Important Definitions Relevant to $n : m$ Synchronization 121
   6.1.1   Poincaré Return Time 121
   6.1.2   Phase of Oscillations 122

Contents xi

6.1.3 Phase of Oscillations via Poincaré Section 122
6.1.4 Poincaré Winding (Rotation) Number 123
6.1.5 Synchronization Order $n:m$ 123
6.2 1:1 Forced Synchronization in Weakly Non-linear Oscillators 123
6.2.1 3:1 Phase (Frequency) Locking 128
6.2.2 3:1 Suppression of Natural Dynamics 131
6.3 $n:m$ Synchronization in Strongly Non-linear Oscillators with
Spiky Forcing... 133
6.3.1 2:3 Phase (Frequency) Locking 136
6.3.2 The Route to 2:3 Suppression 138
6.4 Circle Map: Derivation 138
6.4.1 Amplitude and Phase of Oscillations 139
6.4.2 From Differential to Discrete Equation for Phase 141
6.5 Circle Map: Properties 142
6.6 Arnold Tongues 144
6.7 $n:m$ Synchronization: Experiment 144
6.8 Summary 147

**7 1:1 Forced Synchronization of Periodic Oscillations in the Presence
of Noise** 149
7.1 Introductory Comments on Random Processes 150
7.1.1 One-Dimensional Probability Density, Mean and Variance .. 150
7.1.2 Two-Dimensional Probability Density, Correlation and
Covariance 152
7.1.3 Stationary Process 154
7.1.4 Correlation Time 154
7.1.5 Correlation Between Two Different Processes 155
7.1.6 Spectrum of a Wide-Sense Stationary Process 156
7.2 Truncated Equations 158
7.3 Simplification of the Fluctuational Terms in Truncated Equations .. 158
7.4 Probability Density Distribution of the Phase Difference 165
7.4.1 Case of $Q > 0$ 169
7.5 Bessel Functions 170
7.6 Probability Density Distribution of the Phase Difference, Continued 172
7.7 Mean Frequency of Forced Oscillations with Noise 174
7.8 Interpretation of Phase Dynamics 177
7.9 Phase Diffusion 180
7.10 Full-Scale Biological Experiment 183
7.11 Effects of Noise on the Spectrum of a Synchronized System 185
7.11.1 Effect of Noise on the Spectrum of Oscillations
Synchronized by Suppression 189

xii   Contents

8   Chaos Synchronization 191
  8.1   What Is Chaos? 192
    8.1.1   Exponential Divergence of Phase Trajectories 192
    8.1.2   Chaos Properties in Terms of Phase Space 193
    8.1.3   Chaos Properties in Terms of Spectra 197
  8.2   What Does Synchronization of Chaos Encompass? 197
    8.2.1   Chaos Synchronization: Different Manifestations 197
    8.2.2   Chaos Synchronization in a Classical Sense 198
  8.3   Phase and Basic Frequency of Chaotic Oscillations 199
  8.4   Forcing Chaos Periodically: What to Expect? 201
    8.4.1   Phase Locking of Chaos 203
    8.4.2   Suppression of Chaos................................ 204
    8.4.3   Any Other Options? 204
    8.4.4   Interacting Chaotic Systems 205
  8.5   Synchronization of Chaos by Periodic Forcing 205
    8.5.1   Experiment 205
    8.5.2   Numerical Analysis 211
  8.6   Synchronization of Periodic Oscillations by Chaos............... 212
    8.6.1   Spectra... 213
    8.6.2   Poincaré Sections 215
    8.6.3   Phase Difference................................... 216
    8.6.4   Lyapunov Exponents 220
  8.7   Mutual Synchronization of Chaos 222
    8.7.1   Phase/Frequency Locking 222
    8.7.2   Suppression....................................... 223
    8.7.3   Phase Behavior 225
  8.8   Homoclinic Synchronization of Chaos 227
  8.9   Effects of Noise on a Synchronized Chaos.................... 232
    8.9.1   Chaotic System Frequency-Locked by a Harmonic Signal . . 233
    8.9.2   Periodic System Suppressed by Chaotic Forcing 237
  8.10 Summary 237

9   Synchronization of Noise-Induced Oscillations 239
          Stochastic Limit Cycle............................... 241
  9.1   Noise-Induced Oscillations 242
  9.2   Models 243
    9.2.1   Morris–Lecar Model 243
    9.2.2   Monovibrator Circuit 244
  9.3   Coherence Resonance Oscillator 244
  9.4   Frequency and Phase Locking 248
    9.4.1   Frequency Locking: Electronic Experiment 249
    9.4.2   Phase Locking: Coupled Morris–Lecar Models 251
    9.4.3   Phase Dynamics Inside the Synchronization Region:
            Electronic Experiment 253
  9.5   Synchronization via Suppression............................ 255

Contents xiii

10 Conclusions to Part I 259

Part II Case Studies in Synchronization

---

11 Synchronization of Anisochronous Oscillators 265
   11.1 Phase Velocity Field and Coupling Vector 266
   11.2 Effective Coupling Function 268
      11.2.1 Asymptotic Phase 268
      11.2.2 Effective Coupling Function 269
   11.3 Dephasing 270
   11.4 Examples of 2D Anisochronous Oscillators 273
   11.5 Synchronization near the Homoclinic Bifurcation 279
      11.5.1 Weak Coupling Limit 282
      11.5.2 Finite Coupling Strength 285
      11.5.3 Strong Coupling with Moderate μ 288
      11.5.4 Summary on Synchronization near Homoclinic Bifurcation . 289
   11.6 Phase Locking Patterns of Coupled Fast-and-Slow Oscillators 290
      11.6.1 Antiphase Locking in Coupled FitzHugh–Nagumo Models . 290
      11.6.2 Out-of-phase Synchronization via Slow Channels 293
   11.7 Synchronous Patterns in Coupled Morris–Lecar Models 296
      11.7.1 Model 296
      11.7.2 Overview of the Dynamics 298
      11.7.3 Structure of Arnold Tongue for Antiphase Solution 300
          Chaotic Bursting and Torus Breakdown 306
      11.7.4 Crises at the Boundary of Quasiperiodic Regions 308
      11.7.5 Transition to In-phase Synchronization 311
      11.7.6 Mechanism of Torus Folding in the Vicinity of Unstable
          Orbit 312
      11.7.7 Remarks on Synchronization in Morris–Lecar Systems 314
   11.8 Summary 314

12 Phase Multistability 317
   12.1 Period-Doubling Oscillations 318
      12.1.1 Dynamics of Coupled Rössler Systems 320
      12.1.2 Mapping Approach to Multistability 330
   12.2 Self-Modulated Oscillations 335
      12.2.1 Methods of Analysis 335
      12.2.2 Phase Dynamics of Coupled Oscillators 337
   12.3 Bursting Dynamics 339
      12.3.1 Simple Qualitative Approach to Phase Multistability 342
      12.3.2 Dynamics of Coupled Bursters 344
      12.3.3 Multistability Induced by Dephasing 349
   12.4 Summary 352

xiv   Contents

13   Synchronization in Systems with Complex Multimode Dynamics 353
   13.1 Synchronization of Chaotic Systems with Fast and Slow
      Time Scales 355
      13.1.1 Single System with Two Time Scales 355
      13.1.2 Coupled Systems with Two Mode Dynamics 360
      13.1.3 Conclusions... 363
   13.2 Generation and Synchronization of Oscillations with Several
      Noise-Induced Modes 363
      13.2.1 Description of Experiment 364
      13.2.2 Characterizing Collective Response by Spectra 364
      13.2.3 Mutually Coupled Excitable Units 365
      13.2.4 Three Coupled Excitable Units 369
      13.2.5 Two Mutually Coupled Excitable Units with Inhibitory
          Coupling 369
   13.3 Synchronization of Chaotic Systems with Denumerable Set of
      Equilibrium States 371
   13.4 Summary 376

14   Synchronization of Systems with Resource Mediated Coupling 377
   14.1 Neural Synchronization via Potassium Signaling 379
      14.1.1 Model.. 380
      14.1.2 Identical Cells: Competing In-phase and Antiphase
          Synchronization 383
      14.1.3 Heterogeneous Cells: Dynamical Patterns 386
   14.2 Multimode Dynamics in Linear Array of Electronic Oscillators 388
      14.2.1 Model.. 388
      14.2.2 Clustering 390
      14.2.3 Intracluster Synchronization 393
   14.3 Cascaded Microbiological Oscillators......................... 395
      14.3.1 Model.. 396
      14.3.2 Spatial Dynamics 397
   14.4 Synchronization Patterns in Kidney Autoregulation 401
      14.4.1 Vascular-Nephron Model.............................. 402
      14.4.2 Coupling-Induced Inhomogeneity 405
   14.5 Summary 408

15   Conclusions to Part II 411
      And finally... 412

References... 413

Index 423

1 Introduction

It would not be too much of an exaggeration to say that oscillations are one of the main forms of motion. They range from the periodic motion of planets to random openings of ion channels in cell membranes. They are observed at various levels of organization, have various origins and various properties. Since Newton's crack at the three-body problem and until just a few decades ago, the range of phenomena regarded as oscillations were limited to damped, periodic and quasiperiodic oscillations at best. A significant achievement of the second half of the 20th century is the admission of deterministic chaos and noise-induced rhythms as equals into the oscillation family.

Nature is not based on isolated individual systems. It is rich in connections, interactions and communications of different kinds that are complex beyond belief. With this, synchronization is the most fundamental phenomenon associated with oscillations. It is a direct and widely spread consequence of the interaction of different systems with each other. In most general terms, synchronization means that different systems adjust the time scales of their oscillations due to interaction, but there is a large variety of its manifestations and of the accompanying fascinating phenomena.

Anyone writing a book on synchronization is faced with two problems: on one hand, one has to deal with a huge amount of material on the particular aspects and effects; and on the other hand, there is a need to formulate a universal approach that would embrace all the particular cases. Fortunately, an essential contribution to the second problem has been made by Pikovsky, Rosenblum, and Kurths in their recent book [214], that has provided a contemporary view on synchronization as a universal phenomenon that manifests itself in the entrainment of rhythms of interacting self-sustained systems. This viewpoint is in agreement with the approach developed since the time of Huygens, and is completely shared by ourselves. In writing the present book we were motivated by the following considerations:

- Recently, a large variety of new synchronization phenomena were discovered that are inherent in complex (chaotic) systems, but do not occur in simple periodic oscillators. With the modern fascination for the beauty and the complexity of the new effects, there is a tendency to forget about the basic phenomena and theoretical results associated with "simply" periodic oscillations. This is largely due to the fact that not all involved in the studies of these phenomena, and especially younger researchers and students, have the respective education. It turns

2   1 Introduction

out to be difficult to recommend a book, which would consistently present, equation after equation, the most fundamental theoretical results on synchronization. Without such background, it is problematic to analyze the synchronization of irregular oscillations from the general viewpoint, and to avoid discovering "new" effects that often appear to be merely manifestations of the general principles in a particular situation.

- There is a number of fascinating aspects of synchronization (phase multistability, dephasing, self-modulation, etc.), that are observed in a variety of systems and with various types of interaction, that have not been discussed yet in the framework of the general concept of synchronization.

In order to cover the above problems, our book contains two parts. The first part is a consistent and detailed description of the classical approach to forced and mutual synchronization that is based on frequency/phase locking and suppression of natural dynamics. It is oriented to the people not familiar with the fundamental results of synchronization theory obtained by a number of physicists and mathematicians, such as B. van der Pol, A.A. Andronov, A.A. Vitt, M.L. Cartwright, A.W. Gillies, P.J. Holmes, D.A. Rand, R.L. Stratonovich, V.I. Tikhonov, P.S. Landa, D.G. Aronson and co-authors, and published in their original works. It was our aim:

- To reproduce in every detail the derivations of the most fundamental results, which until now were given only schematically and presented a significant challenge for beginners because of the traditional brevity typical of the scientific works of the beginning and middle of the 20th century. We have made every effort to make the reading easy for non-experts, to reduce to the minimum the need to refer to other literature when following the calculations or the description of geometrical effects, and to exclude expressions like "It is easy to show." As a result, the lengths of the respective sections have increased substantially as compared to those in the original books and papers, but we believe it was worth doing this and hope that the readers will find this material helpful.
- To describe the same phenomena using different languages: the ones of physics and of mathematics. In the early experiments on synchronization, the latter was detected by means of listening to the volume of sound (organ pipes), visually observing the positions of pendulums (clocks), and later Lissajous figures and Fourier power spectra on the oscilloscopes (electric circuits). Thus, synchronization can be naturally understood in physical terms like power, frequency or phase. On the other hand, the systems that synchronize can be described by non-linear mathematical equations. Transitions that occur in coupled systems when their parameters change, can be described in mathematical terms of bifurcation and stability theory. In this book we will analyze the phenomena of synchronization and the associated effects using both languages and making a clear connection between these different means of description.
- To generalize theoretical results to complex oscillations. An important achievement of modern oscillations theory is the recognition of the role of irregular oscillations that can be either deterministic or stochastic. We start by considering synchronization in simple periodic oscillators. Then we move to chaotic

and stochastic oscillations and show that in spite of their complexity, they can synchronize according to the same mechanisms as periodic ones.

We will deem to have achieved our goal, if after reading this part the reader will be convinced that very different types of oscillations obey the same mechanisms of synchronization, although the particular manifestations can be different.

The second part is devoted to the general mechanisms and principles of synchronization, describing them with regard to the non-linear properties of the particular classes of systems and couplings. We discuss synchronization of anisochronous oscillations, when fast and slow motions along the trajectory give arise to additional phase-shifted coexisting regimes and thus change the bifurcational structure of the synchronization region. A separate chapter is devoted to the concept of phase multistability and its development in the systems that oscillate with complex waveform (essential for period-doubling and self-modulated oscillations) and have a particular structure of their phase space. The latter might include regions of fast and slow motion, closeness of the trajectories to some singular points, etc. (essential for bursting behavior). The concept of synchronization is extended to the systems with several time scales of either deterministic, or stochastic origin. Finally, we consider cooperative behavior of systems with a particular type of coupling through the primary resource supply and discuss their applications.

Part I

General Mechanisms of Synchronization

"Begin at the beginning," the King said, very gravely, "and go on till you come to the end: then stop."
Lewis Carroll, "Alice in Wonderland"

You have to learn the rules of the game. And then you have to play better than anyone else.
Albert Einstein

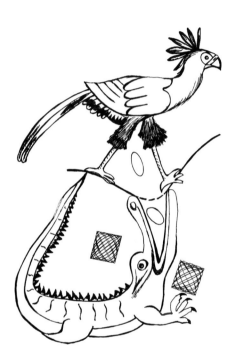

This part offers a tutorial description of the mechanisms of synchronization. We start from the beginning: periodic oscillations and analytical approaches. Then we proceed with irregular oscillations, either chaotic or stochastic, and generalize the classical results. Then we stop.

2 General Remarks

2.1 What Are We Going to Talk About?

"Synchronization, of course, but what is it and why should I bother?" you might ask.

Look, everything around us is moving. As René Descartes used to say [71], "Give me the matter and motion and I will construct the universe." Others say "motion is the mode of existence of matter" [274]. How exactly is the matter moving? One very popular possibility is the motion that demonstrates a certain degree of repetition, this would be an *oscillation*. Your heart is an oscillator, can you hear how it beats?

Now consider several oscillators and let them feel each other's motion, no matter how exactly—the scientists would say "couple them." Most likely, the coupling will not go unnoticed by any of these systems: all of them will change their behavior to this or that extent. In fact, this is going on in your body right now: you inhale and exhale repetitively, and thus influence the way your heart beats without knowing it perhaps. The basic features of oscillations are their amplitude and shape, but when we talk about repetition of anything, a natural question is "how often?" With this question arises the concept of a characteristic time scale of oscillations.

What does coupling have to do with all this? Well, because of coupling all aspects of the system's behavior would generally change, sometimes most drastically. So, before you couple anything that oscillates, it would be good to know the possible consequences in advance, wouldn't it? We can tell you right now that a lot of things can happen. For example, oscillations can stop altogether, which might be good sometimes, but occasionally disastrous. Or they could become totally unpredictable—but you might like it nevertheless because it looks beautiful. But the phenomenon which is most often associated with synchronization is the change of the *time scales* of interacting systems: if you couple the systems cleverly, they can start to oscillate "syn-chronously," which means "sharing the same time" [214]. For your heart and breathing this can mean that, say, while you are breathing once, your heart makes three beats exactly.

One can say that synchronization is the most fundamental phenomenon that occurs in oscillating processes. In most general terms, synchronization can be defined

10   2 General Remarks

as follows:

> *Synchronization is an adjustment of the time scales of oscillations due to interaction between the oscillating processes.*

2.2 Topics to Consider

In Part I we consider the general types of non-damped oscillations that can occur in real-life systems, and introduce the three mechanisms by which all of them can be synchronized. More precisely:

- In Chap. 3 we describe the phenomenon of 1 : 1 forced synchronization which can occur in self-sustained periodic oscillators, i.e., in systems that, without being influenced externally, demonstrate purely periodic oscillations—a definition of these systems is given in Sect. 2.3. Real-life examples of such systems are clocks, either mechanical or electronic, generators of electromagnetic waves, drills, metronomes, a string of a violin while being bowed, etc. If periodic forcing is applied to such systems, and if the frequency of forcing is close to, but slightly different from, the frequency of self-oscillations, the forcing can entrain both their frequency and phase. Two classical mechanisms of forced synchronization are introduced: phase (frequency) locking and suppression of natural dynamics.
- In Chap. 4 the interaction between two periodic oscillators is described, which are coupled to each other bidirectionally or mutually. If the frequencies of uncoupled oscillators are sufficiently close, then depending on the kind of coupling between them, a number of phenomena can occur. One possibility is 1 : 1 mutual synchronization, when both subsystems start to oscillate periodically with the same frequency which is not equal to either of their natural frequencies. Another kind of response induced by coupling is the simultaneous death of oscillations in both subsystems. Also, one can expect the more complicated phenomenon of phase multistability: one out of two (or even out of a larger number) oscillating patterns can be realized at exactly the same set of control parameters, depending on the choice of the initial conditions. However, the mechanisms of synchronization in mutually coupled weakly non-linear oscillators of general type are the same as in forced systems, namely, phase (frequency) locking and suppression.
- Chapter 5 considers the third mechanism of synchronization of periodic oscillations via homoclinic bifurcation, which is different from locking or suppression, and which involves global restructuring of the phase space of interacting systems. This mechanism is less general, but nevertheless can be expected in quite a large class of self-oscillators whose autonomous oscillations are highly inhomogeneous in time. The examples of such systems are populations of microorganisms, neuron systems, lasers, etc. We give mathematical definitions that are essential for the description of this synchronization scenario; discuss in detail

the changes in the phase space that accompany the onset of synchronization via homoclinic bifurcation, and reveal the phenomena associated with this mechanism.

- Chapter 6 is devoted to a more generic case of $n:m$ synchronization, when, as a result of interaction, the ratio of the time scales of the coupled systems becomes equal to $n:m$, where n and m are arbitrary integers. Such a situation typically occurs when the natural time scales of the interacting systems are not close to each other. We describe how the main synchronization mechanisms for this particular case are realized, and derive a simple discrete map called the circle map to analyze this type of synchronization. We illustrate $n:m$ synchronization on an example of cardiorespiratory interaction in humans.
- In Chap. 7 we discuss the general effects of noise on synchronization of periodic oscillations. We demonstrate that noise, which is inevitably present in all real systems, can evoke very non-trivial phenomena in the dynamics of synchronized self-oscillators. Different theoretical approaches for the description of noise-induced phenomena are discussed. For an example of forced 1 : 1 synchronization, we analytically study phase and frequency properties of synchronous oscillations in the presence of noise. The theoretical results are illustrated by experiments with electronic self-oscillators and with the cardiovascular system of humans.
- Chapter 8 describes the mechanisms of synchronization of chaotic oscillations. The latter was found to be typical dynamical regimes in many real systems. Examples of systems with chaotic dynamics are fluid and gas flows, electrical circuits, semiconductors devices, populations of animals, biological objects, and many others. The chapter starts with an explanation of the origin of the dynamical chaos. We discuss different manifestations of synchronization of irregular chaotic oscillations. The concept of phase for a non-periodic process is introduced. We describe the synchronization of chaos in terms of phases and frequencies of chaotic oscillations, and also in terms of saddle periodic orbits embedded into chaotic attractors. Forced and mutual synchronization of chaos is discussed. The main mechanisms of chaos synchronization are revealed, and the effects of noise on them are considered. Some results are illustrated by experiments with an electronic circuit.
- In Chap. 9 synchronization is considered in systems where oscillations are induced merely by external random fluctuations. We discuss different classes of dynamical systems where noise alone is able to induce highly regular oscillations with the properties similar to the properties of deterministic self-oscillations. We show that the mechanisms of synchronization characteristic of purely deterministic systems are also valid for noise-induced oscillations. We discuss the peculiarities of synchronization in stochastic systems and illustrate these results on electronic circuits and on the models of neurons.

2.3 Self-Sustained Oscillations: A Key Concept in Synchronization Theory

Before we start talking about any synchronization at all, we need to outline more precisely the class of systems and processes in which we can expect it to occur. Systems that oscillate in principle are usually called oscillators. But the systems we are interested in should be capable of demonstrating oscillations that are self-sustained, or self-oscillations. The concept of self-oscillations was first proposed by Andronov, Khaikin and Vitt[1] in 1937 [12] (for the English version see [14][2]). Self-oscillations form a special, but rather broad class of all oscillating processes and are characterized by the following features.

2.3.1 Features of Self-Oscillations

Below we list the features of self-oscillations.

- First and foremost, they do not damp, i.e., the repetitive motion of the system does not stop with the course of time, and does not show the tendency to stop.[3]
- Second and equally important, they oscillate "by themselves," i.e., not because they are repetitively kicked from outside.
- The third feature is perhaps the most intriguing and fascinating: the shape, amplitude and time scale of these oscillations are chosen by the oscillating system alone.[4] An outsider cannot easily change them, e.g., by setting different initial conditions.[5]

Examples of self-oscillators are a grandfather pendulum clock, a whistle, your throat when you sing a musical note, as well as many musical instruments, your heart and many other biological systems, a bottle of water with a narrow neck that is put vertically with its neck down (water will come out in pulses). In order to prevent possible confusion, we would like to give just one example of an oscillator which is not a self-sustained one.

Counterexample. Consider a famous bob pendulum consisting of a load on a rope, whose other end is fixed. If we give the load an initial kick, it will start to oscillate,

---

[1] Another popular spelling is "Witt" which is widely used in literature.

[2] As explained in "Preface to the second Russian edition" of [14], the name of Vitt was "by an unfortunate mistake not included on the title page as one of the authors" of [12], but he has contributed equally with the two other authors.

[3] To be more precise, until the power source lasts, as will be explained below; so they are not perpetuum mobile.

[4] In p. 162 of [14] it is said: "The amplitude of these oscillations is determined by the properties of the system and not by the initial conditions. ... Whatever the initial conditions, undamped oscillations are established and (they are) stable."

[5] Except in the case of multistability which will be discussed in Chap. 12. But even then the number of options is usually quite limited and is anyway offered by the same self-oscillating system.

2.3 Self-Sustained Oscillations: A Key Concept in Synchronization Theory

but if we leave it alone, the oscillations will decay and eventually stop due to friction of the whole construction with air, and also at the point of the rope attachment. Of course, a repetitive kicking will resume the oscillations of the pendulum, but these will not be *self*-sustained because they would damp without the kicks. What if there were no friction in the system? Then the oscillations would not damp, but would that make them self-sustained? No, because the properties of these oscillations would be completely defined by the direction and strength of the initial kick made by an outsider who would wish to launch them: the harder one kicks, the larger the swing will be. This would contradict the third feature of self-oscillations.

2.3.2 Features of Self-Oscillating Systems

For self-oscillations to occur, the oscillating system must be designed in a special way—which is quite a popular design, we haste to say. The following three features of the self-oscillating *systems* are most essential: they must be non-linear systems, there must be dissipation in them, and there must be a source of power.

Dissipation

Dissipation is a mechanism due to which energy is being lost by the system while it changes its state, i.e., performs a motion. It has to be said that most macroscopic systems are dissipative anyway, since there is always some sort of friction in it. For example, mechanical systems lose energy because their details experience friction with other details or surrounding air. In electronic systems elementary particles bump into other particles, the elements of the circuits heat and thus lose energy. This list can be continued, but the main idea is clear: dissipation is everywhere.

It would be pertinent to emphasize again that the systems without (or almost without) dissipation are not self-oscillators. The oscillations in such systems are usually associated with the motion of either very small (microscopic) particles like electrons in an atom, or of very large (megascopic) objects like stars and planets. They do oscillate (rotate around their centers) eternally, but just because the energy of their oscillations is not wasted on friction.

Power Source

Having established that dissipation is ubiquitous, a natural line of thought occurs:

- Oscillations have amplitude A. Which, roughly speaking, is half the difference between the maximal and the minimal values of an oscillating quantity. Note that if oscillations are decaying, their amplitude decreases. If the oscillations are on the contrary expanding, A grows with the time course. For self-oscillations the amplitude A should not change in time.[6]

---

[6] At least if the oscillations are periodic. If self-oscillations are not periodic, their amplitude will itself oscillate around some average value, neither growing unboundedly, nor tending to zero, like, e.g., in chaotic oscillations described in Chap. 8.

14   2 General Remarks

- Any oscillations have power O. Which is the energy per time unit, and monotonously depends on the amplitude A.
- Therefore, in order to maintain non-damped oscillations with a constant amplitude, the system performing them must keep its power at a certain sufficient level all the time.
- But how can the system do that, if dissipation persistently pumps the power out of it?

The answer is obvious: The system should simply find the way to feed on some source of power in order to compensate for its losses.[7] Thus we have deduced the need for the source of power in self-oscillating systems.

Non-linearity

First of all, *what* is non-linearity? Suppose we have a system about which we would like to find out whether it is linear or not. Apply some perturbation x_1 to it and record its response y_1. Then apply another perturbation x_2 and record the response y_2 to that. Then apply perturbation equal to $(x_1 + x_2)$ and calculate the response y_3. Then calculate the sum $(y_1 + y_2)$ and compare the two quantities

$$(y_1 + y_2) \quad \text{and} \quad y_3. \tag{2.1}$$

Are they equal for any chosen x_1 and x_2? If yes, then the system is linear. If they are not equal, the system is non-linear. Graphically, linearity can be illustrated as a straight line on the graph of response y as a function of input x. Anything different from a straight line would represent a non-linear system.

Let us come back to the system which wants to self-oscillate, i.e., to decide for itself how to behave and to hold its ground by being resistive to at least minor influences from the outside world. In order to do this, the system must take power from the available source in a proper way.

Suppose the system does not oscillate at all, i.e., its amplitude A is zero. At this state it does not spend power on oscillations, and does not need to compensate for it. Therefore, the amount of power S taken from the source per time unit should be zero. They say that the system is in equilibrium. Now let the initial conditions be such that the amplitude of oscillations is finite $A > 0$. Then the system is not in equilibrium and should take power. How?

It is convenient to express powers O and S as functions of A^2 rather than A. The reason is that quite often the power O spent on oscillations is proportional to A^2, i.e., $O = kA^2$ with k being some proportionality constant. Consider this case for the start.

The amount of power S that enters the system from the source is a function of A^2, and this function can be either linear or non-linear. A few possibilities are illustrated in Fig. 2.1. In order to maintain oscillations with a certain amplitude A_0,

---

[7] In [14] it is said: "A self-oscillating system is an apparatus which produces a periodic process at the expense of a non-periodic source of energy."

2.3 Self-Sustained Oscillations: A Key Concept in Synchronization Theory

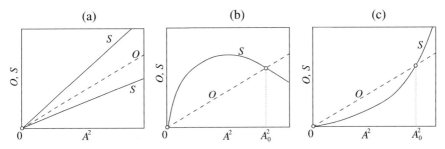

Fig. 2.1. Powers in the system as functions of the square of oscillations amplitude A^2. *Dashed line*: power O spent on oscillations; *solid line*: power S supplied into the system. The approximation $O = kA^2$ is used. **a** S is a linear function of A^2, no non-damped oscillations can occur; **b** S is a non-linear function of A^2, oscillations with A_0 are stable; **c** S is a non-linear function of A^2, oscillations with A_0 are unstable

the supplied power S must compensate the dissipated power O. If S is a linear function of A^2 as shown in (a), then S and O can intersect only at one zero point, which is equivalent to the absence of oscillations. If S is non-linear as illustrated in (b), then two intersections are possible: at zero and at a certain A_0^2. This means that if the amplitude A of oscillations reaches the value of A_0, the lost power is being compensated. With this, if $A > A_0$, the supplied power S is not enough to compensate for the power loss O, and the amplitude of oscillations will decay automatically until it reaches A_0. Similar considerations show that from A_0 the system tends to establish the amplitude A_0 as well. The system that demonstrates the given character of a non-linearity is capable of self-oscillations. In (c) an example of a non-linear function S is given at which oscillations with $A = A_0$ can occur, but they will be unstable. Indeed, setting $A > A_0$ leads to more power entering the system than being lost, and A is pushed to grow further. Setting $A < A_0$ leads to the power spent is not compensated, and consequently to the decrease of A towards zero amplitude, i.e., towards no oscillations. Although this last system can oscillate with a non-zero amplitude A_0, it is not a self-sustained system, because such oscillatory regime is not stable: a tiny perturbation will ruin it.

If the power of oscillations O is not a linear function of A^2, the picture would qualitatively look as in Fig. 2.2. It is qualitatively similar to the one in Fig. 2.1(b), so the same principles apply. Based on these simple considerations, it can be concluded that it is an interplay between the non-linear power supply and dissipation that makes self-oscillations possible. Thus, a self-sustained system must be non-linear.

Note, that the figures above schematically illustrate the requirements for the simplest periodic self-oscillations to arise, but would not be sufficient to explain the origin of more complex self-oscillations whose amplitude is not constant. However, the fundamental physical principles explained here remain valid for all self-sustained systems, provided the modern developments are taken into account that are discussed in the next paragraph.

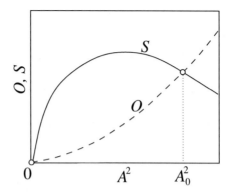

Fig. 2.2. Powers in the system as functions of the square of oscillations amplitude A^2. *Dashed line*: power O spent on oscillations; *solid line*: power S supplied into the system. Self-oscillations can occur

2.3.3 Modern Revisions of the Definition of a Self-Sustained System

Andronov et al. [12, 14] have defined a self-sustained system as periodic, but nowadays a family of self-oscillations has expanded considerably to include quasiperiodic and irregular oscillations, so this requirement is obviously out of date.

Also, the original definition required the self-sustained system to be autonomous, which in fact means that the power available from the source should be constant and not depend on time explicitly.[8] This definition was revised by Landa [161, 162] in view of modern developments of oscillations theory. She excluded the word "autonomous" and has thus allowed the source of energy to change in time. This addition made alone would immediately include oscillations that exist only because of rhythmic external forcing, i.e., forced oscillations which are not self-sustained. However, forced oscillations would have the same or similar time scales as the forcing itself. So in order to exclude forced oscillations, Landa adds a requirement that reads "The complete or partial independence of the frequency spectrum of oscillations from the spectrum of the energy (power) source" [161]. This means that at least a part of the spectrum of oscillations does not come as a result of the transformation of the spectrum of the source of power, i.e., the frequency components are not harmonics or subharmonics of those of the spectrum of the source. At least a part of the spectrum of oscillations must be defined by the intrinsic properties of the system itself.

The relaxation of the condition on the constancy of the power source has an important consequence: it allows one to include into the family of self-sustained oscillations the ones that are induced merely by random perturbations and would not occur without them. In Chap. 9 it will be demonstrated that this classification of noise-induced oscillations is justified, and that they do behave like self-sustained systems in many respects, and in particular can be synchronized.

---

[8] Although the amount of power actually taken from the source at the given time instant does depend on the stage of oscillations, and thus depends on time *implicitly*.

2.3.4 Self-Sustained Oscillations and Attractors

A distinctive feature of self-oscillating systems is their ability to self-organize. When we launch a process in such a system by, e.g., switching it on, the initial conditions can be chosen at random in a wide range. In general, the time course of a process thus launched can depend on the initial settings quite substantially. However, a self-oscillator is very confident about what it is ought to do, and after some transient (relaxation) time passes by, it arrives at the same regime of oscillations from a large range of initial conditions. In mathematical terms, such regimes are characterized by the attractors in the phase space. Sometimes, certain systems can have a choice of the possible attracting regimes to which they can go, depending on the initial conditions provided, and this is called multistability. Nevertheless, self-sustained systems are generally quite firm in their decisions on how to behave, and are resistant to weak attempts to distract them from their course. A mathematical term for this property is robustness.

2.3.5 Synchronization as a Control Tool

In various applications it might become necessary to amend the conduct of a self-sustained system either slightly or substantially. One might even want to stop all oscillations in it. However, this might not be a straightforward and easy task, given the above-mentioned stubbornness of self-oscillators. In this respect, our book shows you the possible ways to control the behavior of self-oscillating systems by means of clever and inexpensive perturbations. But before one is able to choose the best way to tame the particular system, it is necessary to classify it, to learn about the temper and habits of the systems from the given family, and to arm oneself with the full range of the available taming tools. We wish our reader good luck in this exciting journey.

2.4 Duality of the Description of Synchronization

Synchronization of oscillations is a phenomenon that was originally discovered by Christian Huygens in 1665 in a mechanical system: two pendulum "grandfather" clocks hanging on the same beam [125]. The interaction between the organ pipes was studied by Rayleigh [243]. The first observations of synchronization in a electronic tube generators were done by Eccles [58, 75] in relation to the problem of creating a precision clock and the transmission of naval signals. Almost at the same time experiments with electric circuits were performed by Appleton [28] and by van der Pol [292, 293] while they were studying the reception of radio signals with electric circuits with triodes. The same authors developed the first theoretical approaches that were able to explain their results to some extent. However, the first *non-linear* mathematical theory of synchronization which was able to capture the phenomena observed much more accurately, was created in the Soviet Union by Andronov and Witt, also with regard to a very practical problem: stabilization of

18   2 General Remarks

the frequency of a powerful generator of electromagnetic waves by energy-efficient weak external forcing [13, 299].

In the experiments synchronization was detected by observing Lissajous figures on the screens of oscilloscopes that provided one with information on the phase shifts, amplitudes and frequencies. Thus, synchronization can be naturally described in physical terms like power, frequency or phase. On the other hand, the systems which can demonstrate synchronization, can be described by non-linear mathematical equations. Transitions that occur in coupled systems when their parameters change, can be described in terms of dynamical systems theory including bifurcation theory. We emphasize that the same phenomena can be described using different languages, the language of physics or the language of mathematics. But whatever approach we choose, the underlying phenomena remain the same. In this book we will analyze the phenomenon of synchronization and the associated effects using both languages and making a clear connection between these different levels of description.

2.5 Oscillations Helping Each Other Out

A reader who has reached this point in the book might be already thinking: "First, they were talking about my heartbeats, whistles, clocks and bottles, then about some electronic experiments and organ pipes. In between they promised me something exciting to arise out of the coupling of various devices, and also gave a definition of some imaginary self-sustained system. These look like all different things to me, having nothing to do with each other. Even if they are saying that two clocks can be synchronized, so what? How does it help me to understand what happens to organ pipes? And above all, what does it have to do with my heart?"

This is a fundamental question which we would be delighted to receive and to answer.

We need to make a short excursion into the past. Before the beginning of the 20th century, non-linearity was perceived as an annoying misfortune that could be encountered in this or that physical phenomenon. Every physical problem seemed to contain some non-linearity, but it would be perceived as *its own* non-linearity specific to the given problem [256], just like you might have suggested above. In the early 1930s Soviet physicist Leonid Mandelstam was the first to recognize the burning need to develop a unique approach to non-linearity and proposed the ideas of *non-linear thinking*. In addition to that, in 1944 in one of his lectures he made an observation that starting from Kepler laws, most fundamental discoveries made in physics were in fact *oscillatory* in this or that way. He also observed that oscillations were a key element that was common in all traditional subdivisions of physics: optics, electricity, acoustics, etc. Now we know that oscillations are common in biology, chemistry, geology, finances and social sciences as well, and this list can be continued. His ideas of *commonness of oscillations* and *oscillations' mutual aid* consisted in that there are the same fundamental laws of nature that lie behind oscillations of all kinds. An understanding of the principles behind oscillations in one system would help one to understand oscillations in the other systems.

The statement above might not sound immediately obvious, so we continue. Already at the end of the 19th century it was clear that if one considers *small* oscillations in acoustics and in electricity, and consistently, from the first principles, derives mathematical equations describing them, the resulting equations will be *the same* [243]! Moreover, it was shown that the same equations are valid for small oscillations in mechanical systems. Is it a coincidence?

Let us go further. Later on, when deriving from the first principles the differential equations underlying the *non-small* oscillations in the systems of all kinds (chemical, biological, physical) it was noticed that quite often these equations appear *equivalent* in the sense of topology. The latter means that a change of variables would reduce one set of equations to another, i.e., that there is no real difference between them from the viewpoint of mathematics!

At present these ideas are quite well established and might even be occasionally regarded as trivial. But, as Mandelstam once said "It is the triviality of this, which is non-trivial" [256]. Thus, the theory of oscillations serves as a common language that can be spoken by different disciplines. The laws of the theory of oscillations are common between oscillations of the same class, regardless of the nature of the particular system demonstrating them.

Coming back to the question in the beginning of this section, we are safe in saying that all seemingly different systems and phenomena that were mentioned here, and a huge lot of those not mentioned, just because there are too many of them, obey the same fundamental principles. If we state that self-oscillations can be synchronized, this means that *all* self-oscillations can do that, no matter where they are found. In the remainder of Part I of this book the *simplest* paradigmatic models will be discussed, that describe periodic, chaotic, noisy and noise-induced oscillations. The mathematical results will predict that certain interesting things can happen to them. But then qualitatively the same phenomena can occur even in much more complex systems, provided that their oscillatory properties are equivalent to those described by simple equations.

Therefore, when learning about, say, phase locking in van der Pol oscillator, one learns about phase locking in a general periodically self-oscillating system.

2.6 Terms of Bifurcations Theory

In the next chapters we use a number of terms that belong to the theory of differential equations, including bifurcation theory. It is not possible to give a detailed introduction into differential equations here, and anyway this is very well done by other authors before us. Just a few useful sources are [65, 101, 156], and we would like to mention separately [2] for those who feel that they need to start from a very basic level. For an excellent historical introduction to dynamical chaos we would mention [3].

3 1 : 1 Forced Synchronization of Periodic Oscillations

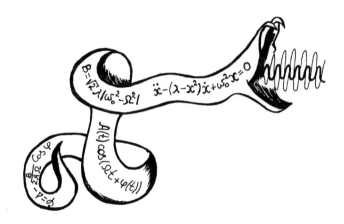

In this chapter we study the simplest case of synchronization: synchronization of unidirectionally coupled periodic oscillators. Another name for this phenomenon is forced synchronization, which reflects the fact that one system influences the other, but does not experience any influence from the other system in return. Another simplifying assumption employed here is that the frequency of external stimulus is sufficiently close to the frequency of natural, i.e., unforced, oscillations. Provided that the strength and frequency of forcing satisfy certain conditions, a remarkable effect can take place: the system that experiences only weak external perturbation can start to oscillate with the frequency equal to the one of this perturbation. They say that the phenomenon of 1 : 1 phase (frequency) locking, or entrainment, occurs. This is a special case of a more general phenomenon of $n:m$ synchronization that can be observed when the forcing frequency f_f is not close to the natural frequency f_0 of oscillations in the forced system, but instead is close to a value $\frac{n}{m}f_0$. $n:m$ synchronization will be considered in Chap. 6.

We will start with considering a periodic weakly non-linear oscillator that is forced harmonically. As a particular example we use a famous paradigm for periodic self-sustained oscillations, the van der Pol equation, which has been used to describe a variety of oscillatory phenomena including oscillations of current in electric circuit [292], signal of electrocardiogram [294], dynamics of semiconductor lasers [49],

22   3 1:1 Forced Synchronization of Periodic Oscillations

generation of relativistic magnetrons [168], and the activity of a single neuron [197]. The equation reads

$$\ddot{x} - \left(\lambda - x^2\right)\dot{x} + \omega_0^2 x = 0. \tag{3.1}$$

Here, dots over the variables denote derivatives over time t, λ is the non-linearity parameter and also the bifurcation parameter: at $\lambda < 0$ there are no self-oscillations, and the only stable solution of the system is a stable fixed point at the origin. At $\lambda = 0$, Andronov–Hopf bifurcation occurs, as a result of which the fixed point becomes unstable, and a stable limit cycle is born. At the moment of birth, oscillations on the limit cycle are harmonic, and their frequency is exactly equal to $\omega_0 > 0$, which is also called *eigenfrequency*. If λ is positive and small, i.e., $0 < \lambda \ll 1$, the periodic self-sustained oscillations remain almost harmonic, and their frequency remains approximately equal to the value of ω_0. The solution to (3.1) for large t, i.e., after the system has relaxed to the limit cycle from the arbitrarily chosen initial conditions, can be approximately described by

$$x(t) = A\cos(\omega_0 t + \varphi_0), \quad A = \text{const}, \ \omega_0 = \text{const}, \ \varphi_0 = \text{const}, \tag{3.2}$$

where A is the amplitude, ω_0 is the frequency, and φ_0 is the initial phase of oscillations. The respective phase portrait on the plane $(\dot{x}, x)$ and the realization of $x(t)$ are shown in Fig. 3.1 by a black line. This is the case of the so-called *weakly non-linear oscillator*,[1] which can be analyzed analytically by means of the approximate methods of the theory of oscillations.

Generally, by a weakly non-linear oscillator we understand a system with a limit cycle, whose control parameters are just above the values corresponding to a supercritical Andronov–Hopf bifurcation.[2] Note that when the non-linearity λ in (3.1) is no longer small, the oscillations, although remaining periodic, are no longer close to harmonic, their amplitude grows and the frequency is less than ω_0 (Fig. 3.1, grey line). The larger the λ, the slower the oscillations, and the bigger their amplitude is.

Now, let us introduce external periodic forcing into the system in its simplest harmonic form as follows:

$$\ddot{x} - \left(\lambda - x^2\right)\dot{x} + \omega_0^2 x = B\cos(\Omega t). \tag{3.3}$$

Here, B and Ω are the strength (amplitude) and frequency of the external forcing, respectively. The solutions of (3.3) at $\omega_0 = 1$, fixed small value of $B = 0.01$ and four different values of Ω close to 1 are illustrated in Fig. 3.2. The external forcing

$$F(t) = B\cos(\Omega t) \tag{3.4}$$

is shown by black in the fourth column together with the solution $x(t)$. Note that the amplitude B of forcing here is much smaller than the amplitude of x. That is why, in order to allow the reader to compare the details of the behavior of both x and F, in the last column of Fig. 3.2 we show not F, but $10F$.

---

[1] Or nearly sinusoidal, as they are called in [12, 14].

[2] There are two forms of Andronov–Hopf bifurcation, a supercritical and a subcritical one. The former is encountered more often, therefore in what follows we will call it simply "Andronov–Hopf bifurcation" for brevity.

3 1:1 Forced Synchronization of Periodic Oscillations

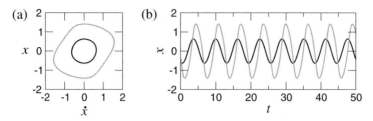

Fig. 3.1. **a** Phase portraits and **b** realizations of the autonomous van der Pol oscillator (3.1) at $\omega_0 = 1$ and two different values of non-linearity λ: $\lambda = 0.1$ (*black*) and $\lambda = 0.5$ (*grey*)

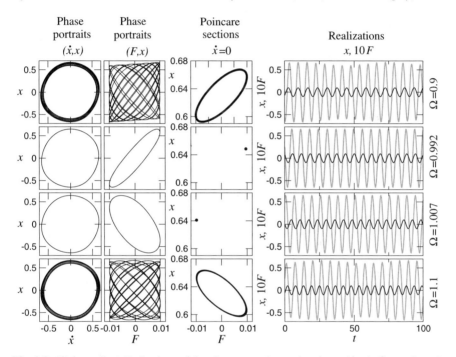

Fig. 3.2. (Color online) Projections of the phase portraits on the planes $(\dot{x}, x)$ (**first column**) and (F, x) (**second column**), Poincaré sections on the plane (F, x) (**third column**), and realizations $x(t)$ and $F(t) = B\cos(\Omega t)$ (**fourth column**) of the forced van der Pol oscillator (3.3) at $\lambda = 0.1$, $B = 0.01$ and different values of Ω: $\Omega = 0.9$, $\Omega = 0.992$, $\Omega = 1.007$, $\Omega = 1.1$

One can see that if the forcing frequency Ω is sufficiently close to the natural frequency $\omega_0 = 1$ ($\Omega = 0.992$ and $\Omega = 1.007$), which means that the frequency detuning between the systems is small, the forced oscillations $x(t)$ are periodic. Namely, the phase trajectories tend to the stable limit cycle, and the Poincaré section is a fixed point. Each time F takes its maximal value, x tends to be at the same "stage" of its oscillations. This is phase synchronization of oscillations by external

24   3 1 : 1 Forced Synchronization of Periodic Oscillations

forcing. Note that synchronized oscillations have constant amplitude, and the values of x at the local maxima are the same from one oscillation to another.

However, when the forcing frequency Ω is not close enough to ω_0 ($\Omega = 0.9$ and $\Omega = 1.1$), i.e., frequency detuning is not small, an interesting phenomenon occurs: the amplitude of oscillations oscillates itself. This is called *beating*. The instantaneous amplitudes, that are roughly half distances between the closest maxima and minima $x(t)$, oscillate periodically with a certain beat frequency. To some extent, this can be visible in the Poincaré section, defined by $\dot{x} = 0$, $\ddot{x} < 0$, that shows the maxima against the values of the forcing F taken at the same instants (Fig. 3.2, third column). Later we will consider the beat frequency in more detail and make some theoretical analyses.

Perhaps more importantly, when the amplitude of oscillations is not constant the forced oscillations are not synchronous with the forcing: when F takes its maximal values, x can take any value. The projection (F, x) is very informative: it is clearly visible that x and F move independently of each other. In more rigorous terms, the oscillations are quasiperiodic: the phase trajectories lie on the two-dimensional invariant tori whose Poincaré sections are closed curves. This regime corresponds to the absence of synchronization between the system and the forcing.

In Fig. 3.2 the 1 : 1 synchronization phenomenon is illustrated numerically. However, for the weakly non-linear oscillator (3.3) considered, synchronization also allows for analytical treatment, which will be illustrated in the next sections.

3.1 Phase of Quasiharmonic Oscillations

Here, we need to introduce an important idea closely associated with the phenomenon of synchronization—the idea of phase. When discussing the synchronization illustrated in Fig. 3.2, we mentioned the "stage" of oscillations, which is the current position of the system inside the given cycle of oscillations, e.g., the beginning, the first quarter, the middle, the end, etc. We need a quantity characterizing the "stage" of oscillations at any given time moment t: call it phase $\psi(t)$. For a purely harmonic function of time like in (3.2) the phase can be introduced uniquely as an argument of the cosine or sine, which in the given case will be $\psi(t) = \omega_0 t + \varphi_0$. In Fig. 3.3(a) a harmonic function $\cos(1.005t + 0.5)$ is shown by a solid line that represents harmonic forcing. The period of the cosine function is 2π, so if we start to observe the cosine at some time moment t, the onset of the nth full oscillation cycle ($n = 1, 2, \ldots$) can be characterized by the values of $\psi(t)$ equal to $\varphi_0 + 2\pi(n-1)$, and the end of it by $\varphi_0 + 2\pi n$. Thus, the first oscillation cycle will be within $\psi(t) \in [\varphi_0; \varphi_0 + 2\pi]$, and in terms of time inside the interval $t \in [0; 2\pi/\omega]$. *For harmonic oscillations, phase is a linear function of time.* In Fig. 3.3(c) the phase $\psi(t) = (1.005t + 0.5)$ of the signal in (a) is shown with filled circles. In (a) filled circles indicate the values of the signal at the instants $t = 2\pi n/1.005$, i.e., when phase $\psi(t)$ changes by 2π.

In this chapter we will deal with oscillations in a forced weakly non-linear system. Such oscillations, generally not being harmonic, can be viewed as almost har-

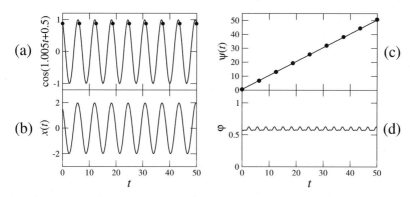

Fig. 3.3. Illustration of phase of quasiharmonic oscillations. **a** Harmonic signal cos(1.005t + 0.5) that represents forcing; **b** quasiharmonic signal x(t) that represents response of the forced system; **c** phases of a harmonic forcing (*filled circles*) and quasiharmonic response (*solid line*); **d** phase difference φ(t) between the response and the forcing that oscillates slightly around some constant and is thus an evidence of 1 : 1 phase synchronization

monic, or "quasiharmonic." This term means that the oscillations can be described as cosine (or sine) whose argument is not a linear function of time, but is close to being linear, and whose amplitude is not a constant, but changes either slowly, or slightly. For quasiharmonic oscillations phase can be also introduced as an argument of cosine. An example of quasiharmonic oscillations is given in Fig. 3.3(b), and its phase in (c) by solid line.

It has to be mentioned that purely harmonic or quasiharmonic oscillations are rare in real life. Unfortunately, even if oscillations are periodic, but non-harmonic, there is no unique way to introduce a phase. In more detail, the problem of introduction of a phase for non-periodic oscillations of complex shape, including chaotic ones, will be discussed in Sect. 8.3. Because in this chapter we are not going to consider any other oscillations besides the weakly non-linear ones, we do not need to be bothered with the more difficult cases right now.

Note that phase itself can serve as a useful instrument for describing the oscillations. However, in relation to the synchronization problem, phase represents a convenient tool for detection whether two oscillations are synchronized or not. Namely, one can introduce phases for the two oscillations and consider their difference φ that is usually referred to as "phase difference." If the phase difference happens to be a constant or to oscillate slightly around a constant[3] this would usually imply that two oscillations are 1 : 1 synchronized. An illustration of this is given in Fig. 3.3(d) that shows the phase difference between the forcing in (a) and the response in (b). If the phase difference grows or decreases monotonously in time, there is no 1 : 1 synchronization.

---

[3] If phase difference oscillates around some constant, it does not necessarily mean synchronization. For more detail, see the description of Fig. 3.9.

3.2 Derivation of Truncated Equations for Phase Difference and Amplitude

It should be noted that the *exact* oscillating solution of (3.3) for the arbitrary values of parameters λ, B and Ω cannot be found analytically with the mathematical tools available so far. However, for a certain range of values of these parameters, one can analytically find the approximate solutions that would describe the unknown exact solutions with a certain degree of accuracy. This would be quite sufficient from the practical viewpoint, since whatever is measured in a real experiment is anyway measured with a certain error. Hence, approximate theory can be good enough when one compares it with an experiment. The idea of calculations similar to the ones presented here first occurred to Andronov and Witt in the 1920s [13, 299]. The significance of their results is that they were one of the first to successfully analyze the *non-linear* systems by the approximate methods, and to provide an accurate explanation of the earlier experimental observations of synchronization in electric circuits. The analysis of equations of the form similar to that of (3.3), i.e., of non-linear, dissipative ordinary differential equations of the second order with weak non-linearity and with periodic excitation, was also done by Cartwright [61, 62], Gillies [88], Holmes and Rand [117], Arrowsmith [36]. An introduction into this analysis was made in [179] and [110] with more references.

In what follows, we will restrict our analysis to the small values of λ, for which without the forcing ($B = 0$) the solution is almost harmonic (3.2). The addition of forcing ($B \neq 0$) will obviously change the solution. However, let us assume that the forcing is not too strong, i.e., B is not large as compared to the amplitude A_0 of unperturbed self-oscillations, and the forcing frequency Ω is only slightly different from ω_0. Then the solution of (3.3) can be approximately described as a quasiharmonic oscillation, i.e., oscillation in the form of (3.2), whose amplitude A and the argument ψ (phase) of the cosine are perturbed by the forcing.

Since the system, (3.3), under study is close to being linear with small λ, it is natural to suppose that its response to an external forcing at the frequency Ω contains the frequency component Ω. We will thus be looking for a solution in the form of a quasiharmonic function of time, namely,

$$x(t) = A(t)\cos(\Omega t + \varphi(t)). \tag{3.5}$$

Here, $A(t)$ is the envelope of the oscillations $x(t)$ illustrated by Fig. 3.2 (fourth column). A does not change in time when synchronization occurs, and oscillates slowly when beating starts.

We assume that both $A(t)$ and $\varphi(t)$ are slow functions of time compared to the function $\cos(\Omega t)$. Mathematically, this condition can be written as

$$\dot{A}(t) \ll \Omega A(t), \qquad |\dot{\varphi}(t)| \ll \Omega. \tag{3.6}$$

The full phase $\psi(t)$ of the forced oscillations is

$$\psi(t) = \Omega t + \varphi(t), \tag{3.7}$$

3.2 Derivation of Truncated Equations for Phase Difference and Amplitude   27

while the phase of forcing $\psi_f(t)$ is

$$\psi_f(t) = \Omega t.$$

Hence, $\varphi(t)$ is the phase difference between the forcing and the forced oscillations. When $\varphi(t)$ is a constant, oscillations $x(t)$ in the system are $1:1$ synchronized by the external forcing and are harmonic with frequency Ω. When $\varphi(t)$ changes in time, there is no $1:1$ synchronization. Thus, in order to reveal the synchronization conditions, we need to formulate the explicit equations describing the evolution of φ and A in time. Synchronization will mean that there is/are stable fixed point/s in these equations, so we will have to find the conditions for these points to exist and to be stable.

To derive the equations for φ and A, one can use the method of averaging, also known as the Krylov–Bogoliubov[4] method [150]. In the following, for brevity we will omit the brackets "(t)" denoting the explicit dependence on time of A and φ. If we calculate the time derivative of $x(t)$ rigorously, we obtain

$$\dot{x} = \dot{A}\cos(\Omega t + \varphi) - A\Omega\sin(\Omega t + \varphi) - A\dot{\varphi}\sin(\Omega t + \varphi). \tag{3.8}$$

By representing the solution in the form of (3.5), instead of one independent phase variable $x(t)$ we introduce two phase variables: $A(t)$ and $\varphi(t)$. Thus, an ambiguity is introduced into the system. In order to remove the introduced ambiguity, we have to specify an additional condition that $A(t)$ and $\varphi(t)$ should satisfy. It is convenient to set such a condition that the derivative of $x(t)$ is a simple expression in the form

$$\dot{x} = -A\Omega\sin(\Omega t + \varphi), \tag{3.9}$$

which would immediately imply

$$\dot{A}\cos(\Omega t + \varphi) - A\dot{\varphi}\sin(\Omega t + \varphi) = 0. \tag{3.10}$$

Next, we need to find $\ddot{x}$. The calculations can be continued using the expressions above that contain sines and cosines, but it is usually more convenient to operate with exponential functions. Thus, we want to express all trigonometric functions in $x(t)$, $\dot{x}$ and $\ddot{x}$ in terms of exponents of complex arguments. We start from reformulating the solution, (3.5),

$$x = A\cos(\Omega t + \varphi) = A\frac{e^{i(\Omega t+\varphi)} + e^{-i(\Omega t+\varphi)}}{2} = \frac{e^{i\Omega t}Ae^{i\varphi} + e^{-i\Omega t}Ae^{-i\varphi}}{2}.$$

Let us introduce a complex function of time a, such that

$$a = Ae^{i\varphi}, \qquad a^* = Ae^{-i\varphi}, \tag{3.11}$$

where the asterisk denotes the complex conjugate. Then $x(t)$ can be represented through a as

---

[4] Bogoliubov is also sometimes spelled as Bogolyubov or Bogolioubov in literature.

3 1:1 Forced Synchronization of Periodic Oscillations

$$x = \frac{1}{2}\left(ae^{i\Omega t} + a^*e^{-i\Omega t}\right).$$ (3.12)

We can call a a complex amplitude of oscillations. The condition (3.10) can be rewritten as

$$\dot{A}\frac{e^{i(\Omega t+\varphi)} + e^{-i(\Omega t+\varphi)}}{2} - A\dot{\varphi}\frac{e^{i(\Omega t+\varphi)} - e^{-i(\Omega t+\varphi)}}{2i}$$

$$= \frac{e^{i\Omega t}\dot{A}e^{i\varphi} + e^{-i\Omega t}\dot{A}e^{-i\varphi}}{2} - \frac{e^{i\Omega t}A\dot{\varphi}e^{i\varphi} - e^{-i\Omega t}A\dot{\varphi}e^{-i\varphi}}{2i}$$

$$= \frac{1}{2}e^{i\Omega t}\left(\dot{A}e^{i\varphi} + iA\dot{\varphi}e^{i\varphi}\right) + \frac{1}{2}e^{-i\Omega t}\left(\dot{A}e^{-i\varphi} - iA\dot{\varphi}e^{-i\varphi}\right) = 0.$$

With the account of the following:

$$\dot{a} = \dot{A}e^{i\varphi} + Ai\dot{\varphi}e^{i\varphi} \quad \text{and} \quad \dot{a}^* = \dot{A}e^{-i\varphi} - Ai\dot{\varphi}e^{-i\varphi},$$

the condition (3.10) turns into

$$\dot{a}e^{i\Omega t} + \dot{a}^*e^{-i\Omega t} = 0.$$ (3.13)

Now, consider $\dot{x}$ and rewrite (3.9) as

$$\dot{x} = -A\Omega\frac{e^{i(\Omega t+\varphi)} - e^{-i(\Omega t+\varphi)}}{2i} = \frac{i\Omega}{2}\left(ae^{i\Omega t} - a^*e^{-i\Omega t}\right).$$ (3.14)

Consider $\ddot{x}$ as a derivative of (3.14)

$$\ddot{x} = \frac{i\Omega}{2}\left(\dot{a}e^{i\Omega t} + ai\Omega e^{i\Omega t} - \dot{a}^*e^{-i\Omega t} + a^*i\Omega e^{-i\Omega t}\right)$$

$$= \frac{i\Omega}{2}\dot{a}e^{i\Omega t} - \frac{\Omega^2}{2}ae^{i\Omega t} - \frac{i\Omega}{2}\dot{a}^*e^{-i\Omega t} - \frac{\Omega^2}{2}a^*e^{-i\Omega t}.$$

Add and subtract $\frac{i\Omega}{2}\dot{a}e^{i\Omega t}$ and regroup terms

$$\ddot{x} = \left(i\Omega\dot{a}e^{i\Omega t} - \frac{i\Omega}{2}\dot{a}e^{i\Omega t}\right) - \frac{\Omega^2}{2}ae^{i\Omega t} - \frac{i\Omega}{2}\dot{a}^*e^{-i\Omega t} - \frac{\Omega^2}{2}a^*e^{-i\Omega t}.$$

The sum of the second and the fourth terms satisfies the condition (3.13) and is equal to zero. Hence

$$\ddot{x} = i\Omega\dot{a}e^{i\Omega t} - \Omega^2\frac{1}{2}\left(ae^{i\Omega t} + a^*e^{-i\Omega t}\right).$$ (3.15)

Substitute x, $\dot{x}$ and $\ddot{x}$ ((3.12), (3.14), (3.15), respectively) into (3.3)

$$i\Omega\dot{a}e^{i\Omega t} - \frac{\Omega^2}{2}\left(ae^{i\Omega t} + a^*e^{-i\Omega t}\right)$$

$$- \left(\lambda - \frac{1}{4}\left(ae^{i\Omega t} + a^*e^{-i\Omega t}\right)^2\right)\frac{i\Omega}{2}\left(ae^{i\Omega t} - a^*e^{-i\Omega t}\right)$$

$$+ \frac{\omega_0^2}{2}\left(ae^{i\Omega t} + a^*e^{-i\Omega t}\right)$$

$$= B\frac{e^{i\Omega t} + e^{-i\Omega t}}{2}.$$

3.2 Derivation of Truncated Equations for Phase Difference and Amplitude 29

Regroup terms

$$
i\Omega \dot{a} e^{i\Omega t} + \frac{(\omega_0^2 - \Omega^2)}{2}\left(a e^{i\Omega t} + a^* e^{-i\Omega t}\right) - \lambda \frac{i\Omega}{2} a e^{i\Omega t} + \lambda \frac{i\Omega}{2} a^* e^{-i\Omega t}
$$
$$
+ \frac{1}{4}\left(a^2 e^{i2\Omega t} + a^{*2} e^{-i2\Omega t} + 2aa^*\right)\frac{i\Omega}{2}\left(a e^{i\Omega t} - a^* e^{-i\Omega t}\right)
$$
$$
= B\frac{e^{i\Omega t} + e^{-i\Omega t}}{2}.
$$

Open the brackets

$$
i\Omega \dot{a} e^{i\Omega t} + \frac{(\omega_0^2 - \Omega^2)}{2}\left(a e^{i\Omega t} + a^* e^{-i\Omega t}\right) - \lambda \frac{i\Omega}{2} a e^{i\Omega t} + \lambda \frac{i\Omega}{2} a^* e^{-i\Omega t}
$$
$$
+ \frac{i\Omega}{8} a^3 e^{i3\Omega t} - \frac{i\Omega}{8} a^2 a^* e^{i\Omega t} + \frac{i\Omega}{8} aa^{*2} e^{-i\Omega t} - \frac{i\Omega}{8} a^{*3} e^{-i3\Omega t}
$$
$$
+ \frac{i\Omega}{4} a^2 a^* e^{i\Omega t} - \frac{i\Omega}{4} aa^{*2} e^{-i\Omega t}
$$
$$
= B\frac{e^{i\Omega t} + e^{-i\Omega t}}{2}. \tag{3.16}
$$

Collect similar terms and multiply the whole equation by $e^{-i\Omega t}/(i\Omega)$,

$$
\dot{a} + \frac{(\omega_0^2 - \Omega^2)}{2i\Omega}\left(a + a^* e^{-i2\Omega t}\right) - \frac{\lambda}{2} a + \frac{\lambda}{2} a^* e^{-i2\Omega t} + \frac{1}{8} a^3 e^{i2\Omega t} + \frac{1}{8} a^2 a^*
$$
$$
- \frac{1}{8} aa^{*2} e^{-i2\Omega t} - \frac{1}{8} a^{*3} e^{-i4\Omega t}
$$
$$
= \frac{B}{2i\Omega}\left(1 + e^{-i2\Omega t}\right). \tag{3.17}
$$

We remind you, that the aim of our calculations is to write down the equations describing the evolution in time of the complex amplitude $a(t)$, and then to solve them. Knowing $a(t)$, we will know A and φ, and thus we will know the approximate solution $x(t)$ of van der Pol equation (3.3). However, (3.17) is not simpler than the original (3.3), and it is not easier to find the amplitude a from it than it was to find x from (3.3). To simplify the problem we can make more use of the fact of slowness of A and φ and exploit the method of averages by Krylov and Bogoliubov [150]. Note that a, $\dot{a}$ and a^* are slow functions of time as compared to the functions $e^{\pm n\Omega t}$, n being an integer number. This means that they almost do not change during one period of fast oscillations with the frequency Ω. If we average the whole equation over one period $T = 2\pi/\Omega$ of fast oscillations, we can get rid of the fast terms, and only the slow terms will remain in the equation. The time average $\bar{f}^T$ of a smooth function $f(t)$ over the time interval T is defined as follows:

$$
\bar{f}^T = \frac{1}{T}\int_{t_0}^{t_0+T} f(t)\,dt. \tag{3.18}
$$

30 3 1:1 Forced Synchronization of Periodic Oscillations

Consider terms in (3.17) containing $e^{-i2\Omega t}$. The time average of the second such term is equal to

$$\frac{1}{T}\int_{t_0}^{t_0+T}\frac{\lambda}{2}a^*e^{-i2\Omega t}\,dt$$

$$\approx \frac{\lambda}{2}a^*\frac{\Omega}{2\pi}\int_{t_0}^{t_0+2\pi/\Omega}e^{-i2\Omega t}\,dt = \frac{\lambda}{2}a^*\frac{\Omega}{2\pi}\left(\frac{1}{-2i\Omega}\right)e^{-i2\Omega t}\bigg|_{t_0}^{t_0+2\pi/\Omega}$$

$$= \frac{\lambda}{2}a^*\frac{\Omega}{2\pi}\frac{1}{-2i\Omega}\Big[\underbrace{\cos(2\Omega t)\big|_{t_0}^{t_0+2\pi/\Omega}}_{=0} - i\underbrace{\sin(2\Omega t)\big|_{t_0}^{t_0+2\pi/\Omega}}_{=0}\Big] = 0.$$

By analogy, it is easy to show that the average values of terms containing $e^{i2\Omega t}$ and $e^{-i4\Omega t}$ are equal to zero as well. Thus, we obtain the time-averaged equations which, with account of $a^2 a^* = a(aa^*) = a|a|^2$, read

$$\dot{a} + \frac{(\omega_0^2 - \Omega^2)}{2i\Omega}a - \frac{\lambda}{2}a + \frac{1}{8}a|a|^2 = -i\frac{B}{2\Omega}. \tag{3.19}$$

Recall that $a = Ae^{i\varphi}$ and substitute it into (3.19)

$$\dot{A}e^{i\varphi} + Ai\dot{\varphi}e^{i\varphi} - i\frac{(\omega_0^2 - \Omega^2)}{2\Omega}Ae^{i\varphi} - \frac{\lambda}{2}Ae^{i\varphi} + \frac{1}{8}A^3e^{i\varphi} = -i\frac{B}{2\Omega}.$$

Multiply everything by $e^{-i\varphi}$

$$\dot{A} + Ai\dot{\varphi} - i\frac{(\omega_0^2 - \Omega^2)}{2\Omega}A - \frac{\lambda}{2}A + \frac{1}{8}A^3 = -i\frac{B}{2\Omega}e^{-i\varphi}.$$

Introduce the frequency detuning Δ between the unperturbed system and the forcing

$$\Delta = \frac{(\omega_0^2 - \Omega^2)}{2\Omega} \approx (\omega_0 - \Omega), \tag{3.20}$$

the latter approximation being valid when the forcing frequency Ω is close to the natural frequency of unperturbed oscillations ω_0 ($\Omega \approx \omega_0$). Represent $e^{-i\varphi}$ through $\cos\varphi$ and $\sin\varphi$

$$\dot{A} + iA\dot{\varphi} - iA\Delta - \frac{\lambda}{2}A + \frac{1}{8}A^3 = -i\frac{B}{2\Omega}e^{-i\varphi} = -i\frac{B}{2\Omega}(\cos\varphi - i\sin\varphi).$$

Separate the real and imaginary parts of the equation

$$\dot{A} - \frac{\lambda}{2}A + \frac{1}{8}A^3 = -\frac{B}{2\Omega}\sin\varphi,$$

$$A\dot{\varphi} - A\Delta = -\frac{B}{2\Omega}\cos\varphi.$$

3.3 Amplitude of Unperturbed Oscillations at Small Non-linearity 31

Finally, we obtain

$$\dot{A} = \frac{\lambda}{2}A - \frac{1}{8}A^3 - \frac{B}{2\Omega}\sin\varphi, \qquad (3.21)$$

$$\dot{\varphi} = \Delta - \frac{B}{2A\Omega}\cos\varphi. \qquad (3.22)$$

These are the famous *truncated equations* for the amplitude A of forced oscillations and for the phase difference φ between the latter and the external forcing. These equations have a fundamental importance in the theory of synchronization. Their significance is due to the fact that the analysis of a van der Pol equation (3.3) that is non-autonomous, i.e., depends on time explicitly, is reduced to the analysis of the *autonomous* system of equations (3.21)–(3.22). In terms of bifurcation theory, instead of analyzing periodic orbits of (3.3), we can analyze the fixed points in (3.21)–(3.22), which is obviously much easier.

The fixed points of (3.21)–(3.22) mean that the phase difference between the system and external forcing does not change in time ($\varphi = $ const), i.e., the external forcing has synchronized the system, and the oscillations are periodic with constant amplitude A and the frequency of external forcing Ω. Thus, finding of the conditions when these fixed points are stable will mean finding the conditions at which $1:1$ forced synchronization occurs.

3.3 Amplitude of Unperturbed Oscillations at Small Non-linearity

It is clearly seen that both truncated equations (3.21)–(3.22) are non-linear, A and φ influencing each other in the presence of forcing ($B \neq 0$).

If there is no forcing ($B = 0$), one can estimate the stationary amplitude of natural self-oscillations, i.e., the amplitude that the oscillations will have after the sufficiently long relaxation time will pass. In order to do this, in (3.21) one should set $\dot{A} = 0$ and solve the algebraic equation

$$f_A(A) = \frac{\lambda}{2}A - \frac{1}{8}A^3 = 0, \qquad \begin{aligned} A_0 &= 0, \\ A_0 &= 2\sqrt{\lambda}. \end{aligned} \qquad (3.23)$$

Solution $A_0 = 0$ corresponds to the absence of oscillations, i.e., to the fixed point. Strictly speaking, there are two roots corresponding to the non-zero solution, but only the positive one makes sense, since the amplitude is supposed to be a positive value by definition. The stability of the fixed points is determined by the sign of $\partial f_A(A)/\partial A$: if it is negative (positive), the point is stable (unstable):

$$\left.\frac{\partial f_A(A)}{\partial A}\right|_{A=0} = \left.\frac{\lambda}{2} - \frac{3}{8}A^2\right|_{A=0} = \frac{\lambda}{2} > 0,$$

$$\left.\frac{\partial f_A(A)}{\partial A}\right|_{A=2\sqrt{\lambda}} = \left.\frac{\lambda}{2} - \frac{3}{8}A^2\right|_{A=2\sqrt{\lambda}} = \frac{\lambda}{2} - \frac{3}{8}4\lambda = -\lambda < 0.$$

32   3 1 : 1 Forced Synchronization of Periodic Oscillations

Thus, the non-oscillatory solution is unstable, and the oscillatory one is stable. In what follows let us denote the amplitude of natural (unperturbed) self-oscillations as A_0. Hence, A_0 is proportional to the square root of the non-linearity parameter λ while the latter remains small.

3.4 Analysis of Truncated Equations for Weak Forcing

Consider the non-zero forcing. The analysis of these equations for arbitrary values of B and Ω is difficult. However, a few special cases can be considered that allow for approximate analytical solutions. In the simplest case when the strength B of forcing can be regarded as very small

$$B \ll \varepsilon A_0, \tag{3.24}$$

the amplitude of the perturbed oscillations is not very different from A_0. In the equation for φ we can set $A = A_0$ as in (3.23), and then it becomes independent of A

$$\dot{\varphi} = \Delta - \frac{B}{4\sqrt{\lambda}\Omega} \cos \varphi = f_\varphi(\varphi). \tag{3.25}$$

The fixed points of this equation that correspond to $\dot{\varphi} = 0$ can be found by solving a non-linear algebraic equation

$$\cos \varphi = \frac{4\sqrt{\lambda}\Omega \Delta}{B}, \tag{3.26}$$

which is illustrated in Fig. 3.4. One can see that $\cos \varphi$ can intersect the horizontal line twice, thus there can be two solutions

$$\varphi_1 = \cos^{-1} \frac{4\sqrt{\lambda}\Omega \Delta}{B}, \qquad \varphi_2 = 2\pi - \cos^{-1} \frac{4\sqrt{\lambda}\Omega \Delta}{B},$$

which exist provided that

$$4\sqrt{\lambda}\Omega |\Delta| \le B. \tag{3.27}$$

Their stability is determined by the sign of $\partial f_\varphi(\varphi)/\partial \varphi$ in (3.25). Namely,

$$\begin{aligned}
\frac{\partial f_\varphi(\varphi)}{\partial \varphi}\bigg|_{\varphi_1} &= \frac{B}{4\sqrt{\lambda}\Omega} \sin\left(\cos^{-1} \frac{4\sqrt{\lambda}\Omega \Delta}{B}\right) \\
&= \frac{B}{2A_0\Omega} \sqrt{1 - \cos^2\left(\cos^{-1} \frac{4\sqrt{\lambda}\Omega \Delta}{B}\right)} \\
&= \frac{B}{4\sqrt{\lambda}\Omega} \sqrt{1 - \left(\frac{4\sqrt{\lambda}\Omega \Delta}{B}\right)^2} \ge 0
\end{aligned}$$

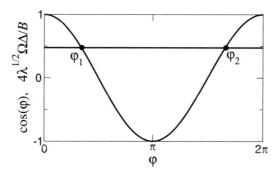

Fig. 3.4. Graphical illustration of the solution of the non-linear equation (3.26)

and

$$\left.\frac{\partial f_\varphi(\varphi)}{\partial \varphi}\right|_{\varphi_2} = \frac{B}{4\sqrt{\lambda\Omega}} \sin\left(2\pi - \cos^{-1}\frac{4\sqrt{\lambda\Omega}\Delta}{B}\right)$$

$$= \frac{B}{4\sqrt{\lambda\Omega}}\left[\sin 2\pi \times \cos\left(\cos^{-1}\frac{4\sqrt{\lambda\Omega}\Delta}{B}\right) - \cos 2\pi \times \sin\left(\cos^{-1}\frac{4\sqrt{\lambda\Omega}\Delta}{B}\right)\right]$$

$$= -\frac{B}{4\sqrt{\lambda\Omega}}\sqrt{1 - \cos^2\left(\cos^{-1}\frac{4\sqrt{\lambda\Omega}\Delta}{B}\right)}$$

$$= -\frac{B}{4\sqrt{\lambda\Omega}}\sqrt{1 - \left(\frac{4\sqrt{\lambda\Omega}\Delta}{B}\right)^2} \leq 0,$$

which means that the fixed point φ_2 is stable and φ_1 is unstable. When the strict equality in (3.27) is satisfied, two fixed points merge, and when (3.27) is no longer valid, the pair of fixed points disappear via saddle-node bifurcation. This means that there is no longer a constant phase difference between the forcing and the response in (3.3). Hence, the equation

$$B = 4\sqrt{\lambda\Omega}|\Delta| \tag{3.28}$$

describes the borderline of 1:1 synchronization region at very small strengths of forcing B.

At $\lambda = 0.1$ and $\omega_0 = 1$, and with the approximation in (3.20) for Δ, the synchronization region defined by (3.28) is outlined by the shaded area in Fig. 3.5 on the plane of forcing parameters (Ω, B). We see that it has the characteristic shape of a tongue with a tip at $\Omega \approx \omega_0 = 1$. The solid lines show the numerically estimated lines of saddle-node bifurcations of a stable and a saddle periodic orbits of the original non-autonomous equation (3.3) for the same parameters. We see that the approximation for the synchronization region border given by (3.28) is quite accurate for a significant range of B.

34   3 1:1 Forced Synchronization of Periodic Oscillations

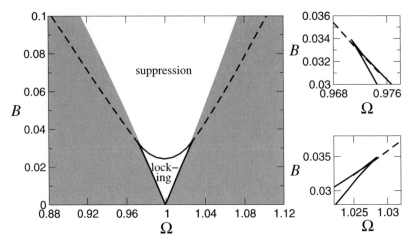

Fig. 3.5. (Color online) 1 : 1 synchronization tongue for the forced van der Pol system (3.3) on the plane of forcing parameters Ω and B, at $\lambda = 0.1$, $\omega_0 = 1$. Lines correspond to bifurcations of periodic solutions: *solid lines* mark saddle-node bifurcations, *dashed lines* mark torus birth (Neimark–Sacker) bifurcations, both obtained numerically from the direct analysis of (3.3). *Shaded area* shows analytical prediction of the locking region according to (3.28). *Insets* to the right show the connections between the above types of lines in more detail (compare with insets in Fig. 3.7)

3.5 Derivation of Truncated Equations in Descartes Coordinates

If the amplitude B of forcing is not vanishingly small, the truncated equations (3.21)–(3.22) for the amplitude A and phase φ of the complex amplitude a that modulates periodic oscillations according to (3.12) can have up to three fixed points. To reveal the borderlines of the synchronization region, one could find the fixed points, analyze their stability and find the lines in the parameter plane on which bifurcations occur. However, although the equations look quite compact, their analysis is quite involved. It appears that if the same equations are rewritten in Descartes coordinates instead of the polar ones, their analysis becomes less cumbersome. In this section we show how to obtain the Descartes form of the truncated equations.

At the arbitrary values of B, A can no longer be regarded as a constant approximately equal to the amplitude of the unperturbed oscillations A_0, and the two equations cannot be separated. Perform the following variable substitution:

$$\bar{u}(t) = A(t)\cos\varphi(t), \qquad \bar{v}(t) = A(t)\sin\varphi(t). \tag{3.29}$$

The time derivatives of the new variables can be expressed through A and φ as

$$\dot{\bar{u}} = \dot{A}\cos\varphi - A\dot{\varphi}\sin\varphi,$$
$$\dot{\bar{v}} = \dot{A}\sin\varphi + A\dot{\varphi}\cos\varphi,$$

and further with account of (3.21)–(3.22) as

3.5 Derivation of Truncated Equations in Descartes Coordinates 35

$$\dot{u} = \cos\varphi \left[\frac{\lambda}{2} A - \frac{1}{8} A^3 - \frac{B}{2\Omega} \sin\varphi \right] - A \sin\varphi \left[\Delta - \frac{B}{2A\Omega} \cos\varphi \right],$$

$$\dot{v} = \sin\varphi \left[\frac{\lambda}{2} A - \frac{1}{8} A^3 - \frac{B}{2\Omega} \sin\varphi \right] + A \cos\varphi \left[\Delta - \frac{B}{2A\Omega} \cos\varphi \right].$$

$$(3.30)$$

Note that

$$A = \sqrt{\bar{u}^2 + \bar{v}^2}, \qquad \cos\varphi = \frac{\bar{u}}{\sqrt{\bar{u}^2 + \bar{v}^2}}, \qquad \sin\varphi = \frac{\bar{v}}{\sqrt{\bar{u}^2 + \bar{v}^2}},$$

and substitute this into (3.30),

$$\begin{aligned}
\dot{u} &= \frac{\bar{u}}{\sqrt{\bar{u}^2 + \bar{v}^2}} \left[\frac{\lambda}{2} \sqrt{\bar{u}^2 + \bar{v}^2} - \frac{1}{8} (\bar{u}^2 + \bar{v}^2)^{3/2} - \frac{B}{2\Omega} \frac{\bar{v}}{\sqrt{\bar{u}^2 + \bar{v}^2}} \right] \\
&\quad - \sqrt{\bar{u}^2 + \bar{v}^2} \frac{\bar{v}}{\sqrt{\bar{u}^2 + \bar{v}^2}} \left[\Delta - \frac{B}{2\Omega\sqrt{\bar{u}^2 + \bar{v}^2}} \frac{\bar{u}}{\sqrt{\bar{u}^2 + \bar{v}^2}} \right] \\
&= \bar{u}\frac{\lambda}{2} - \frac{\bar{u}}{8}(\bar{u}^2 + \bar{v}^2) - \frac{B\bar{u}\bar{v}}{2\Omega(\bar{u}^2 + \bar{v}^2)} - \bar{v}\Delta + \frac{B\bar{u}\bar{v}}{2\Omega(\bar{u}^2 + \bar{v}^2)},
\end{aligned}$$

$$\begin{aligned}
\dot{v} &= \frac{\bar{v}}{\sqrt{\bar{u}^2 + \bar{v}^2}} \left[\frac{\lambda}{2} \sqrt{\bar{u}^2 + \bar{v}^2} - \frac{1}{8} (\bar{u}^2 + \bar{v}^2)^{3/2} - \frac{B}{2\Omega} \frac{\bar{v}}{\sqrt{\bar{u}^2 + \bar{v}^2}} \right] \\
&\quad + \sqrt{\bar{u}^2 + \bar{v}^2} \frac{\bar{u}}{\sqrt{\bar{u}^2 + \bar{v}^2}} \left[\Delta - \frac{B}{2\Omega\sqrt{\bar{u}^2 + \bar{v}^2}} \frac{\bar{u}}{\sqrt{\bar{u}^2 + \bar{v}^2}} \right] \\
&= \bar{v}\frac{\lambda}{2} - \frac{\bar{v}}{8}(\bar{u}^2 + \bar{v}^2) - \frac{B\bar{v}^2}{2\Omega(\bar{u}^2 + \bar{v}^2)} + \bar{u}\Delta - \frac{B\bar{u}^2}{2\Omega(\bar{u}^2 + \bar{v}^2)}.
\end{aligned}$$

The third and the fifth terms of the right-hand part of the equation for $\dot{u}$ cancel each other. Collect similar terms and rewrite the set of equations for $\dot{u}$ and $\dot{v}$,

$$\dot{u} = \frac{\bar{u}}{2} \left[\lambda - \frac{(\bar{u}^2 + \bar{v}^2)}{4} \right] - \bar{v}\Delta, \tag{3.31}$$

$$\dot{v} = \frac{\bar{v}}{2} \left[\lambda - \frac{(\bar{u}^2 + \bar{v}^2)}{4} \right] + \bar{u}\Delta - \frac{B}{2\Omega}. \tag{3.32}$$

Note that (3.31)–(3.32) are completely equivalent to (3.21)–(3.22) and describe exactly the same kind of behavior, only in different coordinates. Typical phase portraits of these two types of equations are shown in Fig. 3.6 for parameter values $\lambda = 0.1$, $\Omega = 1.005$, $B = 0.01$ (for reference, see the bifurcation diagram in Fig. 3.5): in polar coordinates in Fig. 3.6(a), and in Descartes coordinates in Fig. 3.6(b). The phase difference φ is shown by the modulus of 2π.

At these parameters phase locking takes place, and there are two fixed points in the truncated equations: one saddle (empty circle) and one stable (filled circle). Note that the closed curve formed by the manifolds of the saddle point in Fig. 3.6(b) is almost a perfect circle. One might think that for any value of φ the amplitude A should take the same values, but as seen from Fig. 3.6(a), this is not the case. The

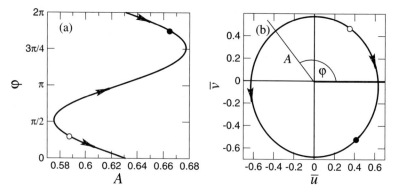

Fig. 3.6. Phase portraits of the truncated equations for the amplitude A and phase difference φ corresponding to the forced van der Pol oscillator: **a** in polar coordinates (A, φ) from (3.21)–(3.22); **b** in Descartes coordinates $(\bar{u}, \bar{v})$ from (3.31)–(3.32). Parameters are $\lambda = 0.1$, $\Omega = 1.005$, $B = 0.01$ and correspond to the phase locking. *Filled* (*empty*) *circles* show stable (saddle) point. *Solid lines* show the unstable manifolds of the saddle points

reason is that the amplitude A is the distance between the phase point and the origin in Fig. 3.6(b). With this, the center of this circle is not at the origin, so at different values of φ, the distance between the origin and the phase point is different.

It is not convenient to analyze (3.31)–(3.32) in their present form, since they contain too many control parameters. For the further analysis, let us try to make them look simpler. Denote

$$u = \frac{\bar{u}}{2\sqrt{\lambda}}, \qquad v = \frac{\bar{v}}{2\sqrt{\lambda}}, \tag{3.33}$$

substitute into (3.31)–(3.32) and divide the result by $2\sqrt{\lambda}$

$$\begin{aligned}
\frac{du}{dt} &= \lambda \frac{u}{2}\left[1 - (u^2 + v^2)\right] - v\Delta, \\
\frac{dv}{dt} &= \lambda \frac{v}{2}\left[1 - (u^2 + v^2)\right] + u\Delta - \frac{B}{4\sqrt{\lambda}\Omega}.
\end{aligned} \tag{3.34}$$

Introduce new independent argument τ,

$$\tau = \frac{\lambda t}{2},$$

and divide both equations by $\lambda/2$,

$$\begin{aligned}
\frac{du}{d\tau} &= u\left[1 - (u^2 + v^2)\right] - v\frac{2\Delta}{\lambda}, \\
\frac{dv}{d\tau} &= v\left[1 - (u^2 + v^2)\right] + u\frac{2\Delta}{\lambda} - \frac{B}{2\Omega\lambda\sqrt{\lambda}}.
\end{aligned}$$

3.6 Analysis of Truncated Equations in Descartes Coordinates 37

Denote

$$\delta = \frac{2\Delta}{\lambda}, \qquad F = \frac{B}{2\Omega\lambda\sqrt{\lambda}} \tag{3.35}$$

and rewrite the last equations as

$$\frac{du}{d\tau} = u\left[1 - \left(u^2 + v^2\right)\right] - \delta v = f(u, v), \tag{3.36}$$

$$\frac{dv}{d\tau} = v\left[1 - \left(u^2 + v^2\right)\right] + \delta u - F = g(u, v). \tag{3.37}$$

These are the equations for a non-linear system that is potentially able to demonstrate self-sustained periodic oscillations.

3.6 Analysis of Truncated Equations in Descartes Coordinates

We will now analyze the stability of the fixed points of (3.36)–(3.37) without making any simplifying assumptions on the values of parameters δ and F. The fixed points are defined by

$$\dot{u} = f(u, v) = 0, \qquad \dot{v} = g(u, v) = 0. \tag{3.38}$$

From (3.36) we obtain

$$\delta v = u\left(1 - \left(u^2 + v^2\right)\right), \tag{3.39}$$

and from (3.37)

$$1 - \left(u^2 + v^2\right) = \frac{F - \delta u}{v}. \tag{3.40}$$

Substitute (3.40) into (3.39)

$$v^2 = \frac{u(F - \delta u)}{\delta}. \tag{3.41}$$

Substitute v^2 from (3.41) into (3.40)

$$(F - \delta u) = v\left(1 - u^2 - \frac{Fu}{\delta} + u^2\right) = v\left(1 - \frac{Fu}{\delta}\right), \tag{3.42}$$

$$v = \delta\frac{F - \delta u}{\delta - Fu}. \tag{3.43}$$

Take a square of the last expression to obtain

$$v^2 = \delta^2\frac{(F - \delta u)^2}{(\delta - Fu)^2}. \tag{3.44}$$

Equating (3.44) and (3.41) leads to the following equation that only includes the u-variable:

$$\delta^2\frac{(F - \delta u)^2}{(\delta - Fu)^2} = \frac{u(F - \delta u)}{\delta}. \tag{3.45}$$

38   3 1:1 Forced Synchronization of Periodic Oscillations

$(F - \delta u) = 0$ could be a root of (3.45). But due to (3.43), it would lead to $v = 0$, which is not a root of (3.38). Thus, $(F - \delta u) \neq 0$, and we can safely divide by it both parts of (3.45)

$$\frac{\delta^3(F - \delta u)}{(\delta - Fu)^2} = u.$$

Simple transformation leads to the cubic equation for u

$$F^2 u^3 - 2F\delta u^2 + (\delta^2 + \delta^4)u - \delta^3 F = 0. \tag{3.46}$$

We need to solve this equation. Denote

$$\tilde{a} = F^2, \qquad \tilde{b} = -2F\delta, \qquad \tilde{c} = (\delta^2 + \delta^4), \qquad \tilde{d} = -\delta^3 F. \tag{3.47}$$

The first step in solving a cubic equation is obtaining a "depressed cubic equation," i.e., equation without a quadratic term, by making the following variable substitution:

$$u = u_* - \frac{\tilde{b}}{3\tilde{a}}. \tag{3.48}$$

Substitute u in the above form into (3.46) and obtain

$$u_*^3 + \left(\frac{\tilde{c}}{\tilde{a}} - \frac{\tilde{b}^2}{3\tilde{a}^2}\right)u_* + \left(\frac{2\tilde{b}^3}{27\tilde{a}^3} - \frac{\tilde{b}\tilde{c}}{3\tilde{a}^2} + \frac{\tilde{d}}{\tilde{a}}\right) = 0. \tag{3.49}$$

Denote

$$C = \left(\frac{\tilde{c}}{\tilde{a}} - \frac{\tilde{b}^2}{3\tilde{a}^2}\right), \qquad D = \left(\frac{2\tilde{b}^3}{27\tilde{a}^3} - \frac{\tilde{b}\tilde{c}}{3\tilde{a}^2} + \frac{\tilde{d}}{\tilde{a}}\right), \tag{3.50}$$

so that (3.49) becomes

$$u_*^3 + Cu_* + D = 0. \tag{3.51}$$

Express C and D through δ and F using (3.47)

$$C = \frac{\delta^2 + \delta^4}{F^2} - \frac{4F^2\delta^2}{3F^4} = \frac{3\delta^4 - \delta^2}{3F^2}, \tag{3.52}$$

$$D = \frac{2 \times (-2F\delta)^3}{27F^6} - \frac{(-2F\delta) \times (\delta^2 + \delta^4)}{3F^4} + \frac{(-\delta^3 F)}{F^2}$$

$$= \frac{\delta^3}{27F^3}\left(18\delta^2 - 27F^2 + 2\right). \tag{3.53}$$

A cubic equation (3.51) has either three real roots, or one real and two complex-conjugate roots. With this, either all three, or two of three, roots can coincide, but there is always at least one real root u_{*1}. If we find this real root, we can divide (3.51) by $(u_* - u_{*1})$, obtain a quadratic equation and then find the two remaining roots. u_{*1} can be found from (3.51) by substitution

$$u_{*1} = s - t \tag{3.54}$$

3.6 Analysis of Truncated Equations in Descartes Coordinates 39

with s and t such that

$$3st = C, \tag{3.55}$$

$$s^3 - t^3 = -D. \tag{3.56}$$

If we express s through t from the first equation and substitute into the second one, we obtain an equation of the sixth order with respect to t

$$t^6 - Dt^3 - \frac{C^3}{27} = 0.$$

Denoting $z = t^3$, we obtain a quadratic equation with respect to z

$$z^2 - Dz - \frac{C^3}{27} = 0. \tag{3.57}$$

The solution of this equation is

$$z = \frac{1}{2}\left(D \pm \sqrt{D^2 + \frac{4C^3}{27}}\right) = \frac{1}{2}(D \pm \sqrt{R}), \tag{3.58}$$

where R is

$$R = D^2 + \frac{4C^3}{27}. \tag{3.59}$$

Because of the sign "$\pm$," there are two solutions for $z = t^3$, but we can take either one of them: whatever sign we choose before $\sqrt{R}$, the final solution u_{*1} will not depend on this choice. The properties of the cubic equations are such that if $R < 0$, there are three real roots in (3.46); if $R > 0$, there is only one real root in (3.46), and the other two are complex and thus not "physical." When $R = 0$, two of the three real roots u_{*1}, u_{*2} and u_{*3} coincide. We remind you that u_{*j}, $j = 1, 2, 3$ are the shifted and rescaled components of the amplitudes A_j modulating periodic oscillations in the forced van der Pol equation (see (3.29)). If at certain values of the normalized detuning δ and normalized forcing strength F, two of these amplitudes coincide, this means that a saddle-node bifurcation occurs to the fixed points of the system (3.36)–(3.37) and to the periodic orbits of the original forced van der Pol system (3.3). Thus, the equation

$$R = D^2 + \frac{4C^3}{27} = 0 \tag{3.60}$$

is the condition of a saddle-node bifurcation. Let us reveal the equation of the line of saddle-node bifurcation on the plane of parameters (δ, F) by substituting into (3.60) the expressions for C and D from (3.52) and (3.53)

$$R = \frac{\delta^6}{27^2 F^6}\left(18\delta^2 - 27F^2 + 2\right)^2 + \frac{4}{27}\left(\frac{3\delta^4 - \delta^2}{3F^2}\right)^3 \tag{3.61}$$

$$= \frac{4\delta^6}{F^6}\left[\frac{F^4}{4} - \frac{F^2}{27}(1 + 9\delta^2) + \frac{\delta^2}{27}(\delta^2 + 1)^2\right]. \tag{3.62}$$

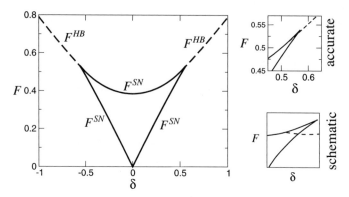

Fig. 3.7. 1 : 1 forced synchronization tongue on the plane (δ, F) of the parameters of the truncated equations in the Descartes coordinates (3.36)–(3.37). Saddle-node (*solid line*) and Andronov–Hopf (*dashed line*) bifurcation lines of the fixed points are calculated exactly and are described by (3.64) and (3.72), respectively. *Insets* to the right show the connections between two types of bifurcation lines in more detail: the upper one is accurate, and the lower one is schematic which is included in order to emphasize the configuration of the curves

Thus, the curve defined by the condition

$$\frac{F^4}{4} - \frac{F^2}{27}(1 + 9\delta^2) + \frac{\delta^2}{27}(\delta^2 + 1)^2 = 0 \tag{3.63}$$

is the line of saddle-node bifurcation. We can plot this line denoted by F^{SN} by noting that the condition above defines a quadratic equation with respect to F^2. We can solve this equation and find two branches $F_1^{SN}(\delta)$ and $F_2^{SN}(\delta)$ of the saddle-node line

$$F_{1,2}^{SN}(\delta) = \sqrt{2}\left(\frac{(1 + 9\delta^2)}{27} \pm \sqrt{\frac{(1 + 9\delta^2)^2}{27^2} - \frac{\delta^2}{27}(\delta^2 + 1)^2}\right)^{1/2}, \tag{3.64}$$

taking the positive values of $F_{1,2}^{SN}$. F^{SN} is symmetric with respect to $\delta = 0$, and is shown in Fig. 3.7 by a solid line.

One can see that F^{SN} outlines a closed region in the (δ, F) plane. It can be easily shown, e.g., by trying just one point inside this area, that R is negative there. Thus there are three real roots of (3.46), implying three fixed points in (3.36)–(3.37), and hence there are three possible values of fixed amplitude of periodic oscillations of the original forced van der Pol oscillator (3.3). The analysis of the eigenvalues of the fixed points of (3.36)–(3.37) reveals that only one point is stable. Outside the region bounded by the saddle-node bifurcation line, R is positive, hence there is only one fixed point in (3.36)–(3.37).

There is also a special point in the (δ, F) parameter plane, corresponding to $C = D = 0$, at which all three roots of (3.51) coincide and equal to zero. This means that three fixed points of (3.36)–(3.37), and the three respective periodic orbits of (3.3), merge. From (3.52) it follows that

3.6 Analysis of Truncated Equations in Descartes Coordinates 41

$$3\delta^2 = 1, \qquad \delta = \pm\frac{1}{\sqrt{3}},$$

and the respective value of F from (3.53) is

$$F = \sqrt{\frac{18\delta^2 + 2}{27}} \bigg|_{\delta=1/\sqrt{3}} = \sqrt{\frac{8}{27}}.$$

Thus, the coordinates of this point are $(\delta, F) = (\pm 1/\sqrt{3}, \sqrt{8/27})$ for positive F. At these points the two branches of the saddle-node bifurcation F^{SN} line defined by (3.64) meet.

Derive the stability conditions for the fixed points of (3.36)–(3.37). First, calculate the derivatives of the functions f and g

$$f_u = \frac{\partial f}{\partial u} = 1 - 3u^2 - v^2, \qquad f_v = \frac{\partial f}{\partial v} = -\delta - 2uv,$$

$$g_u = \frac{\partial g}{\partial u} = \delta - 2uv, \qquad g_v = \frac{\partial g}{\partial v} = 1 - u^2 - 3v^2.$$

The characteristic equation for the eigenvalues $\mu_{1,2}$ of a fixed point is

$$\begin{vmatrix} (f_u - \mu) & f_v \\ g_u & (g_v - \mu) \end{vmatrix} = 0.$$

The eigenvalues $\mu_{1,2}$ are then expressed as

$$\mu_{1,2} = \frac{1}{2}\left[(f_u + g_v) \pm \sqrt{\tilde{D}}\right]. \tag{3.65}$$

Here,

$$\tilde{D} = (f_u - g_v)^2 + 4g_u f_v = 4(u^2 + v^2)^2 - 4\delta^2, \tag{3.66}$$

$$f_u + g_v = 2 - 4u^2 - 4v^2. \tag{3.67}$$

At point $(\delta, F) = (1/\sqrt{3}, \sqrt{8/27})$, $\mu_1 = 0$, $\mu_2 = -2/3$. Since one of the eigenvalues is indeed zero, this confirms that a saddle-node bifurcation occurs at this point.

Now, find the conditions for Andronov–Hopf bifurcation. We remind you that as a result of Andronov–Hopf bifurcation occurring to the (rescaled) components of the amplitude of periodic oscillations u and v of (3.36)–(3.37), the original forced van der Pol system (3.3) undergoes the bifurcation of a birth of a torus from a limit cycle, i.e., oscillations become quasiperiodic and synchronization is lost. When Andronov–Hopf bifurcation occurs, R in (3.58) is no longer zero. First, we have to express the solution (u, v) through u_{*1}, which is now the only real root of the cubic equation (3.51). From (3.58) it follows that

$$t^3 = \frac{1}{2}(D + \sqrt{R}).$$

42   3 1:1 Forced Synchronization of Periodic Oscillations

Then, from (3.56) we find s^3

$$s^3 = -D + t^3 = \frac{1}{2}(-D + \sqrt{R}).$$

Thus,

$$t = \frac{1}{\sqrt[3]{2}}(D + \sqrt{R})^{1/3}, \qquad s = \frac{1}{\sqrt[3]{2}}(-D + \sqrt{R})^{1/3}.$$

From (3.48) and (3.54) we find

$$u = \frac{1}{\sqrt[3]{2}}\left[(-D + \sqrt{R})^{1/3} - (D + \sqrt{R})^{1/3}\right] + \frac{2\delta}{3F}, \qquad (3.68)$$

D being defined by (3.53) and R by (3.59). Andronov–Hopf bifurcation occurs when the eigenvalues $\mu_{1,2}$ of the fixed point (u, v) become purely imaginary, which with account (3.66)–(3.67) means

$$f_u + g_v = 0, \qquad \tilde{D} < 0.$$

For convenience, express v^2 through u using (3.41) and substitute into (3.67) to obtain

$$f_u + g_v = 2 - \frac{4F}{\delta}u = 0,$$
$$u = \frac{\delta}{2F}. \qquad (3.69)$$

In (3.52)–(3.53) denote

$$D_1 = 18\delta^2 - 27F^2 + 2, \qquad C_1 = \left(3\delta^2 - 1\right), \qquad (3.70)$$

so that

$$D = \frac{\delta^3}{27F^3}D_1, \qquad C = \frac{\delta^2}{3F^2}C_1.$$

Then from (3.59)

$$R = \frac{\delta^6}{27^2 F^6}\left(D_1^2 + 4C_1^3\right).$$

Express u in (3.68) through D_1 and C_1 and substitute into (3.69)

$$\frac{1}{\sqrt[3]{2}}\frac{\delta}{3F}\left[(-D_1 + \sqrt{D_1^2 + 4C_1^3})^{1/3} - (D_1 + \sqrt{D_1^2 + 4C_1^3})^{1/3}\right] + \frac{2\delta}{3F} = \frac{\delta}{2F}.$$

Simplification gives

$$\left(D_1 + \sqrt{D_1^2 + 4C_1^3}\right)^{1/3} - \left(-D_1 + \sqrt{D_1^2 + 4C_1^3}\right)^{1/3} = \frac{\sqrt[3]{2}}{2}. \qquad (3.71)$$

3.6 Analysis of Truncated Equations in Descartes Coordinates

Take a cube of both parts

$$\left(\sqrt{D_1^2 + 4C_1^3} + D_1\right) - 3\left(\sqrt{D_1^2 + 4C_1^3} + D_1\right)^{2/3}\left(\sqrt{D_1^2 + 4C_1^3} - D_1\right)^{1/3}$$
$$+ 3\left(\sqrt{D_1^2 + 4C_1^3} + D_1\right)^{1/3}\left(\sqrt{D_1^2 + 4C_1^3} - D_1\right)^{2/3} - \left(\sqrt{D_1^2 + 4C_1^3} - D_1\right)$$
$$= \frac{1}{4}.$$

Simplify

$$2D_1 + 3\left(D_1^2 + 4C_1^3 - D_1^2\right)^{1/3}$$
$$\times \left[\left(\sqrt{D_1^2 + 4C_1^3} - D_1\right)^{1/3} - \left(\sqrt{D_1^2 + 4C_1^3} + D_1\right)^{1/3}\right] = \frac{1}{4}.$$

Substitute (3.71) to obtain

$$2D_1 - 3\sqrt[3]{4}C_1 \times \frac{\sqrt[3]{2}}{2} = \frac{1}{4}$$

or

$$2D_1 - 3C_1 = \frac{1}{4}.$$

Substitute D_1 and C_1 from (3.70)

$$2\left(18\delta^2 - 27F^2 + 2\right) - 3\left(3\delta^2 - 1\right) = \frac{1}{4},$$
$$36\delta^2 - 54F^2 + 4 - 9\delta^2 + 3 = \frac{1}{4},$$
$$27\delta^2 - 54F^2 = -\frac{27}{4},$$
$$4\delta^2 + 1 = 8F^2,$$
$$F^{\mathrm{HB}} = \sqrt{\frac{\delta^2}{2} + \frac{1}{8}}, \tag{3.72}$$

where F^{HB} denotes the Andronov–Hopf bifurcation line. With this, $\tilde{D}$ from (3.66) should be negative to enable eigenvalues $\mu_{1,2}$ being imaginary. Substitute v^2 from (3.41) into (3.66) for $\tilde{D}$

$$\tilde{D} = 4\left[\frac{F^2}{\delta^2}u^2 - \delta^2\right] < 0,$$

leading to

$$\delta^2 > |Fu|. \tag{3.73}$$

Substituting u from (3.69) into (3.73) gives

$$\delta^2 > \left|\frac{\delta}{2}\right| \quad \text{or} \quad |\delta| > \frac{1}{2}. \tag{3.74}$$

With account of (3.74), the curve F^{HB} defined by (3.72) has two branches at positive F: for $\delta > 0.5$ and for $\delta < -0.5$. F^{HB} is shown in Fig. 3.7 by dashed lines.

By the present point in the book, we have made a long journey: we started from the full equations for a forced van der Pol oscillator (3.3), arrived at the truncated equations for the amplitude and phase difference in Descartes coordinates (3.36)–(3.37), and then to (3.64) and (3.72), the latter equations forming the result we were looking for: the borderlines of the synchronization region. Remember that the truncated equations (3.36)–(3.37) were supposed to describe the dynamics of the nonautonomous van der Pol oscillator (3.3) around the 1 : 1 synchronization region. Figure 3.8 shows the synchronization tongues obtained numerically by applying continuation methods [73] to the analysis of bifurcations of periodic orbits in (3.3). In the same figure grey lines are the lines of bifurcations of the fixed points in (3.36)–(3.37). Graphs are plotted on the plane of parameters (Ω, B) of (3.3). Parameters (δ, F) were renormalized to (Ω, B) using (3.35). Figure 3.8(a) shows the graphs for non-linearity $\lambda = 0.1$, and (b) for $\lambda = 0.5$ of (3.3). The agreement between the numerical and analytical graphs is remarkable.

Note that truncated equations (3.36)–(3.37) are valid as long as the quasiharmonic approximation (3.5) chosen for the forced system is satisfactory. This approximation works better, the smaller the non-linearity λ is. Indeed, at $\lambda = 0.1 \ll 1$, the analytical graphs coincide with the numerical ones with a very high accuracy. $\lambda = 0.5$ is no longer much less than 1, the coincidence between the analytical and numerical graphs is somewhat less accurate, but is still remarkably good.

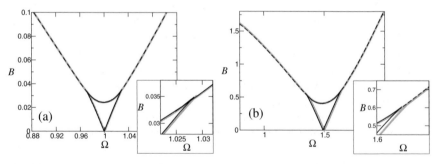

Fig. 3.8. 1 : 1 forced synchronization tongue on the plane (Ω, B) of (3.3). Lines of bifurcations of periodic orbits of (3.3) (numerical methods) are compared with those of the fixed points of (3.36)–(3.37) (analytical methods). Designations: saddle-node (*solid black line*) and torus birth (*dashed black line*) lines in (3.3). *Grey lines* show bifurcations in (3.36)–(3.37). *Insets* show enlarged segments of the diagrams in the vicinity of the special point $(\delta, F) = (1/\sqrt{3}, \sqrt{8/27})$. **a** $\lambda = 0.1$, $\omega_0 = 1$; **b** $\lambda = 0.5$, $\omega_0 = 1.5$

3.7 Synchronization Region from the Truncated Equations: Non-bifurcational Approach

In Sect. 3.6 the synchronization conditions were derived as a result of the analysis of relevant bifurcations. In experiment when one normally observes a single realization of a process, it is not straightforward to detect the occurrence of a bifurcation. Historically, experimentalists used to characterize the processes being observed by their frequencies. In this and subsequent sections, Fourier spectra of the realizations of the forced van der Pol system (3.3) will be estimated analytically by analyzing the truncated equations in the form (3.34).

One might ask "What other analysis do we need if in Sect. 3.6 the exact solutions of the truncated equations were already obtained?" Indeed, from the u and v found, one can go back to the polar form and obtain the amplitude A and phase difference φ explicitly using (3.29). Then A and φ can be substituted into (3.5) to obtain the approximate solution $x(t)$ to (3.3) that will be accurate enough for the range of parameters for which the truncated equations remain valid. Then if function $x(t)$ is known explicitly, one can introduce Fourier power spectral density. However, the problem is that the explicit accurate solution $x(t)$ will be described by a bulky function of time which is difficult to manipulate. And even if we finally obtain an expression for its spectrum, it will be cumbersome and quite difficult to interpret. Hence, the whole purpose of these calculations will be lost.

In this and subsequent sections a different approach will be used. Namely, when solving the truncated equations, further physically motivated approximations will be used in order to obtain a solution in a form that can be interpreted easily. Obviously, further approximations will introduce further inaccuracies into the solution. However, as long as the latter is capable of describing the main physical effects, it will be regarded as satisfactory. We will also compare the approximate solutions of the truncated equations with the results of direct numerical simulation of (3.3). Since the comparison of *exact* solutions with numerical simulations was already made in Fig. 3.8, this will allow one to assess the difference between the exact and the approximate solutions as well.

Approximate analytic solutions of truncated equations for large values of forcing strength B were presented by Landa in her book [160], where a special form of a weakly non-linear self-oscillator on a phase plane was considered. In that book, several special cases were treated analytically, including phase locking at B not vanishingly small, and suppression of natural dynamics. The calculations given in [160], although being quite sufficient to be understood by an expert, might still be lacking detail if considered by a beginner. Moreover, unfortunately, this book is not translated from Russian so far, and is thus unavailable for the non-Russian readers. We regard these results as being fundamental and crucial for the understanding of the phenomenon of phase synchronization, so here we give the full details of this analysis for the van der Pol oscillator (3.3), which is similar to the version of the oscillator considered in [160] but is written here in a simpler form for the convenience of presentation.

46   3 1:1 Forced Synchronization of Periodic Oscillations

We start from (3.34), where we denote

$$\tilde{F} = \frac{B}{4\sqrt{\lambda\Omega}},\tag{3.75}$$

i.e.,

$$\dot{u} = \frac{\lambda}{2}u\left[1 - \left(u^2 + v^2\right)\right] - v\Delta,$$
$$\dot{v} = \frac{\lambda}{2}v\left[1 - \left(u^2 + v^2\right)\right] + u\Delta - \tilde{F}.\tag{3.76}$$

These equations describe a non-linear system that is potentially able to demonstrate self-sustained periodic oscillations. In (3.76), λ is explicitly multiplied by the non-linear terms in the equations. If we assume that

$$2\Delta \gg \lambda \quad \text{and} \quad \tilde{F} \gg \lambda\tag{3.77}$$

then the non-linearity is weak, and λ becomes a "small parameter" of these equations. Then we can apply the approximate methods for the analysis of weakly non-linear oscillators, like the Bogoliubov–Krylov method. Note that the assumption (3.77) implies that both the detuning and strength of forcing must be sufficiently large and ultimately cannot be regarded as vanishingly small. In the null approximation, we can neglect the term with λ and analyze the linear equations

$$\dot{u} = -\Delta v,$$
$$\dot{v} = \Delta u - \tilde{F},$$

or

$$\ddot{u} + \Delta^2 u = \tilde{F}\Delta.\tag{3.78}$$

This is a linear ordinary differential equation (ODE) of the second order, which is non-homogeneous due to the non-zero right-hand part. A general solution of a linear non-homogeneous ODE can be found as a general solution u_h of a respective homogeneous ODE (by setting the right-hand part to zero) plus any particular solution u_n of the original non-homogeneous equation. The general solution of the homogeneous equation can be written down as

$$u_h(t) = \tilde{C}_1 e^{i\Delta t} + \tilde{C}_2 e^{-i\Delta t}.$$

The constants $\tilde{C}_1$ and $\tilde{C}_2$ could be found from initial conditions, and can generally be complex numbers. If we set

$$\tilde{C}_1 = \frac{C}{2i}e^{i\Psi}, \qquad \tilde{C}_2 = -\frac{C}{2i}e^{-i\Psi},$$

u_h can be simpler written as

$$u_h(t) = C\sin(\Delta t + \Psi).$$

3.7 Synchronization Region from the Truncated Equations: Non-bifurcational Approach 47

Next, we have to find a particular solution u_n for the non-homogeneous equation (3.78) It is enough to find the simplest solution, so we try a constant $u_n(t) = D$. By substituting D into the equation, we find that

$$u_n(t) = \frac{\tilde{F}}{\Delta}.$$

Hence, the general solution of (3.76) is

$$u(t) = C\sin(\Delta t + \Psi) + \frac{\tilde{F}}{\Delta}. \tag{3.79}$$

The variable v can be found as $v = -\frac{1}{\Delta}\dot{u}$, i.e.,

$$v(t) = -C\cos(\Delta t + \Psi). \tag{3.80}$$

Here, C and Ψ are constants depending on initial conditions. Thus, in the null approximation, (3.76) describes non-damped oscillations around a center point $(u_0, v_0) = (\tilde{F}/\Delta, 0)$ with amplitude C and frequency Δ. However, we are not happy with the null approximation, since λ is not zero, but a small number. Using the smallness of λ, we will be looking for a solution in the form of quasiharmonic oscillations with amplitude $C(t)$ and phase $\Psi(t)$, both functions changing slowly in time t as compared to $\cos(\Delta t)$, i.e.,

$$|\dot{C}(t)| \ll C(t)\Delta, \qquad \dot{\Psi}(t) \ll \Delta.$$

In order to find the solutions u and v that will describe the complex amplitude of forced oscillations, we need to write down the equations for C and Ψ and to solve them.

Then the derivatives of u and v are

$$\dot{u} = \dot{C}\sin(\Delta t + \Psi) + C\cos(\Delta t + \Psi)(\dot{\Psi} + \Delta),$$
$$\dot{v} = -\dot{C}\cos(\Delta t + \Psi) + C\sin(\Delta t + \Psi)(\dot{\Psi} + \Delta).$$

Note that

$$u^2 + v^2 = C^2\sin^2(\Delta t + \Psi) + \frac{\tilde{F}^2}{\Delta^2} + \frac{2C\tilde{F}}{\Delta}\sin(\Delta t + \Psi) + C^2\cos^2(\Delta t + \Psi)$$

$$= C^2 + \frac{\tilde{F}^2}{\Delta^2} + \frac{2C\tilde{F}}{\Delta}\sin(\Delta t + \Psi).$$

Substitute u, v, $\dot{u}$ and $\dot{v}$ into (3.76)

$$\dot{C}\sin(\Delta t + \Psi) + C\cos(\Delta t + \Psi)(\dot{\Psi} + \Delta)$$

$$= \frac{\lambda}{2}\left[1 - C^2 - \frac{\tilde{F}^2}{\Delta^2} - \frac{2C\tilde{F}}{\Delta}\sin(\Delta t + \Psi)\right]$$

$$\times \left[C\sin(\Delta t + \Psi) + \frac{\tilde{F}}{\Delta}\right] + \Delta C\cos(\Delta t + \Psi), \tag{3.81}$$

3 1:1 Forced Synchronization of Periodic Oscillations

$$-\dot{C}\cos(\Delta t + \Psi) + C\sin(\Delta t + \Psi)(\dot{\Psi} + \Delta)$$
$$= \frac{\lambda}{2}\left[1 - C^2 - \frac{\tilde{F}^2}{\Delta^2} - \frac{2C\tilde{F}}{\Delta}\sin(\Delta t + \Psi)\right]$$
$$\times [-C\cos(\Delta t + \Psi)] + \Delta\left[C\sin(\Delta t + \Psi) + \frac{\tilde{F}}{\Delta}\right] - \tilde{F}$$
$$= -\frac{\lambda}{2}\left[1 - C^2 - \frac{\tilde{F}^2}{\Delta^2} - \frac{2C\tilde{F}}{\Delta}\sin(\Delta t + \Psi)\right]$$
$$\times C\cos(\Delta t + \Psi) + \Delta C\sin(\Delta t + \Psi). \tag{3.82}$$

Multiply (3.81) by $\sin(\Delta t + \Psi)$

$$\dot{C}\sin^2(\Delta t + \Psi) + C\cos(\Delta t + \Psi)\sin(\Delta t + \Psi)(\dot{\Psi} + \Delta)$$
$$= \frac{\lambda}{2}\left[1 - C^2 - \frac{\tilde{F}^2}{\Delta^2} - \frac{2C\tilde{F}}{\Delta}\sin(\Delta t + \Psi)\right]$$
$$\times \left[C\sin^2(\Delta t + \Psi) + \frac{\tilde{F}}{\Delta}\sin(\Delta t + \Psi)\right]$$
$$+ \Delta C\sin(\Delta t + \Psi)\cos(\Delta t + \Psi),$$

and (3.82) by $\cos(\Delta t + \Psi)$

$$-\dot{C}\cos^2(\Delta t + \Psi) + C\sin(\Delta t + \Psi)\cos(\Delta t + \Psi)(\dot{\Psi} + \Delta)$$
$$= -\frac{\lambda}{2}\left[1 - C^2 - \frac{\tilde{F}^2}{\Delta^2} - \frac{2C\tilde{F}}{\Delta}\sin(\Delta t + \Psi)\right]C\cos^2(\Delta t + \Psi)$$
$$+ \Delta C\sin(\Delta t + \Psi)\cos(\Delta t + \Psi).$$

Subtract the latter equation from the previous one to obtain

$$\dot{C} = \frac{\lambda}{2}\left[1 - C^2 - \frac{\tilde{F}^2}{\Delta^2} - \frac{2C\tilde{F}}{\Delta}\sin(\Delta t + \Psi)\right]\left[C + \frac{\tilde{F}}{\Delta}\sin(\Delta t + \Psi)\right].$$

Next, multiply (3.81) by $\cos(\Delta t + \Psi)$

$$\dot{C}\sin(\Delta t + \Psi)\cos(\Delta t + \Psi) + C\cos^2(\Delta t + \Psi)(\dot{\Psi} + \Delta)$$
$$= \frac{\lambda}{2}\left[C\sin(\Delta t + \Psi)\cos(\Delta t + \Psi) + \frac{\tilde{F}}{\Delta}\cos(\Delta t + \Psi)\right]$$
$$\times \left[1 - C^2 - \frac{\tilde{F}^2}{\Delta^2} - \frac{2C\tilde{F}}{\Delta}\sin(\Delta t + \Psi)\right] + \Delta C\cos^2(\Delta t + \Psi),$$

and (3.82) by $\sin(\Delta t + \Psi)$

$$-\dot{C}\cos(\Delta t + \Psi)\sin(\Delta t + \Psi) + C\sin^2(\Delta t + \Psi)(\dot{\Psi} + \Delta)$$
$$= -\frac{\lambda}{2}\left[1 - C^2 - \frac{\tilde{F}^2}{\Delta^2} - \frac{2C\tilde{F}}{\Delta}\sin(\Delta t + \Psi)\right]C\cos(\Delta t + \Psi)\sin(\Delta \tau + \Psi)$$
$$+ \Delta C\sin^2(\Delta t + \Psi).$$

3.7 Synchronization Region from the Truncated Equations: Non-bifurcational Approach 49

Add the previous and the latter equations together to obtain

$$C(\dot{\Psi} + \Delta) = \frac{\lambda}{2}\left[1 - C^2 - \frac{\tilde{F}^2}{\Delta^2} - \frac{2C\tilde{F}}{\Delta}\sin(\Delta t + \Psi)\right]\frac{\tilde{F}}{\Delta}\cos(\Delta t + \Psi) + \Delta C.$$

We thus obtain a system of two equations for C and Ψ

$$\dot{C} = \frac{\lambda}{2}\left[1 - C^2 - \frac{\tilde{F}^2}{\Delta^2} - \frac{2C\tilde{F}}{\Delta}\sin(\Delta t + \Psi)\right]\left[C + \frac{\tilde{F}}{\Delta}\sin(\Delta t + \Psi)\right],$$

$$\dot{\Psi} = \frac{\lambda}{2}\left[1 - C^2 - \frac{\tilde{F}^2}{\Delta^2} - \frac{2C\tilde{F}}{\Delta}\sin(\Delta t + \Psi)\right]\frac{\tilde{F}}{\Delta C}\cos(\Delta t + \Psi). \tag{3.83}$$

These are still quite complicated equations that are not easier to analyze than (3.76) or (3.21)–(3.22). However, here we can use the fact that $C(t)$ and $\Psi(t)$ are slow functions of t and to average these equations over one period $T = 2\pi/\Delta$ of fast oscillations, similarly to what we did when deriving the truncated equations. Regrouping the terms in (3.83) gives

$$\dot{C} = \frac{\lambda C}{2}\left[1 - C^2 - \frac{\tilde{F}^2}{\Delta^2}\right] + \frac{\lambda C^2 \tilde{F}}{\Delta}\sin(\Delta t + \Psi)$$

$$+ \frac{\lambda \tilde{F}}{2\Delta}\left[1 - C^2 - \frac{\tilde{F}^2}{\Delta^2}\right]\sin(\Delta t + \Psi) - \frac{\lambda C\tilde{F}^2}{2\Delta^2}(1 - \cos(2(\Delta t + \Psi)))$$

and

$$\dot{\Psi} = \frac{\lambda \tilde{F}}{2\Delta C}\left[1 - C^2 - \frac{\tilde{F}^2}{\Delta^2}\right]\cos(\Delta t + \Psi) - \frac{\lambda \tilde{F}^2}{2\Delta^2}\sin(2(\Delta t + \Psi)).$$

It is clear that averages over $T = 2\pi/\Delta$ of all terms containing $\sin(\Delta t + \Psi)$, $\cos(\Delta t + \Psi)$, $\sin(2(\Delta t + \Psi))$ and $\cos(2(\Delta t + \Psi))$ are equal to zero. The resulting averaged equations for C and Ψ read

$$\dot{C} = \frac{\lambda C}{2}\left[1 - C^2 - \frac{\tilde{F}^2}{\Delta^2}\right] - \frac{\lambda C\tilde{F}^2}{2\Delta^2} = f(C), \tag{3.84}$$

$$\dot{\Psi} = 0. \tag{3.85}$$

The fixed points of the system above are

$$C_1 = 0, \qquad C_2 = \sqrt{1 - \frac{2\tilde{F}^2}{\Delta^2}}.$$

C_2 exists when $\Delta^2 \geq 2\tilde{F}^2$, and it is stable since the derivative

$$\frac{\mathrm{d}f(C)}{\mathrm{d}C}\bigg|_{C_2} = -\lambda\left[1 - \frac{2\tilde{F}^2}{\Delta^2}\right]$$

is negative at this point. Thus, the range of parameters at which the stable amplitude C is non-zero corresponds to the absence of synchronization. The borderline

50 3 1:1 Forced Synchronization of Periodic Oscillations

of synchronization region will be defined by the equality

$$\Delta^2 = 2\tilde{F}^2, \qquad \Delta = \pm\sqrt{2}|\tilde{F}| = \pm\sqrt{2}\frac{B}{4\sqrt{\lambda}\Omega},$$

$$B = |\Delta|2\sqrt{2\lambda}\Omega.$$

With account of (3.20), B becomes

$$B = \sqrt{2\lambda}|\omega_0^2 - \Omega^2|. \tag{3.86}$$

This will be the borderline of synchronization region when the strength of forcing B is not very small.

3.8 Fourier Power Spectra at Strong Forcing

Outside synchronization region the amplitude C of (3.84) is not zero. Recall that oscillations in the forced system are in the form $x = A\cos(\Omega t + \varphi)$, which after simple transformations and using (3.29) and (3.33), reduces to

$$x = A\cos\varphi\cos\Omega t - A\sin\varphi\sin\Omega t = \bar{u}\cos\Omega t - \bar{v}\sin\Omega t$$

$$= 2\sqrt{\lambda}\left(C\sin(\Delta t + \Psi) + \frac{\tilde{F}}{\Delta}\right)\cos\Omega t + 2\sqrt{\lambda}C\cos(\Delta t + \Psi)\sin\Omega t.$$

Use trigonometric identities to express sums of sines and cosines in the calculations above

$$x = 2\sqrt{\lambda}C(\sin\Delta t\cos\Psi + \cos\Delta t\sin\Psi)\cos\Omega t$$

$$+ \frac{2\sqrt{\lambda}\tilde{F}}{\Delta}\cos\Omega t + 2\sqrt{\lambda}C(\cos\Delta t\cos\Psi - \sin\Delta t\sin\Psi)\sin\Omega t.$$

Now, use the approximation $\Delta \approx (\omega_0 - \Omega)$ for Ω sufficiently close to ω_0:

$$x = 2\sqrt{\lambda}C\big[(\sin\omega_0 t\cos\Omega t - \cos\omega_0 t\sin\Omega t)\cos\Psi$$

$$+ (\cos\omega_0 t\cos\Omega t + \sin\omega_0 t\sin\Omega t)\sin\Psi\big]\cos\Omega t$$

$$+ \frac{2\sqrt{\lambda}\tilde{F}}{\Delta}\cos\Omega t + 2\sqrt{\lambda}C\big[(\cos\omega_0 t\cos\Omega t + \sin\omega_0 t\sin\Omega t)\cos\Psi$$

$$- (\sin\omega_0 t\cos\Omega t - \cos\omega_0 t\sin\Omega t)\sin\Psi\big]\sin\Omega t$$

$$= \frac{2\sqrt{\lambda}\tilde{F}}{\Delta}\cos\Omega t$$

$$+ 2\sqrt{\lambda}C\big[\sin\omega_0 t\cos^2\Omega t\cos\Psi - \cos\omega_0 t\sin\Omega t\cos\Psi\cos\Omega t$$

$$+ \cos\omega_0 t\cos^2\Omega t\sin\Psi + \sin\omega_0 t\sin\Omega t\sin\Psi\cos\Omega t$$

$$+ \cos\omega_0 t\cos\Omega t\cos\Psi\sin\Omega t + \sin\omega_0 t\sin^2\Omega t\cos\Psi$$

$$- \sin\omega_0 t\cos\Omega t\sin\Psi\sin\Omega t + \cos\omega_0 t\sin^2\Omega t\sin\Psi\big]$$

$$= \frac{2\sqrt{\lambda}\tilde{F}}{\Delta}\cos\Omega t + 2\sqrt{\lambda}C[\sin\omega_0 t\cos\Psi + \cos\omega_0 t\sin\Psi].$$

3.8 Fourier Power Spectra at Strong Forcing   51

Finally,

$$x = A_1 \cos \Omega t + A_2 \sin(\omega_0 t + \Psi),\tag{3.87}$$

where

$$A_1 = \frac{2\sqrt{\lambda}\tilde{F}}{\Delta} = \frac{B}{2\Delta\Omega},$$

$$A_2 = 2\sqrt{\lambda}C_2 = 2\sqrt{\lambda}\sqrt{1 - \frac{2\tilde{F}^2}{\Delta^2}} = 2\sqrt{\lambda}\sqrt{1 - \frac{2A_1^2}{A_0^2}},$$

where A_0 is the amplitude of unperturbed oscillations $A_0 = 2\sqrt{\lambda}$ (see Sect. 3.3). Thus, the resulting oscillations outside the synchronization region consist of two terms: one at the frequency Ω of forcing, and another at the frequency ω_0 of unperturbed oscillations. These oscillations can be classified as quasiperiodic. It is clear that the larger the strength of forcing B is, the larger the component at Ω, and the smaller the component at ω_0. The component at ω_0 vanishes when the forcing becomes sufficiently strong, i.e., when the following condition is satisfied:

$$\frac{2A_1^2}{A_0^2} = 1, \quad 2\left[\frac{B}{(\omega_0^2 - \Omega^2)}\right]^2 \frac{1}{A_0^2} = 1, \quad B = \frac{A_0}{\sqrt{2}}|\omega_0^2 - \Omega^2|,$$
$$B = \sqrt{2\lambda}|\omega_0^2 - \Omega^2|.\tag{3.88}$$

The latter relationship is the equation defining the borderline of the synchronization region. Note that the original assumption (3.77) under which this equation was derived implies that B is large and thus the forcing is strong. This will be the region of suppression of natural dynamics by the external forcing.

In Fig. 3.9(a), (b) the numerically calculated $1:1$ synchronization tongues (solid and dashed lines) are compared with the estimate of the border of the suppression region given by (3.88) (shaded), for two sets of values of (ω_0, λ): $\omega_0 = 1, \lambda = 0.1$, and $\omega_0 = 1, \lambda = 0.5$. For the convenience of comparison, the analytic line (3.88) is given for the whole range of Ω and B, not only for selected values of B that are large enough. The dashed lines show the borderlines of the synchronization regions formed by the lines of Neimark–Sacker (torus birth) bifurcation. As predicted, the larger the value of B, the better the agreement is between the theoretical prediction of (3.88) and the Neimark–Sacker bifurcation line. Also, as usual with the methods exploiting the weakness of non-linearity, the smaller the value of λ, the better the analytic prediction is.

The lines shown by filled circles in Fig. 3.9 deserve special attention. As one leaves the region of suppression by crossing the torus birth bifurcation line (dashed line), in the original forced system (3.3) a torus is born from the stable limit cycle. In truncated equations this transition is associated with the birth of a limit cycle from a fixed point via Andronov–Hopf bifurcation in coordinates "amplitude A"–"phase difference φ." Importantly, the cycle in (A, φ) is born with zero amplitude which gradually increases as one goes further away from the synchronization region. As soon as this cycle is born, the synchronization is lost, since the behavior of the full

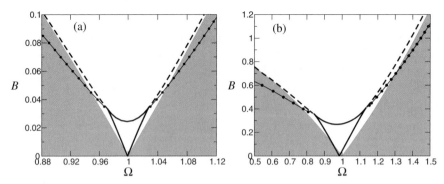

Fig. 3.9. 1 : 1 synchronization tongue for the forced van der Pol system (3.3) on the plane of forcing parameters Ω, B at **a** $\omega_0 = 1$ and $\lambda = 0.1$, **b** $\omega_0 = 1$ and $\lambda = 0.5$. *Solid lines* mark saddle-node bifurcations, *dashed lines* mark torus birth (Neimark–Sacker) bifurcations, *filled circles* mark the line at which the phase φ in truncated equations (3.21)–(3.22) start to grow or decay unboundedly, all lines being estimated numerically. *Shaded area* shows analytical prediction of the suppression region according to (3.88)—in agreement with the assumption, for small B this formula is not very accurate, while it works quite well for larger B

system has become quasiperiodic. However, for a range of Ω outside, but close to, the suppression region the absolute value of phase difference $|\varphi|$ does *not* grow unboundedly, but oscillates around the fixed point, i.e., belongs to a limit cycle. This behavior is called the "phase trapped" solution in [35]. Only on the line marked by filled circles in Fig. 3.9 does this cycle undergo a non-local bifurcation and disappears, as a result of which $|\varphi|$ starts its unbounded growth. According to [35], this is called "phase drift."

For the realizations $x(t)$ of the forced oscillations that at large B are approximately described by (3.87), we can introduce the correlation function $K[X, X_\tau]$ and the power spectral density $S(\omega)$. Strictly speaking, the latter functions are defined for random processes, i.e., processes whose realizations are random functions of time. This means that as the same experiment is repeated several times, each trial produces a different realization of the process. The realization described by (3.87) is a deterministic function of time, and hence does not describe a random process. However, in the future we will consider synchronization between random processes, and it will be convenient to compare the effects in deterministic systems with the ones oscillating randomly. Thus, here we will introduce the basic functions describing random processes.

We bear in mind that a deterministic process can be formally regarded as a special case of a random process. Also, we will assume that the processes we consider are already well settled down, and can be regarded as stationary, at least in the wide sense.[5] Roughly speaking, stationarity means that the statistical characteristics of the process do not change in time; e.g., the mean value that is introduced as an average over the ensemble of all realizations of the same random process takes the same

---

[5] For the introductory remarks into the theory of random processes see Sect. 7.1.

3.8 Fourier Power Spectra at Strong Forcing   53

value at any time moment. Our last assumption is that the processes we consider are in addition ergodic. This means that averaging over the ensemble of realizations can be substituted by averaging over time; e.g., mean value $\langle X(t) \rangle$ of the process $X(t)$ that is defined as an average of all realizations at the given time moment, and is a constant for a stationary process, is equal to the value $\bar{x}$ that is defined as an average over time of a particular realization $x(t)$ of the process. Namely,

$$\langle X \rangle = \bar{x},$$
$$\bar{x} = \lim_{T \to \infty} \frac{1}{2T} \int_{-T}^{T} x(t) \, dt. \tag{3.89}$$

For the realization defined by (3.87), $\bar{x} = 0$. Note that stationarity is a necessary condition for ergodicity, but not a sufficient one.

A correlation function $K[X, X_\tau]$ of a random process $X(t)$ is introduced as

$$K[X, X_\tau] = \int_{-\infty}^{\infty} \int_{-\infty}^{\infty} x x_\tau \, p_2^{XX}(x, t, x_\tau, t + \tau) \, dx \, dx_\tau, \tag{3.90}$$

where x and x_τ are the values of the realizations of the random process at the time moments t and $t + \tau$, respectively. Function $p_2^{XX}(x, t, x_\tau, t + \tau)$ is the two-dimensional probability density distribution describing the probability of the two events taking place simultaneously: that the realization $x(t)$ of the process X takes the value from $[x; x + \Delta x]$ at time t, and the value $[x_\tau; x_\tau + \Delta x_\tau]$ at time $t + \tau$. This is the general definition that is applicable to a random process with arbitrary properties. If the process is wide-sense stationary, the correlation $K[X, X_\tau]$ does not depend on time t: it depends only on τ which is the temporal distance between the two events considered. If the process is in addition ergodic, its correlation can be estimated by means of averaging over time of a single realization $x(t)$ of the process $X(t)$ as follows:

$$K[X, X_\tau] = \lim_{T \to \infty} \frac{1}{2T} \int_{-T}^{T} x x_\tau \, dt. \tag{3.91}$$

We estimate the correlation for the process described by (3.87)

$$K[X, X_\tau] = \lim_{T \to \infty} \frac{1}{2T} \int_{-T}^{T} [A_1 \cos \Omega t + A_2 \sin(\omega_0 t + \Psi)]$$

$$\times [A_1 \cos(\Omega t + \Omega \tau) + A_2 \sin(\omega_0 t + \omega_0 \tau + \Psi)] \, dt$$

$$= \lim_{T \to \infty} \frac{1}{2T} \int_{-T}^{T} [A_1^2 \cos \Omega t \cos(\Omega t + \Omega \tau)$$

$$+ A_1 A_2 \cos \Omega t \sin(\omega_0 t + \omega_0 \tau + \Psi)$$

$$+ A_1 A_2 \sin(\omega_0 t + \Psi) \cos(\Omega t + \Omega \tau)$$

$$+ A_2^2 \sin(\omega_0 t + \Psi) \sin(\omega_0 t + \omega_0 \tau + \Psi)] \, dt.$$

54   3   1:1 Forced Synchronization of Periodic Oscillations

Using trigonometric identities, we represent the products of sines and cosines through the sums of sines and cosines. Continuing from above,

$$
\begin{aligned}
K[X, X_\tau] = \lim_{T \to \infty} \frac{1}{2T} \int_{-T}^{T} \Bigg\{ &\frac{A_1^2}{2}[\cos \Omega\tau + \cos(2\Omega t + \Omega\tau)] \\
&+ \frac{A_1 A_2}{2}[\sin(\Omega t + \omega_0 t + \omega_0 \tau + \Psi) \\
&\qquad - \sin(\Omega t - \omega_0 t - \omega_0 \tau - \Psi)] \\
&+ \frac{A_1 A_2}{2}[\sin(\omega_0 t + \Psi + \Omega t + \Omega\tau) \\
&\qquad + \sin(\omega_0 t + \Psi - \Omega t - \Omega\tau)] \\
&+ \frac{A_2^2}{2}[\cos(\omega_0 \tau) - \cos(2\omega_0 t + \omega_0 \tau + 2\Psi)] \Bigg\} \, dt.
\end{aligned}
$$

After the integration and taking the limit as $T \to \infty$, we obtain the following expression for $K[X, X_\tau]$:

$$
K[X, X_\tau] = \frac{A_1^2}{2} \cos \Omega\tau + \frac{A_2^2}{2} \cos(\omega_0 \tau). \tag{3.92}
$$

Fourier power spectral density (spectrum) $S(\omega)$ of the random process is introduced by Wiener–Khintchine[6] theorem as a Fourier Transform (FT) of its correlation function, i.e.,

$$
S(\omega) = \mathcal{F}\{K[X, X_\tau]\} = \int_{-\infty}^{\infty} K[X, X_\tau]e^{-i\omega\tau} \, d\tau. \tag{3.93}
$$

Substitute (3.92) into (3.93)

$$
\begin{aligned}
\mathcal{F}\{K[X, X_\tau]\} &= \int_{-\infty}^{\infty} \left[\frac{A_1^2}{2} \cos \Omega\tau + \frac{A_2^2}{2} \cos(\omega_0 \tau) \right] e^{-i\omega t} \, d\tau \\
&= \frac{A_1^2}{2} \int_{-\infty}^{\infty} \frac{e^{i\Omega\tau} + e^{-i\Omega\tau}}{2} e^{-i\omega\tau} \, d\tau \\
&\quad + \frac{A_2^2}{2} \int_{-\infty}^{\infty} \frac{e^{i\omega_0\tau} + e^{-i\omega_0\tau}}{2} e^{-i\omega\tau} \, d\tau. \tag{3.94}
\end{aligned}
$$

In the equations above, cosine is represented through exponents using a Euler formula. Before calculating the above FT, we need to introduce the following identity.

---

[6] The name of the Russian scientist Khintchine is also spelled in literature as Khintchin, Khinchin or Hinchin.

3.8 Fourier Power Spectra at Strong Forcing 55

First, consider a Dirac delta-function $\delta(\omega)$ defined as

$$\delta(\omega) = \begin{cases} 0, & \omega \neq 0, \\ \infty, & \omega = 0, \end{cases} \qquad \int_{-\infty}^{\infty} \delta(\omega)\, d\omega = 1. \tag{3.95}$$

The property of this function is

$$\int_{-\infty}^{\infty} f(x)\delta(x-a)\, dx = f(a). \tag{3.96}$$

The inverse FT of delta-function with account of the property (3.96) is

$$\frac{1}{2\pi} \int_{-\infty}^{\infty} \delta(\omega) e^{i\omega\tau}\, d\omega = \frac{1}{2\pi} e^{i\omega\tau} \bigg|_{\omega=0} = \frac{1}{2\pi}. \tag{3.97}$$

Hence, $\delta(\omega)$ is a FT of the function $1/(2\pi)$, i.e.,

$$\int_{-\infty}^{\infty} \frac{1}{2\pi} e^{-i\omega\tau}\, d\tau = \delta(\omega). \tag{3.98}$$

Second, consider a FT of an exponential function $e^{i\Omega\tau}$

$$\mathcal{F}\big[e^{i\Omega\tau}\big] = \int_{-\infty}^{\infty} e^{i\Omega\tau} e^{-i\omega\tau}\, d\tau = \int_{-\infty}^{\infty} e^{-i(\omega-\Omega)\tau}. \tag{3.99}$$

Comparing (3.99) with (3.98), we conclude that

$$\mathcal{F}\big[e^{i\Omega\tau}\big] = 2\pi\delta(\omega - \Omega). \tag{3.100}$$

By analogy, the FT of $e^{i\omega_0\tau}$ is

$$\mathcal{F}\big[e^{i\omega_0\tau}\big] = 2\pi\delta(\omega - \omega_0). \tag{3.101}$$

Thus, with account of (3.100) and (3.101), the spectrum of the process (3.87) is

$$S(\omega) = A_1^2\pi[\delta(\omega - \Omega) + \delta(\omega + \Omega)] + A_2^2\pi[\delta(\omega - \omega_0) + \delta(\omega + \omega_0)]. \tag{3.102}$$

The power spectral densities are symmetric with respect to zero frequency, and thus are normally plotted only for the positive frequencies. Spectra are the tools commonly used in experiments in order to characterize the process. According to the approximate estimates above, the spectrum of forced oscillations with large amplitudes of forcing consists of two delta-peaks placed at the frequency of forcing Ω and at the natural frequency of oscillations ω_0. As the forcing strength B grows, the positions of the peaks do not change, while their heights change: the peak at $\omega = \Omega$ grows, and the peak at $\omega = \omega_0$ decreases and finally vanishes.

In the next section we will illustrate the evolution of spectra and other useful characteristics of the forced oscillations by means of numerical simulations of (3.3). We will associate the bifurcations in the original forced van der Pol system with the spectra of its realizations.

3.9 Phase Locking and Suppression: Numerical Simulation

In this section, we illustrate the two mechanisms for phase (frequency) synchronization using the common experimentally accessible methods: realizations, stroboscopic sections[7] and Fourier power spectral densities (spectra). This illustration will provide the link between the mathematical and the physical languages that can be used to describe the same phenomena of forced synchronization of periodic oscillations.

3.9.1 Phase Locking

We start with considering phase locking mechanism of synchronization. It is realized as one enters the synchronization region at small amplitudes of forcing, and crosses the line of saddle-node bifurcation that generally forms the lower part of synchronization region. In particular, we will consider the $1:1$ synchronization tongue in Fig. 3.10(a) (the same as Fig. 3.9(b)) which describes the case of a forced van der Pol system (3.3) with non-linearity parameter $\lambda = 0.5$. The saddle-node bifurcation line is marked by solid line here. Let us fix the forcing frequency $\Omega = 1.05$, and gradually increase the forcing strength B from 0 to 0.4. The evolution of the stroboscopic section is shown in the first column of Fig. 3.11, the realizations of forcing and of the forced system in the second column, and the spectra in the third column. Each row describes the same forcing strength B, whose value is given to the right of the respective row.

In the absence of forcing $B = 0$, the stroboscopic section is not defined. But bearing in mind that with forcing, stroboscopic section is equivalent to Poincaré section, we would like to artificially extend this analogy to the case when the forcing is absent in order to illustrate the transitions. Without forcing, the Poincaré section would be a fixed point, so we would like to show the fixed point in the stroboscopic section, too. Note that the stroboscopic section consists of the phase points $(\dot{x}, x)$ taken at the time moments t_i when the values of the phase of external forcing $\psi_f(t) = \Omega t$ are equal to

$$\phi_f(t_i) = \psi_f^0 + 2\pi i, \quad i = 0, 1, 2, \ldots, \tag{3.103}$$

i.e., this section depends on the choice of the constant ψ_f^0. At different values of ψ_f^0 we get different sections, although all will be topologically equivalent. To illustrate the situation without forcing, we would not like to place the point arbitrarily, but

---

[7] Stroboscopic section is the set of points of the phase trajectory taken in a period of external forcing, i.e., in the case of (3.3) with the time step $2\pi/\Omega$. It is topologically equivalent to the Poincaré section of the periodically forced system: a periodic orbit is shown as a point, and a torus is shown as a closed curve. In experiments with forced systems, stroboscopic section is often preferred to Poincaré section because one does not have to care about choosing the proper Poincaré secant surface, which makes it easier to introduce. A drawback of the stroboscopic section approach is that in the absence of forcing it is not defined, while the Poincaré section is defined in any case.

3.9 Phase Locking and Suppression: Numerical Simulation

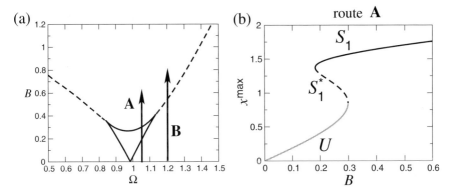

Fig. 3.10. a The vicinity of 1:1 synchronization region of the forced van der Pol oscillator with $\lambda = 0.5$, $\omega_0 = 1$, on the plane of parameters "forcing frequency Ω"–"forcing amplitude B." **b** Evolution of periodic orbits along route **A** is illustrated: the maxima of x are shown versus B. *Solid black line*: stable cycle S; *dashed line*: saddle cycle S^*; *grey line*: unstable cycle U (repeller)

at a position that is meaningful. So, we assume that if the forcing is vanishingly small, the section is going to be very similar to that without forcing. Of course, with forcing, strictly speaking, it is going to be a circle, but of such a small diameter that it is virtually indistinguishable from a point. For all stroboscopic sections of this section, we arbitrarily choose $\psi_f^0 = 4.328$.

The point shown in Fig. 3.11, first row, first column, is the result of computation with vanishingly small B, and it symbolizes the position of the fixed point in the stroboscopic section without forcing. In the first row, second column of Fig. 3.11, the realization of x is given by the solid grey line, without the forcing which is absent. One can see that the oscillations of the system are strictly periodic. Note that since λ is not close to zero here, the system realization is not harmonic which is clearly visible in the figure. In the first row, third column, the spectrum is shown by a solid black line. It contains one peak at the frequency of natural oscillations in the system.[8] The frequency of external periodic forcing is shown by a dashed line. One can see that the frequency of the unforced self-sustained oscillations in the system is different from the frequency of forcing.

As the forcing strength is increased from zero, the oscillations in the system become quasiperiodic ($B = 0.15$, Fig. 3.11, second row). The stroboscopic section is a stable ergodic torus[9] shown by a closed black curve, and the periodic orbit inside

---

[8] The spectra of periodic or quasiperiodic oscillations are discrete, i.e., consist of delta-functions, as shown in Sect. 3.8. However, the spectra shown in this section are estimated numerically from the realizations of finite duration, and because of that are continuous functions of frequency. The positions of the peaks of the numerically estimated spectra coincide with the positions of the respective "true" spectra within numerical accuracy.

[9] The torus densely filled by phase trajectories is often called ergodic. This is opposed to a resonant torus, on whose surface there exist stable and unstable periodic orbits and hence the phase trajectories do not fill the whole torus surface.

58   3 1:1 Forced Synchronization of Periodic Oscillations

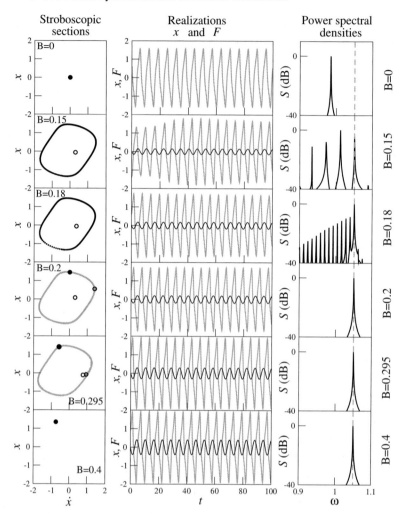

Fig. 3.11. Illustration of 1 : 1 frequency (phase) locking for the forced van der Pol system (3.3) at $\lambda = 0.5$. In Fig. 3.10(a) we move along route **A** defined by $\Omega = 1.05$ as we enter the synchronization tongue via saddle-node bifurcation line. Stroboscopic sections, realizations and power spectral densities (spectra) are shown for each value of B indicated to the right of each row. **First column**: *Black line (circle)*—stable tori (cycles S), *grey line*—resonant tori, *grey circle*—saddle cycles S^*, *white circle*—unstable cycle U (see Fig. 3.10(b) for reference). **Second column**: *Black line*—$x(t)$, *grey line*—force $F(t)$. **Third column**: *Black line*—spectrum of x in (3.3), *vertical dashed line* shows the position of the forcing frequency $\Omega = 1.05$

(white circle) has become unstable. The realization of oscillations has an amplitude that changes in time. The spectrum has several peaks: the highest peak, whose frequency can be called "main frequency" and is associated with the frequency of forced oscillations in the system, the peak at the frequency of forcing (at the dashed

3.9 Phase Locking and Suppression: Numerical Simulation 59

line), and peaks at the combinations of the main frequency and the frequency of forcing (combination frequencies). Note that the main frequency has shifted towards the frequency of forcing as compared to the frequency of the unforced oscillations.

As B is increased to the value 0.18, which is very close to the outer boundary of the phase locking region, the main frequency has coincided with the frequency of forcing (Fig. 3.11, third row). However, the oscillations remain quasiperiodic, which is detected by the presence of spectrum peaks at combination frequencies, by the amplitude modulation of the realization, and by the stroboscopic section in the form of a closed curve. This is not frequency (phase) locking yet.

As B reaches the value of 0.2, the oscillations in the system become periodic with the frequency coinciding with the one of external forcing Ω (Fig. 3.11, fourth row). The stroboscopic section is a stable fixed point (black circle) which lies on the surface of a resonant torus (closed grey line in the section) together with a saddle fixed point (grey circle). This pair of fixed points was born on the torus surface as a result of a saddle-node bifurcation, as was analytically predicted in Sects. 3.4 and 3.6. Inside the torus, there is the same unstable fixed point (white circle) as at $B = 0.15$ and $B = 0.18$. This is the frequency (phase) locked regime.

As B is increased further (Fig. 3.11, fifth row), the stable and saddle fixed points in the stroboscopic section move further away from each other while staying on the surface of the same torus (see row corresponding to $B = 0.295$). At the same time, the unstable fixed point inside the closed curve (white circle) comes closer to the saddle fixed point on the torus surface (grey circle). On the upper line of saddle-node bifurcation in Fig. 3.10(a) they merge and disappear, and this leads to the disappearance of the whole torus surface. As a result of this bifurcation ($B = 0.4$), the stable limit cycle represented with a fixed point (black circle) in the section remains the only object in the phase space of the forced system.

In order to summarize bifurcational transitions as one enters the locking region, let us return to the bifurcation diagram around the $1:1$ synchronization region in Fig. 3.10(a) and consider evolution of all periodic orbits in the system while moving along route **A**, i.e., by fixing $\Omega = 1.05$ and changing B from zero to 0.6. The respective one-dimensional bifurcation diagram is given in Fig. 3.10(b) where the maxima of x for all three periodic orbits are shown depending on B. For more detailed illustrations of the key moments, one can also compare this with Fig. 3.11. At $B = 0$, there is an unstable fixed point in the system which is denoted by U. At $B = 0$, U undergoes Andronov–Hopf bifurcation and an unstable periodic orbit is born from it which we will continue to denote as U (grey line). At small B there are no other periodic orbits, and the full system oscillates quasiperiodically. At $B \approx 0.1825$ a pair of cycles is born via saddle-node bifurcation: one stable S_1 and one saddle S_1^*. This event signifies the entrance to phase (frequency) locking region. As B achieves the value of 0.2988, saddle cycle S_1^* collides with unstable cycle U and both cycles vanish through saddle-node bifurcation. This event marks the transition from the region of locking to the region of suppression of natural dynamics. As B is increased further, there is only one stable cycle S in the system.

3.9.2 Suppression of Natural Dynamics

We continue with considering suppression of natural dynamics mechanism of synchronization. It is realized as one enters the synchronization region at relatively large amplitudes of forcing,[10] and crosses the line of torus birth bifurcation that generally forms the upper part of synchronization region. We will continue considering the $1:1$ synchronization tongue in Fig. 3.10(a), where the torus birth bifurcation line is marked by the dashed line.

Let us fix the forcing frequency $\Omega = 1.2$ (route **B**), which is noticeably bigger than the one chosen for illustration of phase locking. We gradually increase the forcing strength B from 0 to 0.54. The evolution of the stroboscopic section is shown in the first column of Fig. 3.12, the realizations of forcing and of the forced system in the second column, and the spectra in the third column. Each row describes the same forcing strength B whose value is given to the right of the respective row.

In the absence of forcing, the position of the fixed point is determined by the same method as was used to illustrate phase locking at $B = 0$ (Fig. 3.12, first row, first column). Oscillations are strictly periodic, although not harmonic (first row, second column). The spectrum contains just one peak at the natural frequency of oscillations, and the forcing frequency Ω is quite different from that (third column).

As the forcing increases from zero ($B = 0.2$), the oscillations become quasiperiodic (second row): the stroboscopic section is a closed curve, and the fixed point, that was stable without forcing, has become unstable and is shown by the white circle. The spectrum of quasiperiodic oscillations contains the highest peak (corresponds to the main frequency), the peak at the frequency of forcing $\Omega = 1.2$, and the peaks at combination frequencies. This part is very similar to what happened as we were considering phase locking above (compare with Fig. 3.11, second row).

As the forcing strength becomes larger and reaches $B = 0.4$ (third row), so that we approach the torus birth line in Fig. 3.10(a), the picture is qualitatively the same as with $B = 0.2$. However, important quantitative changes can be observed: the diameter of the ergodic torus became smaller, the period of amplitude modulation became bigger, and the spectrum peak at $\omega = \Omega = 1.2$ has grown to become almost as high as the main peak associated with natural dynamics.

As the forcing strength is increased further to reach $B = 0.53$ (fourth row in Fig. 3.12), we have almost touched the torus birth line in Fig. 3.10(a). The torus diameter became drastically smaller than at smaller B, the period of amplitude modulation has increased substantially, and the spectrum peak at the frequency $\omega = \Omega = 1.2$ has become the highest of all peaks. But the oscillations remain quasiperiodic, and this is not synchronization yet.

When $B = 0.54$, the torus birth line is crossed. The stroboscopic section is a single point, the oscillations are strictly periodic and synchronous with the forcing, and the spectrum contains just a single peak at the frequency of forcing. This is

---

[10] Larger than those at which phase locking occurs, but not necessarily large as compared to the amplitude A_0 of natural oscillations in the system. In fact, suppression can be achieved at the forcing strength B considerably less than A_0.

3.9 Phase Locking and Suppression: Numerical Simulation   61

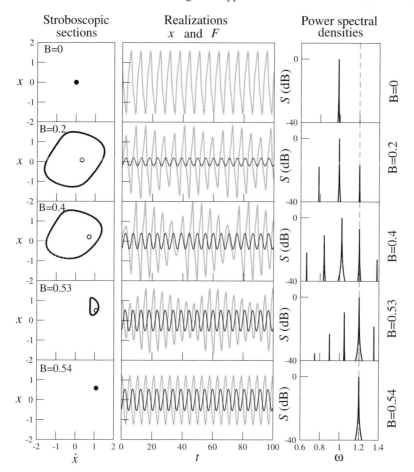

Fig. 3.12. Illustration of suppression of natural dynamics in the forced van der Pol system (3.3) at $\lambda = 0.5$, $\omega_0 = 1$. In Fig. 3.9(b) we move along route **B** defined by $\Omega = 1.2$ as we enter the synchronization tongue via torus birth bifurcation line. Stroboscopic sections, realizations and power spectral densities (spectra) are shown for each value of B indicated to the right of the each row. **First column**: *Black line* (*circle*)—stable torus (cycle S), *grey line*—resonant torus, *grey circle*—saddle cycle S^*, *white circle*—unstable cycle U. **Second column**: *Black line*—$x(t)$, *grey line*—force $F(t)$. **Third column**: *Black line*—spectrum of x in (3.3), *vertical dashed line* shows the position of the forcing frequency $\Omega = 1.2$

the regime of frequency (phase) synchronization. Note that the amplitude of oscillations in the suppression region at $B = 0.54$ is around 1, which is smaller than the amplitude of unforced oscillations $\sqrt{2} \approx 1.41$.

The difference between the two routes to synchronization, phase locking and suppression, manifests itself both in the phase space and in the spectra. The distinctive features of the two synchronization mechanisms are summarized in Table 3.1.

62   3 1 : 1 Forced Synchronization of Periodic Oscillations

Table 3.1. Summary of difference between the two synchronization mechanisms: phase/ frequency locking and suppression of natural dynamics. Manifestations of these mechanisms are given in terms of the phase space and in terms of power spectral densities (spectra)

	Phase locking	Suppression
Changes in phase space	Torus diameter almost does not change. A stable cycle is born on its surface	Torus shrinks to become a stable cycle
Changes in spectra	Peak associated with natural dynamics moves to coincide with forcing frequency	Peak associated with natural dynamics almost does not move, but becomes smaller and finally vanishes

Note that the analytical estimates of the power spectral density (see Sect. 3.8) of the forced oscillations along the suppression route to synchronization describe the real spectra only roughly. First, $\lambda = 0.5$ that was used for numerical simulations, is not close to zero and hence oscillations do not satisfy the assumption of being quasiharmonic. As a result, the real spectrum contains not only two peaks at the frequency of forcing and at the natural frequency, but also peaks at their combinations. Second, the peak associated with the natural oscillations in the system does move slightly towards the forcing frequency, contrary to the estimate of (3.102). However, the analytical calculations predict the fact that the shifting of the peak associated with the natural frequency is negligibly small, as compared to its shifting along the frequency locking route.

3.10 Phase Locking and Suppression: Experiment

In order to convince the reader that the theoretical predictions and the numerical results presented above are not merely the tricks of mathematical theory or of computer simulations, but the descriptions of the real physical phenomena, we present the results obtained from real electronic circuits demonstrating self-sustained oscillations.

In presenting the experimental results it was decided to abandon the modern computer-based interface and to go for an old-fashioned style of the pre-computer experimental techniques. Namely, all signals used for realizations and phase portraits were registered with the help of traditional oscilloscopes, to which they arrived directly from the electric circuits without any pre-processing. All spectra were measured by means of the electronic analogue spectrum analyzers that do not use any numerical techniques. For this reason the spectra of periodic processes have quite broad peaks, as will be seen below. Where possible, we provide additional evidence by showing the snapshots of frequency readings, etc. We hope that it was worth the effort and that the reader will be convinced that different mechanisms of synchronization are something real and not merely the fruits of mathematical imagination.

The electronic implementation used for illustration of forced synchronization is schematically shown in Fig. 3.13. A classical LC-circuit connected to a non-linear

3.10 Phase Locking and Suppression: Experiment

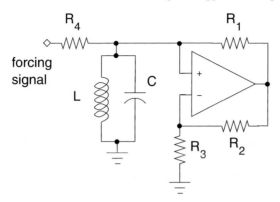

Fig. 3.13. Scheme of an experimental setup used to illustrate the phenomena of forced synchronization. The LC-oscillator is used with the following circuit parameters: $R_1 = R_2 = 1.5\,\text{kOhm}$, $R_3 = 40\,\text{Ohm}$, $R_4 = 50\,\text{kOhm}$, $L = 18\,\text{mH}$, $C = 33\,\text{nF}$. The frequency f_0 of self-sustained oscillations in this circuit was measured to be equal to 6.62 kHz

element with a negative resistance is used as possibly the simplest generator of electromagnetic oscillations. The scheme used is of the same type as the van der Pol oscillator which was studied theoretically and numerically throughout this chapter, although it is not the same. With these experiments we intend to demonstrate that the theoretical results obtained for van der Pol oscillator are sufficiently general and describe a wide class of phenomena in oscillators with similar properties, but possibly described by different equations.

The parameters of the circuit were fixed as indicated in the caption to Fig. 3.13, and the (plain) frequency $f_0 = \omega_0/2\pi$ of self-oscillations was measured to be 6.62 kHz. The external forcing in the form of a periodically oscillating voltage is applied to the circuit, and one can change both amplitude B and frequency $f_\text{f} = \Omega/2\pi$ of forcing in a wide range.

In Fig. 3.14 1 : 1 forced synchronization by phase locking in the electric circuit is illustrated with the snapshots of the screens of oscilloscopes. One can compare this figure with Fig. 3.11 where the same phenomenon is illustrated by numerical simulation. Each row of Fig. 3.14 corresponds to a certain value of the amplitude B of the control force in Volts, which is indicated to the right of the row and grows from top to bottom. The columns show: phase portraits on the plane (y, x) of two voltages in the circuit (first column), voltage x in the circuit (response) and force F versus time (second column), and spectrum of response x (third column). In the bottom of the figure the reading of the forcing frequency $f_\text{f} = \Omega/2\pi$ is given, which is equal to $f_\text{f} = 6.7412\,\text{kHz}$ and is slightly different from the frequency of self-sustained oscillations $f_0 = 6.62\,\text{kHz}$. The arrow at the bottom marks the position of the forcing frequency on the spectra.

64   3 1:1 Forced Synchronization of Periodic Oscillations

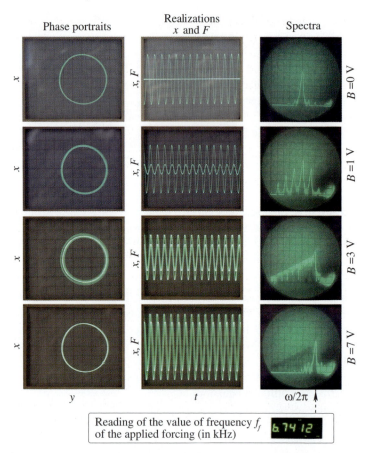

Fig. 3.14. Illustration of 1:1 frequency (phase) locking in an experiment with the forced periodic oscillator whose scheme is given in Fig. 3.13, to be compared with Fig. 3.11. All pictures are photographs of the screens of oscilloscopes on which phase portraits, realizations and spectra are shown. Phase portraits, realizations and power spectral densities (spectra) are shown for each value of B. **First column**: Phase portraits on the plane (y, x) of two voltages in the circuit. **Second column**: Large amplitude—voltage $x(t)$ in the circuit, small amplitude—force $F(t)$. **Third column**: Spectrum of x

3.10.1 Amplitudes from Oscilloscopes

For those who are not familiar with oscilloscopes, we explain in detail how information about the amplitudes of oscillations can be obtained with their help. The screens of all oscilloscopes are covered with grids with square cells, and each cell embraces a certain amount of voltage or spectral power in height, or time or spectral frequency in width. In the phase portraits (first column), the cell height and width are 0.5 V, and from this we can estimate the amplitudes of oscillations in the circuit; e.g. at $B = 1$ V the vertical spread of the phase portrait is approximately 5.5 cells, which

3.10 Phase Locking and Suppression: Experiment 65

means that the amplitude of x is about $5.5/2 \times 0.5$ V $= 1.375$ V. In realizations (second column), the same cell height embraces different amounts of Voltage for x and for F: 0.5 V and 2 V, respectively. Again, at $B = 1$ V x spreads over about 5.5 cells (like in the phase portrait) and its amplitude is about 1.375 V, while F spreads over approximately one cell and hence its amplitude is about 1 V—which coincides with the value of $B = 1$ V. Note that the amplitude of forcing that is illustrated by the oscilloscope here appears to be comparable with the amplitude of oscillations in the circuit, whereas phase locking is expected at much smaller amplitudes of forcing. However, in this experiment forcing F was measured at the point marked by "forcing" in Fig. 3.13, i.e., before the resistor R_4 with a very large resistance. After the signal F passes R_4, its amplitude is decreased considerably by the factor of approximately 35, so that the signal that reaches the self-oscillator is about 35 times smaller than the signal visible in the oscilloscope.

In the experiment, the parameters of the self-oscillating circuit and the forcing frequency were fixed, and the forcing amplitude B was gradually increased from 0 V to 7 V. At $B = 0$ V there is no forcing, the oscillations of the circuit are periodic, and this is clearly visible in the phase portrait in the form of a closed loop, realization with the constant amplitude and the spectrum that has only one peak with frequency that is different from the frequency of forcing. At $B = 1$ V oscillations become quasiperiodic: the phase portrait is no longer a closed loop but a torus (the visible circle becomes slightly thicker as compared to $B = 0$ V), the realization becomes slightly amplitude-modulated, and the spectrum contains peaks not only at the main frequency $f_s = \omega_s/2\pi$, but also at the combination frequencies $n|f_f - f_s|$, where n is integer number. At the same time, the main spectrum peak f_s is shifted against its original position f_0 towards the forcing frequency f_f. At $B = 3$ V the main frequency f_s already coincides with the forcing frequency f_f and the system spends a lot of time near the periodic regime: in the phase portrait the bright closed loop is clearly visible which is a precursor of a periodic regime. However, oscillations are not periodic yet but are still quasiperiodic: in the phase portrait we still see the torus, and the spectrum contains a lot of components besides the main one. Finally, at $B = 7$ V the oscillations become periodic again: phase portrait is a closed loop, realization has constant amplitude and spectrum has only one component. However, this regime is different from the one that existed in the system before forcing was applied: the frequency of oscillations has changed and become equal to the frequency of external forcing. This is how phase (frequency) locking takes place.

With the same experimental setup we now demonstrate the occurrence of $1:1$ forced synchronization by suppression of natural dynamics in the electric circuit with external forcing. Now the forcing frequency is set at $f_f = 7.0767$ kHz as compared to the frequency of natural oscillations $f_0 = 6.62$ kHz, so that the detuning is considerably larger than in the case of locking described above. The experiment on suppression is illustrated in Fig. 3.15, and the designations are the same as in Fig. 3.14. Here, the heights of grid cells in the oscilloscopes are 0.5 V for x and 5 V for F, but realizations are shown during a longer time interval in order to allow one to observe amplitude modulation clearly. In the absence of forcing ($B = 0$ V)

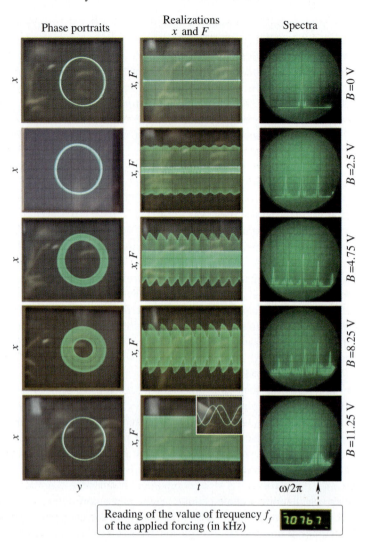

Fig. 3.15. Illustration of 1 : 1 suppression of natural dynamics in an experiment with the forced periodic oscillator whose scheme is given in Fig. 3.13, to be compared with Fig. 3.12. All pictures are photographs of the screens of oscilloscopes on which phase portraits, realizations and spectra are shown. Phase portraits, realizations and power spectral densities (spectra) are shown for each value of B. **First column**: Phase portraits on the plane (y, x) of two voltages in the circuit. **Second column**: Large amplitude—voltage $x(t)$ in the circuit, small amplitude—force $F(t)$. **Third column**: Spectrum of x. *Inset* in fifth row, second column is an enlarged segment of the realizations $x(t)$ and $F(t)$

the oscillations in the circuit are exactly the same as in Fig. 3.14 at $B = 0$ V, i.e., periodic. At $B = 2.5$ V oscillations are quasiperiodic, which can be detected by the "thickened" phase portrait, clearly visible amplitude modulation and the new spectrum components which appear at the frequency f_f of forcing and at combination frequencies. Note that the main spectrum peak stays at the same position as it was at $B = 0$ V, unlike in the case of locking where at a comparable value of $B = 3$ V the position of the main peak has shifted. At $B = 4.75$ V the amplitude modulation of oscillations becomes stronger, new spectrum peaks appear at combination frequencies, and the peak at the f_s is now only slightly lower than the one at f_f. At the same time, it is now clearly visible that in average the amplitude of oscillations in the system has decreased as compared to the amplitude at smaller values of B. Further increase of B up to the value $B = 8.25$ V makes the phase portrait shrink further, while the amplitude modulation becomes even more pronounced. Most importantly, the spectrum peak at forcing frequency f_f is now *higher* than the peak at f_s that corresponds to the natural dynamics of the system! This means that the dominating dynamics is now the one imposed by external forcing. However, this is not synchronization yet, since oscillations are still quasiperiodic and the natural dynamics is present alongside with the one dictated by forcing. Finally, at $B = 11.25$ V only one spectral peak at the forcing frequency is left in the system and the oscillations are periodic again. The natural dynamics is now completely suppressed by the forcing, while the imposed behavior has the frequency of forcing and a slightly different shape and amplitude, as compared to the one that existed in the system before forcing was applied. Note that the "real" amplitude of forcing at which suppression is achieved is about 11.25 V$/35 \approx 0.32$ V which is still considerably smaller than the amplitude of the natural oscillations in the system which is around 1.375 V.

3.11 Beat Frequency: Theory, Simulations and Experiment

In this section we will discuss in detail the beating of oscillations that was mentioned in the introductory part of this chapter. To remind you, when synchronization ceases to exist, the oscillations in the system become modulated by a slow function of time: instantaneous amplitude A. It is interesting to find out what the shape and the frequency of $A(t)$ are, and what the practical meaning of the beat frequency is. The analytical estimates of this section were given in [160], but with less detail.

We remind you that beat frequency $\dot{\varphi}$ is the instantaneous angular velocity with which the phase point on the plane (u, v) rotates around the origin (Fig. 3.6(b)) when there are no stable fixed points in the system (3.31)–(3.32), i.e., no synchronization.

3.11.1 Theory

First, consider weak forcing, i.e., small B. The truncated equation for the phase difference φ can then be approximately written as (3.25) and does not depend on A. Instantaneous beat frequency is given by $\dot{\varphi}$. We are interested in its average $\bar{\dot{\varphi}}$ over time. For this purpose, we first have to find an explicit solution $\varphi(t)$ of this equation as a function of time, and then calculate the average of its derivative. Let us denote

68   3 1:1 Forced Synchronization of Periodic Oscillations

$$A_s = \frac{B}{4\sqrt{\lambda\Omega}}. \tag{3.104}$$

Equation (3.25) can be solved explicitly by separation of variables, namely,

$$\int_0^\varphi \frac{d\varphi}{\Delta - A_s \cos\varphi} = \int_{t_0}^t dt. \tag{3.105}$$

The integral of the same type as in the left-hand side of the above formula can be found, e.g., in [97]. Beats occur outside phase locking region, i.e., when $\Delta > A_s$ (compare with (3.28)). Then the formal condition $\Delta^2 > A_s^2$ is satisfied, and the integral in (3.105) can be written as follows:

$$\frac{2}{\sqrt{\Delta^2 - A_s^2}} \arctan\left\{ \frac{\sqrt{\Delta^2 - A_s^2}\,\tan\frac{\varphi}{2}}{\Delta - A_s} \right\} = t - t_0.$$

In these calculations we assume that at the time moment t_0, φ was zero. We need to express φ through t

$$\arctan\left[\frac{\sqrt{\Delta^2 - A_s^2}\,\tan\frac{\varphi}{2}}{\Delta - A_s} \right] = \frac{\sqrt{\Delta^2 - A_s^2}(t - t_0)}{2}.$$

Take the tangent of both sides:

$$\frac{\sqrt{\Delta^2 - A_s^2}\,\tan\frac{\varphi}{2}}{\Delta - A_s} = \tan\left[\frac{\sqrt{\Delta^2 - A_s^2}(t - t_0)}{2} \right],$$

$$\tan\frac{\varphi}{2} = \frac{\Delta - A_s}{\sqrt{\Delta^2 - A_s^2}} \tan\left[\frac{\sqrt{\Delta^2 - A_s^2}(t - t_0)}{2} \right].$$

Take the arctangent of both sides and multiply by 2:

$$\varphi = 2\arctan\left\{ \frac{\Delta - A_s}{\sqrt{\Delta^2 - A_s^2}} \tan\left[\frac{\sqrt{\Delta^2 - A_s^2}(t - t_0)}{2} \right] \right\}.$$

Instantaneous beat frequency is $\dot\varphi$, so let us find it by differentiating $\varphi(t)$

$$\dot\varphi = \frac{2}{1 + \frac{(\Delta - A_s)^2}{\Delta^2 - A_s^2} \tan^2[\frac{\sqrt{\Delta^2 - A_s^2}(t - t_0)}{2}]} \times \frac{\Delta - A_s}{\sqrt{\Delta^2 - A_s^2}}$$

$$\times \frac{1}{\cos^2[\frac{\sqrt{\Delta^2 - A_s^2}(t - t_0)}{2}]} \times \frac{\sqrt{\Delta^2 - A_s^2}}{2}$$

$$= \frac{\Delta - A_s}{\cos^2[\frac{\sqrt{\Delta^2 - A_s^2}(t - t_0)}{2}] + \frac{(\Delta - A_s)^2}{\Delta^2 - A_s^2} \sin^2[\frac{\sqrt{\Delta^2 - A_s^2}(t - t_0)}{2}]}$$

$$= \frac{\Delta - A_s}{\frac{1}{2}(1 + \cos[\sqrt{\Delta^2 - A_s^2}(t - t_0)]) + \frac{\Delta - A_s}{\Delta + A_s}\frac{1}{2}(1 + \cos[\sqrt{\Delta^2 - A_s^2}(t - t_0)])}$$

$$= \frac{2(\Delta - A_s)}{\frac{2\Delta}{\Delta + A_s} + \frac{2A_s}{\Delta + A_s} \cos[\sqrt{\Delta^2 - A_s^2}(t - t_0)]} = \frac{\Delta^2 - A_s^2}{\Delta + A_s \cos[\sqrt{\Delta^2 - A_s^2}(t - t_0)]}.$$

3.11 Beat Frequency: Theory, Simulations and Experiment 69

Thus, instantaneous beat frequency $\dot{\varphi}$ turns to be a periodic function of time with period $T = 2\pi/\sqrt{\Delta^2 - \Delta_s^2}$. However, we are interested in its average over time. Time average of a periodic function is equal to its average over one period, i.e.,

$$\bar{\dot{\varphi}} = \frac{1}{T} \int_{t_0}^{t_0+T} \dot{\varphi} \, dt.$$

We remind you that φ is an angle of the phase vector in (u, v) plane. Hence, by definition, in one period of oscillations the phase vector rotates by 2π: φ increases by 2π at $\Delta > 0$, and decreases by 2π at $\Delta < 0$. With account of this, we can rewrite

$\Delta > 0$:

$$\bar{\dot{\varphi}} = \frac{1}{T} \int_{t_0}^{t_0+T} \frac{d\varphi}{dt} \, dt = \frac{1}{T} \int_0^{2\pi} d\varphi = \frac{\sqrt{\Delta^2 - \Delta_s^2}}{2\pi} 2\pi = \sqrt{\Delta^2 - \Delta_s^2},$$

$\Delta < 0$: (3.106)

$$\bar{\dot{\varphi}} = \frac{1}{T} \int_0^{-2\pi} d\varphi = -\sqrt{\Delta^2 - \Delta_s^2}.$$

When the detuning Δ approaches Δ_s, the beat frequency smoothly approaches zero. Inside the phase locking region, the amplitude A is a constant, which can formally be associated with an infinite period and zero frequency. Using the definition of Δ from (3.20), we express the beat frequency $\bar{\dot{\varphi}}$ as a function of the forcing frequency Ω

$$\bar{\dot{\varphi}} = \pm\sqrt{(\omega_0 - \Omega)^2 - \frac{B^2}{16\lambda\Omega^2}}.$$

From (3.106) it follows that the beat frequency can be either positive for positive detuning, or negative for negative one. So, for convenience in what follows we will take a modulus of the beat frequency. A typical dependence of $|\bar{\dot{\varphi}}|$ on Ω is given in Fig. 3.16(c) by the solid line for $\lambda = 0.1$, $\omega_0 = 1$, and $B = 0.01$ (see Fig. 3.5 for the theoretical estimate of the *lower* part of the respective synchronization tongue). The same dependence, but for a larger range of Ω, is given in Fig. 3.16(a).

Now, consider large amplitudes of forcing B. In this case, the truncated equations (3.21)–(3.22) cannot be regarded as uncoupled. It is convenient to consider the Descartes coordinates u and v and to introduce the angle φ as

$$\varphi = \arctan \frac{v}{u}.$$

Express u and v through C and Ψ using (3.79)–(3.80) and take a time derivative of φ

$$\dot{\varphi} = \frac{d}{dt} \arctan\left\{ -\frac{C\cos(\Delta t + \Psi)}{C\sin(\Delta t + \Psi) + \frac{\bar{F}}{\Delta}} \right\}$$

$$= \frac{1}{1 + C^2 \cos^2(\Delta t + \Psi)/(C\sin(\Delta t + \Psi) + \frac{\bar{F}}{\Delta})^2}$$

3 1:1 Forced Synchronization of Periodic Oscillations

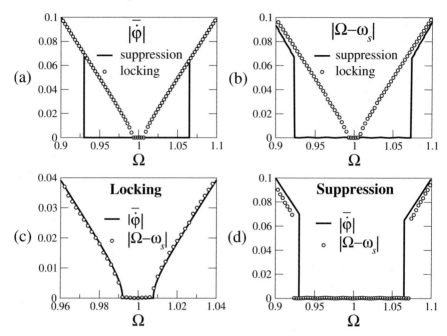

Fig. 3.16. Another illustration of the difference between the locking and suppression mechanisms, see Fig. 3.9(a) for reference. **a** Absolute value of beat frequency $|\bar{\dot\varphi}|$ as in (3.106) for locking ($B = 0.01$), and in (3.107) for suppression ($B = 0.06$). **b** The distance $|\Omega - \omega_S|$ between the frequency ω_S of the highest spectral peak of forced oscillations, and the forcing frequency Ω. **c, d** Comparison between $|\bar{\dot\varphi}|$ and $|\Omega - \omega_S|$ for locking (**c**) and for suppression (**d**). All quantities are shown versus forcing frequency Ω for the forced van der Pol oscillator (3.3) at $\lambda = 0.1$ and $\omega_0 = 1$

$$\times (-) \frac{C(-\sin(\Delta t + \Psi))(C\sin(\Delta t + \Psi) + \frac{\tilde F}{\Delta}) - C^2\cos^2(\Delta t + \Psi)}{(C\sin(\Delta t + \Psi) + \frac{\tilde F}{\Delta})^2}$$

$$\times (\Delta + \dot\Psi)$$

$$= \frac{C^2\sin^2(\Delta t + \Psi) + \frac{C\tilde F}{\Delta}\sin(\Delta t + \Psi) + C^2\cos^2(\Delta t + \Psi)}{C^2\sin^2(\Delta t + \Psi) + \frac{\tilde F^2}{\Delta^2} + \frac{2C\tilde F}{\Delta}\sin(\Delta t + \Psi) + C^2\cos^2(\Delta t + \Psi)} \times (\Delta + \dot\Psi)$$

$$= \frac{C^2 + \frac{C\tilde F}{\Delta}\sin(\Delta t + \Psi)}{C^2 + \frac{\tilde F^2}{\Delta^2} + \frac{2C\tilde F}{\Delta}\sin(\Delta t + \Psi)} \times (\Delta + \dot\Psi)$$

$$= \frac{C^2\Delta^2 + C\tilde F\Delta\sin(\Delta t + \Psi)}{C^2\Delta^2 + \tilde F^2 + 2C\tilde F\Delta\sin(\Delta t + \Psi)} \times (\Delta + \dot\Psi).$$

As was shown previously ((3.84)), in average $\dot\Psi = 0$. Hence, the function $\dot\varphi$ is periodic with period $\frac{2\pi}{\Delta}$.

3.11 Beat Frequency: Theory, Simulations and Experiment 71

By analogy with the phase locking, we calculate the average of $\dot{\varphi}$ over one period T of oscillations

$$\bar{\dot{\varphi}} = \Delta. \tag{3.107}$$

This means that when one approaches the suppression region boundary from outside, the beat frequency drops from the finite value Δ to zero abruptly. A typical dependence of $|\bar{\dot{\varphi}}|$ on Ω is given in Fig. 3.16(d) by the solid line, for $\lambda = 0.1$, $\omega_0 = 1$, $B = 0.06$ (see Fig. 3.9(a) for the theoretical estimate of the *upper* part of the respective synchronization tongue).

In Fig. 3.16(a) the beat frequencies versus Ω are compared for the forced van der Pol oscillator for two different forcing strengths B: $B = 0.01$ corresponding to locking mechanism, and $B = 0.06$ corresponding to suppression. One can see that the crucial difference between the two mechanisms occurs near the boundaries of synchronization region. Namely, a signature of phase locking is the smooth tendency of the beat frequency to zero as the boundary is approached. On the contrary, suppression manifests itself in the almost linear change of the beat frequency as one approaches the synchronization boundary, and its abrupt drop from a final value to zero as the boundary is achieved. However, at a large distance from synchronization border, the beat frequency at small forcing behaves in the same manner as the one at large forcing: the two graphs practically coincide.

3.11.2 Numerical Simulation

One might wonder: "What is the practical use of the beat frequency? From its definition, it seems quite an inconvenient quantity to be estimated from experimental data. How is it related to the more conventional experimentally accessible measures?"

To provide an answer to these questions, let us consider Fourier power spectral density of forced oscillations. Note that outside synchronization region the oscillations are quasiperiodic: an almost periodic signal is amplitude-modulated by $A(t)$. It is known that the Fourier spectrum of such a signal is discrete and consists of frequency components at the main frequency ω_s of oscillations, and at the combinations of this main frequency with the modulating frequency, the latter being the average beat frequency $|\bar{\dot{\varphi}}|$; i.e. the peaks will be placed at

$$\omega_s \pm n|\bar{\dot{\varphi}}|, \tag{3.108}$$

where n is integer number. At the same time, from linear response theory it is known that in the spectrum of forced oscillations, at least with small forcing, one of the components will necessarily be at the forcing frequency Ω. Because we assume that the detuning $(\omega_0 - \Omega)$ is small, it is reasonable to expect that the peak at $\omega_s \pm |\bar{\dot{\varphi}}|$, i.e., the one closest to the main peak, will be the peak corresponding to forcing frequency. Hence, the distance between Ω and ω_s is expected to be the modulating frequency of the signal, i.e., beat frequency $|\bar{\dot{\varphi}}|$.

This assumption can only be checked experimentally or numerically, since it arises from a merely empirical speculation. In order to check its validity, we numerically simulate the forced van der Pol oscillator (3.3) at the same parameters as for

72   3 1:1 Forced Synchronization of Periodic Oscillations

Fig. 3.16(a), and estimate the Fourier power spectral density from its realizations $x(t)$. For each spectrum, the highest peak and its frequency ω_s is found, and the absolute value of the difference between ω_s and Ω is estimated. We fix the forcing strength at two values $B = 0.01$ and $B = 0.06$, and change Ω. Figure 3.16(b) shows the values of $|\Omega - \omega_s|$ depending on Ω and demonstrates a high degree of similarity with Fig. 3.16(a). A closer comparison between the analytical estimates of beat frequencies using (3.106)–(3.107), and the values of $|\Omega - \omega_s|$ is illustrated in Figs. 3.16(c)–(d).

One can see that for the locking mechanism, close to synchronization boundary, the beat frequency changes non-linearly, while at a large distance from the synchronization region its changes are almost linear. In fact, at a large distance from the boundary of synchronization region it is impossible to distinguish between the two routes to synchronization: beat frequencies coincide (Fig. 3.16(a)–(b)).

In experiments that involve high-frequency processes, e.g., in lasers or other semiconductor structures, it is often impossible to record the realizations with sufficiently good time resolution. Hence, it is impossible to extract the envelopes from them and thus to reliably estimate the beat frequencies $\bar{\dot{\varphi}}$ directly using the definition. However, the spectra are normally readily available in such experiments, and are in fact the main tool for the study of such processes. From Fig. 3.16 it can be concluded that the spectral measure $|\Omega - \omega_s|$ serves quite an accurate estimate of the beat frequency $|\bar{\dot{\varphi}}|$ and can be used to distinguish between different synchronization mechanisms.

It should be remembered that the expressions (3.106) and (3.107) for beat frequency are valid only as long as (3.25) and (3.79), (3.80), respectively, remain valid. Hence, when the conditions for the validity of these equations are no longer satisfied, the analytical expressions (3.106)–(3.107) for the beat frequency are not accurate. As an illustration of this, consider the forced van der Pol system (3.3) at the non-linearity $\lambda = 0.5$, which does not satisfy the condition of being much smaller than 1. The respective numerically estimated synchronization region is shown in Fig. 3.9(b). As one fixes the forcing strength at $B = 0.2$ and changes Ω, synchronization is achieved by locking, while at $B = 0.6$ synchronization occurs via suppression. The analytical estimates of the beat frequency using (3.106)–(3.107) are shown in Fig. 3.17, together with their spectral estimates. For suppression, in agreement with analytical prediction illustrated in (a), there is a jump of $|\Omega - \omega_s|$ on the synchronization border as shown in (b). However, outside it, the dependence on Ω is already not strictly linear. For locking, the situation is less clear: synchronization is achieved while $|\Omega - \omega_s|$ tends to zero, but zero is not achieved (see lower part of (b)), and as the locking border is hit, it jumps to zero.

3.11.3 Experiment

The behavior of beat frequency in the vicinity of synchronization region was verified experimentally using the experimental setup whose scheme is shown in Fig. 3.13. The forcing strength B was fixed at a certain value, and the forcing frequency was changed between 5.5 kHz and 8 kHz. Beat frequency was estimated as the distance

3.11 Beat Frequency: Theory, Simulations and Experiment

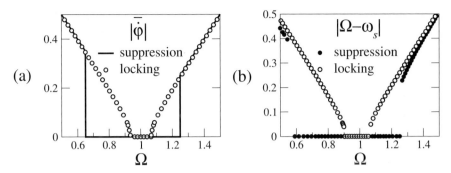

Fig. 3.17. And another illustration of the difference between the locking and suppression mechanisms, see Fig. 3.9(b) for reference. **a** Absolute value of beat frequency $|\dot{\bar{\varphi}}|$ as in (3.106) for locking ($B = 0.2$), and in (3.107) for suppression ($B = 0.6$). **b** The distance $|\Omega - \omega_s|$ between the frequency ω_s of the highest spectral peak of forced oscillations, and the forcing frequency Ω. All quantities are shown versus forcing frequency Ω for the forced van der Pol oscillator (3.3) at $\lambda = 0.5$ and $\omega_0 = 1$

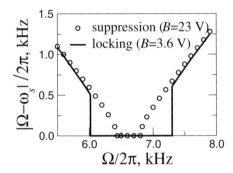

Fig. 3.18. Experimental values of "beat frequency" measured as $|\Omega - \omega_s|/(2\pi)$ (in kHz), versus forcing frequency Ω (compare with Fig. 3.16). Experimental scheme is shown in Fig. 3.13

between the main peak ω_s of the spectrum of the response signal, and the frequency Ω of forcing, i.e., as $|\Omega - \omega_s|$. In Fig. 3.18 beat frequency is shown versus Ω for $B = 3.6\,\text{V}$ at which the phase locking region is being crossed, and at $B = 23\,\text{V}$ at which we cross the suppression region. Remarkably, the experimental graphs demonstrate the same kind of behavior as predicted by the approximate theory illustrated in Fig. 3.16(a), and are in an excellent agreement with numerical simulations illustrated in Fig. 3.16(b). First, the suppression region is proved to be wider than the locking one, as predicted by the theory. Second, we observe that as one approaches the locking region, the beat frequency decreases gradually to become zero, and this dependence is non-linear with the shape very similar to the theoretical one. With this, as one approaches the suppression region, the beat frequency decreases almost linearly and then abruptly jumps to zero.

4 1:1 Mutual Synchronization of Periodic Oscillations

In Chap. 3 we considered the simplest case of two interacting processes, namely, when both processes are periodic, and one influences the other—but not reciprocally. A natural question to ask is "What happens if the other oscillator experiences the influence from the first one in return? Will anything change?" This chapter will try to answer this question.

We consider the case when two periodic self-oscillating systems are coupled mutually, or bidirectionally. One of the earliest observations of mutual synchronization was made by Rayleigh [243], in which he found that two organ pipes whose mouths are sufficiently close can sound in unison. There are a lot of ways one can couple two systems. Generally, both the physical properties of the coupling chain, and the specific features of systems to be coupled, define how the coupling term appears in the model equations. Thus, one can speak about coupling either in terms of its *physical* meaning, or in terms of its *mathematical* description.

The examples of a physical description of coupling are as follows. Two electronic circuits can be coupled through a resistor, a capacitor or an inductor; chemical reactions or living systems can be coupled through the processes of diffusion or directed transport.

76   4 1 : 1 Mutual Synchronization of Periodic Oscillations

At the same time, coupling that is physically exactly the same can be described by different forms of coupling terms in the model equations, depending on the structure of the systems being coupled. One possible way of introducing coupling mathematically is simply to add into one or more of each subsystem's equations terms proportional to the variables of another subsystem or to some functions of them. This would be "direct coupling" according to [35]. This kind of coupling was thoroughly studied, e.g., in [35] and [161].

Another very popular way of coupling is to represent the coupling term as the difference between the coordinates of the interacting systems, which will be considered here. It needs to be noted that the difference between the direct and difference types of coupling becomes important only when one considers the phenomenon of oscillation death (Sect. 4.1), otherwise the synchronization phenomena are similar for both types of coupling.

We again consider the paradigmatic van der Pol oscillators. The model equations read

$$\ddot{x}_1 - \left(\lambda_1 - x_1^2\right)\dot{x}_1 + \omega_1^2 x_1 + B_R(x_1 - x_2) + B_D(\dot{x}_1 - \dot{x}_2) = 0, \qquad (4.1)$$

$$\ddot{x}_2 - \left(\lambda_2 - x_2^2\right)\dot{x}_2 + \omega_2^2 x_2 + B_R(x_2 - x_1) + B_D(\dot{x}_2 - \dot{x}_1) = 0. \qquad (4.2)$$

Here, $\lambda_{1,2}$ are non-linearity parameters, $\omega_{1,2}$ are the eigenfrequencies. B_R and B_D are the strengths of mutual coupling of two different forms. Here we adopt the terminology of [214] where the coupling with $B_D \neq 0$ was called dissipative, and the one with $B_R \neq 0$ was called reactive.[1] Note that when the two systems demonstrate identical oscillations at $\lambda_1 = \lambda_2$ and $\omega_1 = \omega_2$, i.e., are perfectly synchronized, the coupling terms vanish. The more different the oscillations in the two systems, the larger the coupling terms are. Although the forcing applied to a system can be regarded as unidirectional coupling between the two oscillators, there is no direct analogy between the forced oscillator and the oscillator which is diffusively coupled to another one. When external forcing is applied to a system like in (3.3), the forcing term never vanishes whatever the oscillations of the system are.

Mutually coupled van der Pol and some other types of oscillators were studied in [34, 35, 64, 122, 135, 148, 171, 194, 240, 285].

For convenience, we rewrite (4.1)–(4.2) in the form of four first-order ordinary differential equations

$$
\begin{aligned}
\dot{x}_1 &= y_1, \\
\dot{y}_1 &= \left(\lambda_1 - x_1^2\right)y_1 - \omega_1^2 x_1 + B_R(x_2 - x_1) + B_D(y_2 - y_1), \\
\dot{x}_2 &= y_2, \\
\dot{y}_2 &= \left(\lambda_2 - x_2^2\right)y_2 - \omega_2^2 x_2 + B_R(x_1 - x_2) + B_D(y_1 - y_2).
\end{aligned}
\qquad (4.3)
$$

---

[1] In [35] these forms of coupling were referred to as scalar and non-scalar, respectively.

4.1 Truncated Equations for Weakly Non-linear Oscillators

We consider the system of two weakly non-linear van der Pol oscillators (4.1)–(4.2) that are coupled mutually with coupling represented as a difference between the respective variables. We consider both kinds of coupling: dissipative and reactive. The purpose of this section is to derive the truncated equations for the amplitudes in each partial oscillator, and the phase difference between the oscillators.

By analogy with the case of forced oscillations considered in Sect. 3.2, we will be looking for the solution in the form

$$x_{1,2}(t) = A_{1,2}(t)\cos\big(\omega t + \varphi_{1,2}(t)\big) = \frac{1}{2}\big(a_{1,2}(t)e^{i\omega t} + a_{1,2}^*(t)e^{-i\omega t}\big), \qquad (4.4)$$

where the complex amplitudes a_1 and a_2 and their complex-conjugates a_1^* and a_2^* are expressed through real amplitudes A_1 and A_2 and phases φ_1 and φ_2 as

$$a_{1,2} = A_{1,2}e^{i\varphi_{1,2}}, \qquad a_{1,2}^* = A_{1,2}e^{-i\varphi_{1,2}}. \qquad (4.5)$$

In what follows we will omit the brackets "(t)" that emphasize an explicit dependence on time of the variables $A_{1,2}$, $\varphi_{1,2}$ and $a_{1,2}$.

Note that unlike in Sect. 3.2 where we were looking for the solution at the forcing frequency Ω, here we are looking for the solution at some frequency ω which we do not know. Then $\dot{x}_{1,2}$ can be obtained by the direct differentiation of (4.4):

$$\dot{x}_{1,2}(t) = \frac{1}{2}\big(\dot{a}_{1,2}e^{i\omega t} + a_{1,2}i\omega e^{i\omega t} + \dot{a}_{1,2}^* e^{-i\omega t} - a_{1,2}^* i\omega e^{-i\omega t}\big).$$

By analogy with (3.14), we require that

$$\dot{x}_{1,2}(t) = \frac{i\omega}{2}\big(a_{1,2}e^{i\omega t} - a_{1,2}^* e^{-i\omega t}\big). \qquad (4.6)$$

Then the additional condition on the complex amplitudes of the mutually coupled oscillations will read

$$\dot{a}_{1,2}e^{i\omega t} + \dot{a}_{1,2}^* e^{-i\omega t} = 0. \qquad (4.7)$$

$\ddot{x}_{1,2}$ can be obtained by the differentiation of (4.6)

$$\ddot{x}_{1,2}(t) = \frac{i\omega}{2}\big(\dot{a}_{1,2}e^{i\omega t} - \dot{a}_{1,2}^* e^{-i\omega t}\big) - \frac{\omega^2}{2}\big(a_{1,2}e^{i\omega t} + a_{1,2}^* e^{-i\omega t}\big).$$

In the equation above we represent

$$\dot{a}_{1,2}e^{i\omega t} = -\dot{a}_{1,2}e^{i\omega t} + 2\dot{a}_{1,2}e^{i\omega t}. \qquad (4.8)$$

Then, with account of (4.7),

$$\ddot{x}_{1,2}(t) = i\omega\dot{a}_{1,2}e^{i\omega t} - \frac{\omega^2}{2}\big(a_{1,2}e^{i\omega t} + a_{1,2}^* e^{-i\omega t}\big).$$

78   4 1:1 Mutual Synchronization of Periodic Oscillations

Substitute $x_{1,2}$, $\dot{x}_{1,2}$ and $\ddot{x}_{1,2}$ into (4.1)–(4.2)

$$
i\omega\dot{a}_{1,2}e^{i\omega t} - \frac{\omega^2}{2}\left(a_{1,2}e^{i\omega t} + a_{1,2}^*e^{-i\omega t}\right)
$$
$$
- \left(\lambda_{1,2} - \frac{1}{4}\left[a_{1,2}e^{i\omega t} + a_{1,2}^*e^{-i\omega t}\right]^2\right)\frac{i\omega}{2}\left(a_{1,2}e^{i\omega t} - a_{1,2}^*e^{-i\omega t}\right)
$$
$$
+ \frac{\omega_{1,2}^2}{2}\left(a_{1,2}e^{i\omega t} + a_{1,2}^*e^{-i\omega t}\right)
$$
$$
= \frac{B_R}{2}\left(a_{2,1}e^{i\omega t} + a_{2,1}^*e^{-i\omega t} - a_{1,2}e^{i\omega t} - a_{1,2}^*e^{-i\omega t}\right)
$$
$$
+ \frac{B_D i\omega}{2}\left(a_{2,1}e^{i\omega t} - a_{2,1}^*e^{-i\omega t} - a_{1,2}e^{i\omega t} + a_{1,2}^*e^{-i\omega t}\right).
$$

Simplify the left-hand side (l.h.s.):

$$
\text{l.h.s.} = i\omega\dot{a}_{1,2}e^{i\omega t} + \frac{\omega_{1,2}^2 - \omega^2}{2}\left(a_{1,2}e^{i\omega t} + a_{1,2}^*e^{-i\omega t}\right)
$$
$$
- \frac{\lambda_{1,2}i\omega}{2}\left(a_{1,2}e^{i\omega t} - a_{1,2}^*e^{-i\omega t}\right)
$$
$$
+ \frac{i\omega}{8}\left(a_{1,2}^2e^{2i\omega t} + 2a_{1,2}a_{1,2}^* + a_{1,2}^{*2}e^{-2i\omega t}\right)\left(a_{1,2}e^{i\omega t} - a_{1,2}^*e^{-i\omega t}\right)
$$
$$
= i\omega\dot{a}_{1,2}e^{i\omega t} + \frac{\omega_{1,2}^2 - \omega^2}{2}\left(a_{1,2}e^{i\omega t} + a_{1,2}^*e^{-i\omega t}\right)
$$
$$
- \frac{\lambda_{1,2}i\omega}{2}\left(a_{1,2}e^{i\omega t} - a_{1,2}^*e^{-i\omega t}\right) + \frac{i\omega}{8}a_{1,2}^3e^{3i\omega t} - \frac{i\omega}{8}a_{1,2}^2 a_1^* e^{i\omega t}
$$
$$
+ \frac{i\omega}{4}a_{1,2}^2 a_{1,2}^* e^{i\omega t} - \frac{i\omega}{4}a_{1,2}a_{1,2}^{*2}e^{-i\omega t} + \frac{i\omega}{8}a_{1,2}a_{1,2}^{*2}e^{-i\omega t} - \frac{i\omega}{8}a_{1,2}^{*3}e^{-3i\omega t}.
$$

After finding similar terms in the l.h.s., the simplified equation reads

$$
i\omega\dot{a}_{1,2}e^{i\omega t} + \frac{\omega_{1,2}^2 - \omega^2}{2}\left(a_{1,2}e^{i\omega t} + a_{1,2}^*e^{-i\omega t}\right)
$$
$$
- \lambda_{1,2}\frac{i\omega}{2}\left(a_{1,2}e^{i\omega t} - a_{1,2}^*e^{-i\omega t}\right) + \frac{i\omega}{8}a_{1,2}^3e^{3i\omega t} + \frac{i\omega}{8}a_{1,2}^2 a_{1,2}^* e^{i\omega t}
$$
$$
- \frac{i\omega}{8}a_{1,2}a_{1,2}^{*2}e^{-i\omega t} - \frac{i\omega}{8}a_{1,2}^{*3}e^{-3i\omega t}
$$
$$
= \frac{B_R}{2}\left(a_{2,1}e^{i\omega t} + a_{2,1}^*e^{-i\omega t} - a_{1,2}e^{i\omega t} - a_{1,2}^*e^{-i\omega t}\right)
$$
$$
+ \frac{B_D i\omega}{2}\left(a_{2,1}e^{i\omega t} - a_{2,1}^*e^{-i\omega t} - a_{1,2}e^{i\omega t} + a_{1,2}^*e^{-i\omega t}\right).
$$

Multiply both parts by $e^{-i\omega t}/(i\omega)$:

$$
\dot{a}_{1,2} + \frac{\omega_{1,2}^2 - \omega^2}{2i\omega}\left(a_{1,2} + a_{1,2}^*e^{-2i\omega t}\right) - \frac{\lambda_{1,2}}{2}a_{1,2} + \frac{\lambda_{1,2}}{2}a_{1,2}^*e^{-2i\omega t}
$$

$$+ \frac{1}{8} a_{1,2}^3 e^{2i\omega t} + \frac{1}{8} a_{1,2}^2 a_{1,2}^* - \frac{1}{8} a_{1,2} a_{1,2}^{*2} e^{-2i\omega t} - \frac{1}{8} a_{1,2}^{*3} e^{-4i\omega t}$$

$$= \frac{B_R}{2i\omega} \left(a_{2,1} + a_{2,1}^* e^{-2i\omega t} - a_{1,2} - a_{1,2}^* e^{-2i\omega t} \right)$$

$$+ \frac{B_D}{2} \left(a_{2,1} - a_{2,1}^* e^{-2i\omega t} - a_{1,2} + a_{1,2}^* e^{-2i\omega t} \right).$$

By analogy with Sect. 3.2, we assume that $a_{1,2}$ are slow functions of time and almost do not change on a period $T = 2\pi/\omega$ of the frequency ω of fast oscillations. Use the Krylov–Bogoliubov method of averaging, and average the equation above on the period T using the definition (3.18) of a time average of a function $f(t)$:

$$\dot{a}_{1,2} - \frac{\omega_{1,2}^2 - \omega^2}{2\omega} i a_{1,2} - \frac{\lambda_{1,2}}{2} a_{1,2} + \frac{1}{8} a_{1,2}^2 a_{1,2}^* = \left(\frac{B_D}{2} - \frac{B_R}{2\omega} i \right) (a_{2,1} - a_{1,2}).$$

Represent complex amplitudes $a_{1,2}$ through their real amplitudes $A_{1,2}$ and phases $\varphi_{1,2}$ using (4.5) and substitute into the last equation:

$$\dot{A}_{1,2} e^{i\varphi_{1,2}} + A_{1,2} i \dot{\varphi}_{1,2} e^{i\varphi_{1,2}}$$

$$= \frac{\omega_{1,2}^2 - \omega^2}{2\omega} i A_{1,2} e^{i\varphi_{1,2}} + \frac{\lambda_{1,2}}{2} A_{1,2} e^{i\varphi_{1,2}}$$

$$- \frac{1}{8} A_{1,2}^3 e^{i\varphi_{1,2}} + \left(\frac{B_D}{2} - \frac{B_R}{2\omega} i \right) \left(A_{2,1} e^{i\varphi_{2,1}} - A_{1,2} e^{i\varphi_{1,2}} \right).$$

Divide both parts by $e^{i\varphi_{1,2}}$:

$$\dot{A}_{1,2} + A_{1,2} i \dot{\varphi}_{1,2} = \frac{\omega_{1,2}^2 - \omega^2}{2\omega} i A_{1,2} + \frac{\lambda_{1,2}}{2} A_{1,2} - \frac{1}{8} A_{1,2}^3$$

$$+ \left(\frac{B_D}{2} - \frac{B_R}{2\omega} i \right) \left(A_{2,1} e^{i(\varphi_{2,1} - \varphi_{1,2})} - A_{1,2} \right).$$

Represent the exponent through a sine and cosine using the Euler formula

$$\dot{A}_{1,2} + A_{1,2} i \dot{\varphi}_{1,2}$$

$$= \frac{\omega_{1,2}^2 - \omega^2}{2\omega} i A_{1,2} + \frac{\lambda_{1,2}}{2} A_{1,2} - \frac{1}{8} A_{1,2}^3$$

$$+ \left(\frac{B_D}{2} - \frac{B_R}{2\omega} i \right) \left(A_{2,1} \left[\cos(\varphi_{2,1} - \varphi_{1,2}) + i \sin(\varphi_{2,1} - \varphi_{1,2}) \right] - A_{1,2} \right).$$

Separate real and imaginary parts and write the full system of four ordinary differential equations for amplitudes and phases in interacting subsystems:

$$\dot{A}_1 = \frac{\lambda_1}{2} A_1 - \frac{1}{8} A_1^3 + \frac{B_D}{2} \left(A_2 \cos(\varphi_2 - \varphi_1) - A_1 \right) + \frac{B_R}{2\omega} A_2 \sin(\varphi_2 - \varphi_1),$$

$$\dot{\varphi}_1 = \frac{\omega_1^2 - \omega^2}{2\omega} + \frac{B_D}{2} \frac{A_2}{A_1} \sin(\varphi_2 - \varphi_1) - \frac{B_R}{2\omega} \frac{A_2}{A_1} \cos(\varphi_2 - \varphi_1) + \frac{B_R}{2\omega},$$

4 1:1 Mutual Synchronization of Periodic Oscillations

$$\dot{A}_2 = \frac{\lambda_2}{2} A_2 - \frac{1}{8} A_2^3 + \frac{B_D}{2} \left(A_1 \cos(\varphi_1 - \varphi_2) - A_2 \right) + \frac{B_R}{2\omega} A_1 \sin(\varphi_1 - \varphi_2),$$

$$\dot{\varphi}_2 = \frac{\omega_2^2 - \omega^2}{2\omega} + \frac{B_D}{2} \frac{A_1}{A_2} \sin(\varphi_1 - \varphi_2) - \frac{B_R}{2\omega} \frac{A_1}{A_2} \cos(\varphi_1 - \varphi_2) + \frac{B_R}{2\omega}.$$

We note that the right-hand sides of the equations above depend not on φ_1 and φ_2 separately, but on the difference between them. Let us introduce a new variable $\theta = \varphi_2 - \varphi_1$. Also, take into account that $\omega \approx \omega_{1,2}$ and $\omega_1 + \omega_2 \approx 2\omega$, and by Δ denote the detuning which is equal to

$$\Delta = \frac{\omega_2^2 - \omega_1^2}{2\omega} \approx \omega_2 - \omega_1. \tag{4.9}$$

Then we obtain a system of three truncated equations for the amplitudes $A_{1,2}$ and phase difference θ between the oscillators

$$\dot{A}_1 = \frac{\lambda_1}{2} A_1 - \frac{1}{8} A_1^3 + \frac{B_D}{2} \left(A_2 \cos(\theta) - A_1 \right) + \frac{B_R}{2\omega} A_2 \sin(\theta),$$

$$\dot{A}_2 = \frac{\lambda_2}{2} A_2 - \frac{1}{8} A_2^3 + \frac{B_D}{2} \left(A_1 \cos(\theta) - A_2 \right) - \frac{B_R}{2\omega} A_1 \sin(\theta), \tag{4.10}$$

$$\dot{\theta} = \Delta - \frac{B_D}{2} \sin(\theta) \left(\frac{A_2}{A_1} + \frac{A_1}{A_2} \right) + \frac{B_R}{2\omega} \cos(\theta) \left(\frac{A_2}{A_1} - \frac{A_1}{A_2} \right).$$

The equations similar to the above, but in a more general form which is not considered here for the sake of simplicity, were analyzed by Aronson et al. in [34, 35]. Below we will present some of the results of this analysis which we regard as most essential for the understanding of the phenomenon of mutual synchronization of periodic oscillations.

4.2 Periodic Oscillators with Dissipative Coupling

In (4.10) set $B_R = 0$ so that only dissipative coupling is considered. Also, in order to simplify the analysis let us assign $\lambda_1 = \lambda_2 = \lambda$. The truncated equations for amplitudes and phase difference then read

$$\dot{A}_1 = A_1 \left(\frac{\lambda}{2} - \frac{1}{8} A_1^2 - \frac{B_D}{2} \right) + \frac{B_D}{2} A_2 \cos(\theta),$$

$$\dot{A}_2 = A_2 \left(\frac{\lambda}{2} - \frac{1}{8} A_2^2 - \frac{B_D}{2} \right) + \frac{B_D}{2} A_1 \cos(\theta), \tag{4.11}$$

$$\dot{\theta} = \Delta - \frac{B_D}{2} \sin(\theta) \left(\frac{A_2}{A_1} + \frac{A_1}{A_2} \right).$$

Note that the equations above are symmetric with respect to A_1 and A_2: if we swap them, the equations remain the same.

4.2.1 Symmetric Solutions

We will start by looking for the symmetric solutions in the form $A_1 = A_2 = A$ which satisfy

$$\dot{A} = \frac{A}{2}\left(\lambda - B_D - \frac{A^2}{4}\right) + \frac{B_D}{2}A\cos(\theta), \tag{4.12}$$

$$\dot{\theta} = \Delta - B_D\sin(\theta). \tag{4.13}$$

It is clear that the solution (A, θ) to (4.12)–(4.13) generates the solutions (A, A, θ) to (4.11).

A phase-locked symmetric solution corresponds to $\dot{A} = 0$ $(A \neq 0)$ and $\dot{\theta} = 0$. From $(4.13)^2$ which does not depend on A we find θ

$$\Delta - B_D\sin(\theta) = 0, \qquad \sin(\theta) = \frac{\Delta}{B_D},$$

$$\theta_1 = \sin^{-1}\frac{\Delta}{B_D}, \qquad \theta_2 = \pi - \sin^{-1}\frac{\Delta}{B_D}.$$

$\theta_{1,2}$ exist as long as $|\Delta| \leq B_D$. So, in the parameter plane (Δ, B_D) the lines $\Delta = \pm B_D$ are the borderlines of a phase-locking region. In Fig. 4.1 these lines are shown on the plane of parameters B_D and p, where

$$p = \frac{\omega_2}{\omega_1}, \qquad \Delta = \omega_1(1 - p)$$

for $\omega_1 = 1$. One can compare this part of the diagram with the respective part of the similar diagrams for a forced system given in Figs. 3.5 and 3.9 and make sure that the lower boundaries of synchronization regions are qualitatively the same in all cases.

The values of A corresponding to $\theta_{1,2}$ can be found as follows:

$$\frac{A}{2}\left(\lambda - B_D - \frac{A^2}{4}\right) + \frac{B_D}{2}A\cos(\theta) = 0 \quad (A \neq 0). \tag{4.14}$$

Then

$$\left(\lambda - \frac{A^2}{4} - B_D\right) = -B_D\cos(\theta),$$

$$\cos(\theta) = \sqrt{1 - \sin^2(\theta)} = \sqrt{1 - \frac{\Delta^2}{B_D^2}},$$

$$\left(B_D + \frac{A^2}{4} - \lambda\right) = B_D\sqrt{1 - \frac{\Delta^2}{B_D^2}}.$$

---

[2] An equation in the form of (4.13) is sometimes called an Adler equation [5].

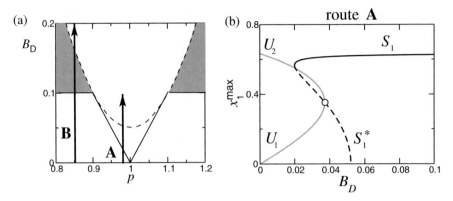

Fig. 4.1. (Color online) **a** The vicinity of 1 : 1 synchronization region of the mutually coupled van der Pol systems (4.3) with *dissipative* coupling, $\lambda_{1,2} = 0.1$, $B_R = 0$ and $B_D > 0$, on the plane of parameters of coupling "detuning p"–"strength of coupling B_D," where $p = \omega_2/\omega_1$, and ω_1 and ω_1 are eigenfrequencies of partial oscillators. *Solid line* marks saddle-node bifurcation, *dashed line* marks Andronov–Hopf bifurcation. *Dash-dotted* lines denote transitions from torus to the fixed point. In the *shaded area* there are no oscillations in the system, i.e., oscillation death occurs. **b** Evolution of periodic orbits along route **A** is illustrated: the maxima of x_1 are shown versus B_D. *Solid black line*: stable cycle S_1; *dashed line*: saddle cycle S_1^*; *grey lines*: twice saddle cycles $U_{1,2}$. *Empty circle* marks transcritical bifurcation

Take the square of both parts of the last equation:

$$B_D^2 + \frac{A^4}{16} + \frac{B_D A^2}{2} + \lambda^2 - 2\lambda B_D - \frac{\lambda A^2}{2} = B_D^2 - \Delta^2.$$

Rearrange terms and reduce to the quadratic equation for A^2:

$$A^4 + A^2 8(B_D - \lambda) + 16(\lambda^2 - 2\lambda B_D + \Delta^2) = 0.$$

Solutions for A^2 are

$$\tilde{A}_1^2 = 4(\lambda - B_D) + 4\sqrt{B_D^2 - \Delta^2}, \qquad \tilde{A}_2^2 = 4(\lambda - B_D) - 4\sqrt{B_D^2 - \Delta^2}.$$

It can be easily checked by an analogy with Sect. 3.4, that $(\tilde{A}_1, \theta_1)$ is a stable solution, and $(\tilde{A}_2, \theta_2)$ is an unstable one. They collide at the region boundary and vanish via saddle-node bifurcation, in full analogy with a forced periodic oscillator.

Let us rewrite (4.12)–(4.13) by making some variable substitutions. First, introduce new "time" τ:

$$t = 2\tau/\lambda: \quad \frac{dA}{d\tau} = A\left(1 - \frac{A^2}{4\lambda} - \frac{1}{\lambda}B_D\right) + \frac{1}{\lambda}B_D A \cos(\theta),$$

$$\frac{d\theta}{d\tau} = \frac{2}{\lambda}(\Delta - B_D \sin(\theta)).$$

4.2 Periodic Oscillators with Dissipative Coupling 83

Next, rescale A by introducing a value ρ such that

$$\rho = \frac{A}{2\sqrt{\lambda}}, \quad A = 2\sqrt{\lambda}\rho,$$

and obtain an equation for ρ:

$$\frac{d\rho}{d\tau} = \rho\left(1 - \rho^2 - \frac{B_D}{\lambda}\right) + \frac{1}{\lambda}B_D\rho\cos(\theta),$$

$$\frac{d\theta}{d\tau} = \frac{2}{\lambda}\Delta - \frac{2}{\lambda}B_D\sin(\theta). \tag{4.15}$$

If we denote

$$\gamma = \frac{B_D}{\lambda}, \quad \delta = \frac{2}{\lambda}\Delta,$$

(4.15) are identical to the equations

$$\dot{\rho} = \rho\left(1 - \gamma - \rho^2\right) + \gamma\rho\cos(\theta), \qquad \dot{\theta} = \delta - 2\gamma\sin(\theta) \tag{4.16}$$

analyzed by Aronson et al. in [35], where the solution to these equations was found
as follows:

$$\rho^2(t) = \frac{a(1 - c^2)}{1 + c\sin[\theta(t) + \psi]}, \tag{4.17}$$

with a, c and ψ expressed as

$$a = (1 - \gamma), \quad c = \frac{\gamma}{\sqrt{a^2 + \delta^2/4}}, \quad \tan\psi = \frac{2a}{\delta}. \tag{4.18}$$

In terms of the parameters of (4.12)–(4.13), a, c and ψ are equal to

$$a = 1 - \frac{B_D}{\lambda}, \quad c = \frac{B_D}{\lambda}\sqrt{a^2 + \frac{\Delta}{\lambda^2}}, \quad \tan\psi = \frac{1 - \frac{B_D}{\lambda}}{\frac{1}{\lambda}\Delta} = \frac{\lambda - B_D}{\Delta}.$$

At small γ, which means that $a > 0$ and $c \in [0; 1)$, (4.17) describes an ellipse in
(ρ^2, θ). The same authors have proved that the solution (4.17) is asymptotically stable, i.e., attracting. An ellipse on the plane (ρ^2, θ) means that the phase trajectories
in the original phase space (x_1, y_1, x_2, y_2) of (4.3) lie on a surface of a torus.

Note that the pair of fixed points $\tilde{A}_{1,2}$ found above are born on the ellipse. Hence,
when these points exist, (4.3) demonstrates the regime of phase locking.

4.2.2 Asymmetric Solutions

In [35], the equilibrium asymmetric solutions to (4.11), such that $A_1 \neq A_2$, were
found as follows:

84   4 1:1 Mutual Synchronization of Periodic Oscillations

$$\dot{A}_1 = 0 \quad \Rightarrow \quad A_1\left(\lambda - \frac{A_1^2}{4} - B_\mathrm{D}\right) = B_\mathrm{D} A_2 \cos(\theta), \qquad (4.19)$$

$$\dot{A}_2 = 0 \quad \Rightarrow \quad A_2\left(\lambda - \frac{A_2^2}{4} - B_\mathrm{D}\right) = B_\mathrm{D} A_1 \cos(\theta), \qquad (4.20)$$

$$\dot{\theta} = 0 \quad \Rightarrow \quad \frac{B_\mathrm{D}}{2} \sin(\theta)\left(\frac{A_2}{A_1} + \frac{A_1}{A_2}\right) = \Delta. \qquad (4.21)$$

The ratio of the first two equations gives

$$\frac{A_1}{A_2} \frac{(\lambda - \frac{A_1^2}{4} - B_\mathrm{D})}{(\lambda - \frac{A_2^2}{4} - B_\mathrm{D})} = \frac{A_2}{A_1}, \quad A_1^2\left(\lambda - \frac{A_1^2}{4} - B_\mathrm{D}\right) = A_2^2\left(\lambda - \frac{A_2^2}{4} - B_\mathrm{D}\right).$$

The last equation is true either when $A_1 = A_2$ which is a symmetric solution studied above, or when

$$A_1^2 = \left(\lambda - \frac{A_2^2}{4} - B_\mathrm{D}\right),$$

which describes an asymmetric solution. One can then use the above expression to reduce (4.19)–(4.21) to a single equation for A_1^2. In [35] it was proved that the asymmetric solutions are always unstable when the coupling is dissipative.

4.2.3 Oscillation Death

Besides the classical phase locking, an interesting phenomenon that can occur in mutually coupled periodic oscillators is oscillation death (quenching) when, due to coupling, oscillations in both systems stop completely. This phenomenon was first discovered experimentally by Rayleigh [243] while he was studying the behavior of coupled organ pipes: he found out that at a certain strength of mutual influence and detuning between the pipes they "may almost reduce one another to silence." In [46] the same effect was discovered in the model of coupled chemical oscillators. There is no analog of this phenomenon in forced oscillations. Mathematically, this is expressed as stabilization of the fixed point at the origin in the original system of coupled oscillators, (4.3). By analyzing the stability of the point $x_1 = x_2 = y_1 = y_2 = 0$, one can outline the region where this point is stable and hence there are no oscillations in the system. The region of oscillator death is marked as the shaded area in Fig. 4.1(a).

4.3 Dissipative Coupling: Numerical Simulation

We now illustrate the changes occurring in the system as one enters synchronization region along two different routes **A** and **B** in Fig. 4.1(a).

4.3.1 Locking

Consider route **A** by fixing the detuning $p = 0.98$, and change the coupling strength B_D from zero to some finite value. Figure 4.2 shows phase portraits on the plane (x_1, x_2) (first column), Poincaré sections on the plane (x_1, x_2) corresponding to the maxima of x_1, i.e., $\dot{x}_1 = 0$, $\ddot{x}_1 < 0$ (second column), realizations of x_1 and of x_2 (third column), and spectra of x_1 and of x_2 (fourth column).

At $B_D = 0$ the oscillators behave independently of each other: although the system as a whole behaves quasiperiodically, the oscillations in each subsystem are periodic. In terms of the phase space, oscillations take place on the surface of a torus which has a "square" shape, see first column. From the third column one can see that as x_1 takes the maximal value (black circles), x_2 can be at any stage of its oscillations (grey circles), and this is also reflected in the Poincaré map in the second column. The spectral peaks of x_1 and of x_2 are well separated as shown in fourth column.

At $B_D = 0.015$ the oscillators start to "feel" each other, and oscillations are now quasiperiodic in each subsystem. The phase trajectory fills the surface of a smooth ergodic torus whose Poincaré section is a closed curve. Note that Poincaré sections shown in Fig. 4.2 reveal only a part of the phase space, while the full structure will

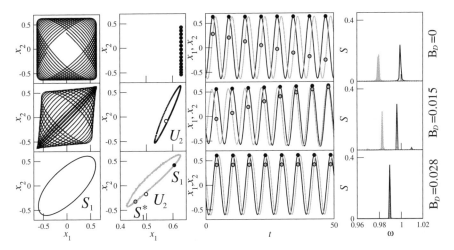

Fig. 4.2. Illustration of 1 : 1 frequency (phase) locking for the mutually coupled van der Pol systems (4.3) with *dissipative* coupling at $\lambda_{1,2} = 0.1$. In Fig. 4.1(a) we move along route **A** corresponding to $p = 0.98$, as we enter the synchronization tongue via saddle-node bifurcation line. Phase portraits, Poincaré sections, realizations and power spectral densities (spectra) are shown for each value of B_D given to the right of the respective row. **First column**: Phase portraits on the plane (x_1, x_2). **Second column**: Poincaré sections on the plane (x_1, x_2). *Black line* (circle S_1)—stable torus (cycle), *grey line*—resonant torus, *grey circle*—saddle cycle S_1^*, *white circle*—twice saddle cycle U_2. **Third column**: *Black line*—$x_1(t)$, *grey line*—$x_2(t)$; *black circles*—maxima of x_1, *grey circles*—values of x_2 when x_1 is at its maxima. **Fourth column**: *Black line*—spectrum of x_1, *grey line*—spectrum of x_2

be discussed below. The absence of phase synchronization is well visible in the realizations: as x_1 takes the maximal value, x_2 can be at any stage of oscillation. There is no frequency synchronization either, since, although the main spectral peaks of the two subsystems are now closer to each other than without coupling, they do not coincide, and the spectra contain combination frequencies.

Finally, at $B_D = 0.028$ the behavior of the system changes drastically: each time x_1 take its maximal value, x_2 is at exactly the same stage (phase) of its oscillations (third column). At the same time, the two spectral peaks coincide and all combination frequencies vanish (fourth column), and oscillations in both systems become strictly periodic (first and third column). This is mutual synchronization via phase (frequency) locking. The structure of the phase space is better visible in Poincaré section in the second column where grey closed curve shows the torus which is now resonant, on whose surface there live a stable limit cycle S_1 (black circle) that is observable in an experiment or simulation, and a saddle periodic orbit S_1^* (grey circle). Outside the torus there is a twice saddle cycle U_2 marked by an empty circle.

Generally, the mechanism of transition to a synchronized state along route **A** is very similar to phase locking in a forced oscillator (compare with Fig. 3.11): the amplitude of oscillations almost does not change, and the major changes occur in frequency. The only difference in terms of spectra is that in case of forcing, the oscillator frequency approaches the forcing frequency to coincide with it, while in case of mutual coupling the frequencies of both oscillators move towards each other to meet at some value which lies in between the original frequencies of uncoupled oscillators.

However, mutual dissipative coupling introduces some changes into the structure of the phase space as compared to forcing, which becomes visible if one compares Poincaré sections in Fig. 3.11 with $B = 0.18$ and $B = 0.2$, and the complete versions of Poincaré sections in Fig. 4.3 at $B = 0.015$ and $B = 0.028$. It becomes immediately obvious that in the vicinity of synchronization region, the dissipatively coupled oscillators possess an additional twice saddle cycle U_2 which did not exist in a forced oscillator.

4.3.2 Bifurcations

In order to better understand what happens in the phase space as the system goes from non-synchronous to synchronous regime, let us consider bifurcations[3] of the special objects involved: of the fixed points and of periodic orbits. We choose to follow route **A** in Fig. 4.1(a) that goes across the locking region and into the suppression region. In this way we illustrate the evolution of all periodic orbits with the change of B_D by displaying the maxima x_1^{max} of their x_1 coordinates (Fig. 4.1(b)). When the two subsystems are uncoupled $B_D = 0$, the phase space contains two significant objects: a twice saddle fixed point at the origin U_1 and a twice saddle

---

[3] All two-parameter and one-parameter bifurcation diagrams given in this and subsequent sections were revealed by means of the free software AUTO2000 [73].

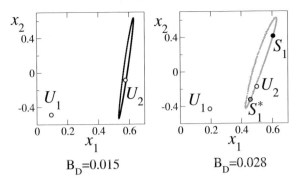

Fig. 4.3. Poincaré sections of mutually coupled van der Pol oscillators (4.3) with *dissipative* coupling at $\lambda_{1,2} = 0.1$, $p = 0.98$ at two different values of coupling strength B_D: **a** $B_D = 0.015$ (no synchronization) and **b** $B_D = 0.028$ (phase locking takes place). This is a complete picture, whose part was given in Fig. 4.2. S_1 is a stable cycle (*black circle*) and S_1^* is a saddle cycle (*grey circle*), both lying on the torus (*closed curve*). The torus is ergodic in **a** which is marked by *black line*, and resonant in **b** which is marked by *grey line*. U_1 and U_2 are twice saddle cycles that lie outside the torus (*empty circles*)

orbit U_2 of finite size, both lying on grey line [285]. At $B_D = 0$ the fixed point U_1 undergoes Andronov–Hopf bifurcation as a result of which an unstable periodic orbit is born which is also labelled as U_1 and is denoted by grey line going from the origin in Fig. 4.1(b).

At $B_D = 0$ a torus is born in the system, which can be either ergodic or resonant. At around $B = 0.02$ a new pair of periodic orbits is born via saddle-node bifurcation: a stable S_1 (black line) and a saddle S_1^* (dashed line). Both newly born orbits lie on the surface of a resonant torus illustrated in Fig. 4.2 at $B = 0.028$. As B grows above the value of 0.02, no changes occur to the stable orbit S_1 in the visible area of the tongue, and the system stays in the synchronous regime associated with the given limit cycle. However, two things do happen to the saddle orbit S_1^*. Namely, at $B = 0.037$ (marked by empty circles) S_1^* meets U_1 and U_2 and undergoes transcritical bifurcation as a result of which $U_{1,2}$ disappear. With the further increase of B_D, the saddle cycle S_1^* shrinks in size, and at $B = 0.052$ vanishes through the inverse Andronov–Hopf bifurcation.

Comparison with Fig. 3.10(b) shows that mutual dissipative coupling causes more complicated bifurcation transitions under the change of parameters, as compared to an applied forcing.

4.3.3 Suppression

Now consider route **B** in Fig. 4.1(a) by setting $p = 0.85$. The phase portraits, Poincaré sections, realizations and spectra for this route can be found in Fig. 4.4. Synchronous regimes can be achieved here in a slightly different manner than in the case of a forced oscillator: via oscillation death. At $B_D = 0$ the behavior of the system is qualitatively the same as with $p = 0.98$ described above, only the

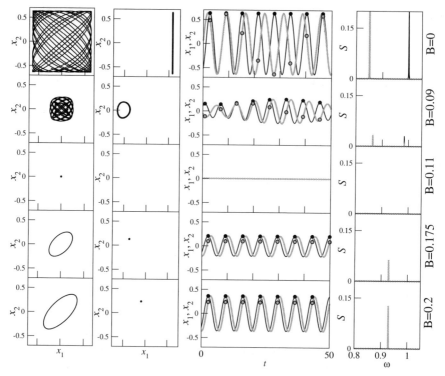

Fig. 4.4. Illustration of arriving at a synchronous regime via oscillation death for the mutually coupled van der Pol systems (4.3) with *dissipative* coupling at $\lambda_{1,2} = 0.1$. In Fig. 4.1(a) we move along route **B** corresponding to $p = 0.85$ as we enter the synchronization tongue after first crossing the oscillation death region. Poincaré sections, realizations and spectra are shown for each value of B_D. Designations are the same as in Fig. 4.2

frequencies of the two systems are better separated. As one increases B_D, the oscillations remain quasiperiodic, i.e., non-synchronous, but their amplitude shrinks (see the case with $B_D = 0.09$). However, the spectral peaks almost do not move. As we cross the boundary of the shaded area, all oscillations cease in both subsystems ($B_D = 0.11$)—this is oscillation death. When we leave the shaded area through its upper boundary ($B_D = 0.17$), the oscillations start again, but with small amplitude and the frequency in between the original frequencies of the uncoupled oscillators. Now the oscillations in both systems are perfectly synchronized: the maxima of x_1 always occur at the same phase of x_2. Further increase of B_D ($B_D = 0.2$) does not change the frequency of oscillations noticeably, but leads to the growth of their amplitude. The part of synchronization region above the Andronov–Hopf bifurcation line can be roughly regarded as a region of synchronization by suppression.

However, the region of oscillation death must not necessarily lie below the borderline of suppression region like in Fig. 4.1(a). In [285] some other coupled oscillators were considered, for which the line of oscillation death lies inside the classical suppression region.

4.4 Reactive Coupling

Now, consider system equations (4.3) at $B_D = 0$, $B_R > 0$, i.e., for reactive coupling only. In [35] interaction of systems was considered with coupling containing both dissipative and reactive terms. It was demonstrated that in the region of synchronization there are four different periodic orbits in the system, that can be roughly classified as: "stable symmetric," "saddle symmetric," "stable asymmetric," "saddle asymmetric." The same solutions can also be called "stable in-phase," "saddle in-phase," "stable anti-phase" and "saddle anti-phase," respectively—and this is how they will be called in the rest of this book. The terms "symmetric" and "asymmetric" (or "in-phase" and "anti-phase") are well justified as one considers two identical oscillators with no detuning. In that case, the existing periodic orbits would satisfy either $x_1 = x_2$ and $y_1 = y_2$, or $x_1 = -x_2$ and $y_1 = -y_2$, which means that in the phase plane (x_1, x_2) or (y_1, y_2) the respective phase portraits would belong to the main diagonal or to the anti-diagonal. At the same time, the realizations x_1 and x_2 will take their maximal values simultaneously for in-phase solutions. For anti-phase solutions, while x_1 will display a maximum, x_2 will have minimum. Of course, when frequency detuning is introduced even between otherwise identical subsystems, their solutions would no longer lie exactly on the diagonals, but could be stretched along them. Also, the maxima and minima in the realizations occur not exactly at the same time, but with some small time (phase) shift. If this is the case, it might still be reasonable to talk about in-phase or anti-phase solutions.

In addition to the four periodic orbits mentioned above, there are two more orbits which are "twice saddle," i.e., have two unstable directions as compared to one unstable direction of the simply "saddle" orbits. Hence, there are six periodic orbits in total, and all of them play their role in synchronization of the reactively coupled systems.

For illustration here, we set $\omega_1 = 1$ and $\omega_2 = p$, while p introduces the frequency detuning between the systems. For $\lambda_1 = \lambda_2 = 0.5$, the $1:1$ synchronization tongue is shown in Fig. 4.5(a) [44], and the shaded area denotes the *absence* of $1:1$ synchronization. Compare this with synchronization tongue for mutual dissipative coupling (Fig. 4.1), and also for forcing (Figs. 3.9 and 3.5). Unlike the dissipative coupling or forcing, reactive coupling makes the structure of the tongue noticeably more complicated. Namely, now we have not one tongue, but essentially two tongues embedded into each other. However, we observe the same bifurcation lines: solid lines mark saddle-node bifurcations, while dashed lines mark torus birth (Neimark–Sacker) bifurcations.

Let us illustrate the two mechanisms of synchronization in reactively coupled oscillators by means of data that one can register in an experiment: realizations, phase portraits, Poincaré sections and spectra.

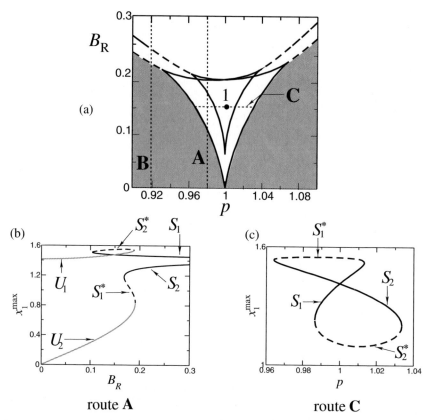

Fig. 4.5. (Color online) **a** 1:1 synchronization tongue for the mutually coupled van der Pol systems (4.3) with *reactive* coupling, $B_D = 0$ and $B_R > 0$, on the plane "detuning p"–"coupling strength B_R," where $p = \omega_2/\omega_1$, and ω_1 and ω_2 are eigenfrequencies of partial oscillators. Identical oscillators are considered with $\lambda_1 = \lambda_2 = 0.5$ and $\omega_1 = 1$. *Solid lines* mark saddle-node bifurcations, *dashed lines* mark torus birth (Neimark–Sacker) bifurcations. **b** Evolution of six periodic orbits along route **A** with $p = 0.98$: the maxima of x_1 of the respective solutions are shown versus B_R. *Solid black line*: stable cycles $S_{1,2}$; *dashed line*: saddle cycles $S^*_{1,2}$; *grey line*: twice saddle cycles $U_{1,2}$. **c** Evolution of four periodic orbits along route **C** with $B_R = 0.15$, designation are as in **b**

4.4.1 Locking

In Fig. 4.5(a) set $p = 0.98$ and increase coupling B_R from zero to 0.3, hence following the route **A** that presumably leads to locking (see Fig. 4.6).

In the absence of coupling $B_R = 0$, the two subsystems have different frequencies which is clearly visible in the spectrum (first row of Fig. 4.6). As we increase coupling B_R two spectrum features change systematically. First, the main spectral peak of the first system marked by △ moves to the right, which means that oscillations in the first subsystem become faster. Second, the main peak of the second

4.4 Reactive Coupling 91

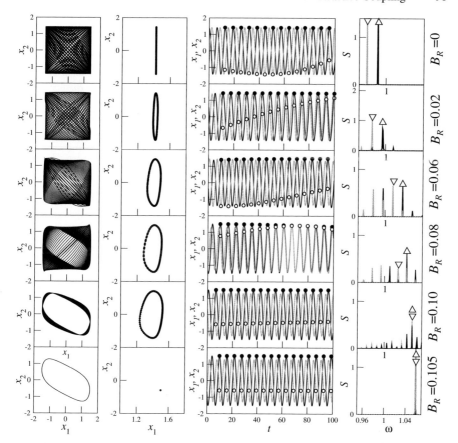

Fig. 4.6. Illustration of 1 : 1 frequency (phase) locking for the mutually coupled van der Pol systems (4.3) with *reactive* coupling at $\lambda_{1,2} = 0.5$. In Fig. 4.5(a) we move along route **A** corresponding to $p = 0.98$, as we enter the synchronization tongue via saddle-node bifurcation line. Phase portraits, stroboscopic sections, realizations and power spectral densities (spectra) are shown for each value of B_R indicated to the right of the respective row. Designations are as in Fig. 4.2. In spectra (**fourth column**) the main peaks of the two subsystems are marked as △ for the first, and ▽ for the second one

system marked by ▽ moves to the right, too. Moreover, the peak ▽ of the second system tends to catch up with peak △, and it finally coincides with the latter at $B_R = 0.10$. However, frequency locking occurs only at $B_R = 0.105$ when oscillations in both subsystems become periodic.

Recall that with dissipative coupling there was no competition between the oscillators and the resulting frequency was settled at a value in between the natural frequencies of two subsystems (Fig. 4.2). However, when coupling is reactive, the two oscillators compete: one of them changes its frequency (speeds up in the given

92   4 1:1 Mutual Synchronization of Periodic Oscillations

example) and pulls the other oscillator with it, so that the final frequency at which they both settle is not in between their natural frequencies.

At the same time with the increase of B_R from zero the phase portrait becomes more stretched along the anti-diagonal, and at $B_R = 0.10$ the trajectory spends a lot of time near the precursor of the stable cycle, which exists at $B_R = 0.15$. With this, the evolution of the Poincaré section is in line with what happens on the way to locking in a forced system and in dissipatively coupled ones: the size of the section does not change near the locking boundary, and the stable cycle is born on the surface of the torus.

The behavior of realizations reflects all the changes described above, and especially the points corresponding to the maxima of x_1 (black circles) are indicative of the phase relationships between the two systems: phase synchronization occurs when each time x_1 is at its maximum, x_2 (empty circles) is at exactly the same stage of its oscillations. The onset of phase locking coincides with the onset of frequency locking.

4.4.2 Suppression

Now consider the route to suppression by setting $p = 0.92$ and increasing B_R from zero, i.e., by moving along the route **B** in Fig. 4.5(a). The respective illustrations are given in Fig. 4.7. As with locking, the spectrum peaks of both subsystems move in the same direction. As coupling becomes stronger, the spectra are enriched by combination frequencies. Also, the oscillators compete, but the first one dominates and tries to suppress the natural dynamics of the second one. At $B_R = 0.2$ the highest peak ∇ of the second system is no longer at the position corresponding to its natural dynamics (compare with case $B_R = 0.18$) but at the position $\triangle$ of the first subsystem. However, this is not synchronization by suppression yet, since oscillations in both systems are quasiperiodic. At $B_R = 0.215$ the main peaks of both subsystems grow above the other peaks, while staying together, but only at $B_R = 0.22$ synchronization occurs and oscillators become periodic.

Transition to suppression is also well visible in the Poincaré section which shrinks just near the boundary of suppression region. Also, the realizations confirm that phase synchronization has occurred only at $B_R = 2.2$ when the maxima of x_1 (black circles) occur at the same stage of x_2 with each oscillatory cycle.

4.4.3 Bifurcations

Now consider bifurcational transitions that occur in the system on its way from non-synchronous to synchronous regime. We will cross the phase locking region in Fig. 4.5(a) in two directions: from below to above by going from "no synchronization" through locking to suppression area (route **A**, see Fig. 4.5(b)); and from left to right and back (route **C**, see Fig. 4.5(c)). This way we are sure to embrace all the objects in the phase space that are involved in the process of synchronization. In all one-parameter bifurcation diagrams the maxima of x_1-coordinate of the periodic orbits are shown against the parameter value.

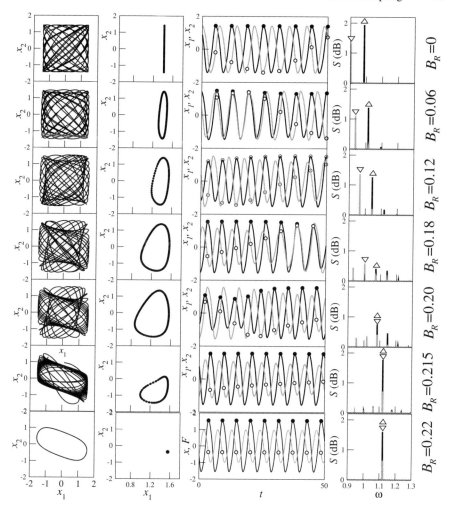

Fig. 4.7. Illustration of 1 : 1 suppression of natural dynamics for the mutually coupled van der Pol systems (4.3) with *reactive* coupling at $\lambda_{1,2} = 0.5$. In Fig. 4.5(a) we move along route **B** corresponding to $p = 0.92$, as we enter the synchronization tongue via saddle-node bifurcation line. Phase portraits, stroboscopic sections, realizations and power spectral densities (spectra) are shown for each value of B_R given to the right of the respective row. Designations are as in Fig. 4.2. In spectra (**fourth column**) the main peaks of the two subsystems are marked as △ for the first, and ▽ for the second one

Route A, Fig. 4.5(b), $p = 0.98$

When considering this route, the reader might find it useful to refer to Poincaré sections in Fig. 4.6. At no coupling $B_R = 0$ the situation is similar to the one with dissipative coupling (compare with Fig. 4.1(b)): there is a twice saddle fixed point U_2 at the origin and a twice saddle cycle U_1 of finite size. At $B_R = 0$ the fixed point U_2 un-

94   4 1:1 Mutual Synchronization of Periodic Oscillations

dergoes Andronov–Hopf bifurcation and a twice saddle cycle is born from it, which we will continue to call U_2. Both $U_{1,2}$ are marked by grey lines. As B_R reaches the value 0.1045, a new pair of periodic orbits is born: a stable S_1 and a saddle S_2^* (black and dashed lines, respectively)—we hit the first line of saddle-node bifurcation in Fig. 4.5(a). At $B_R = 0.16871$ another pair of periodic orbits is born, a stable S_2 and a saddle S_1^*. Then around $B_R \approx 0.2$ both saddle orbits $S_{1,2}^*$ merge with twice saddle orbits $U_{2,1}$, respectively, and disappear via saddle-node bifurcation. This marks the end of locking regions and the beginning of suppression region. As B_R is increased further, we move deeper into suppression region where only a pair of stable cycles $S_{1,2}$ coexist.

Route C, Fig. 4.5(c), $B_R = 0.15$

As we enter the locking region via route **C**, no bifurcations occur to the twice saddle orbits $U_{1,2}$, so they are not shown. The other orbits $S_{1,2}$ and $S_{1,2}^*$ exist only inside the locking region on this route. On the first left border of the locking region a pair of cycles are born via a saddle-node bifurcation, S_2 and S_1^*. While moving deeper into the locking region, we hit the other left saddle-node bifurcation line at $B_R = 0.9864$ on which another pair of cycles appears, S_1 and S_2^*. Within the range of $B_R \in (0.9864; 0.1013)$ there are two coexisting stable limit cycles in the system. At $B_R = 0.1013$ the two cycles S_1 and S_1^* merge and disappear via saddle-node bifurcation, and within $B_R \in (0.9864; 0.10337)$ only one stable cycle exists in the system. At $B_R = 0.10337$ with another saddle-node bifurcation the system gets rid of the remaining pair of orbits S_2 and S_2^*. Note that all four orbits that can exist inside phase locking region are born and evolve on the surface of the same torus.

4.4.4 Phase Multistability

It is important to emphasize that a crucially new phenomenon is induced by reactive coupling, which does not occur in forced or diffusively coupled systems: at exactly the same set of control parameters within the smaller inner tongue in Fig. 4.5(a), there are two stable cycles in the phase space of the system! This means that the system has two different oscillatory regimes to choose from. The oscillations corresponding to these different cycles are different in amplitude and in period. Exactly what regime will be selected depends on the initial conditions. It has significant implications for experiments: since in a typical experiment initial conditions are set rather arbitrarily and are often beyond the control of the experimentalist, it is very hard to predict how the system will behave when the experimental set-up is switched on! The phenomenon of coexistence of two or more stable solutions in the phase space of the system at exactly the same set of control parameters is called multistability.

4.5 Reactive Coupling and the Saddle Torus

In Sect. 4.4 both mechanisms of synchronization, locking and suppression, in reactively coupled van der Pol oscillators were illustrated by experimentally observable data. In addition, bifurcational transitions in the system were considered with regard to these mechanisms. As one carefully considers the latters, one might think of the following:

- Inside the locking region there are two coexisting stable limit cycles, i.e., two attractors.
- We know that any attractor must have a basin of attraction: the set of initial conditions from which the trajectory goes to this particular attractor.
- With two attractors in the phase space, there must be two basins of attraction.
- With two basins, there must be a boundary in the phase space that would separate them from each other.
  Normally, the role of separating boundaries in the phase space is played by the manifolds of various saddle objects, which are also called separatrices.[4] Inside the inner tongue of locking region, there are indeed two saddle cycles, and we can imagine that their manifolds do separate the basins of attraction of the stable cycles.
- However, the same two stable cycles continue to coexist in the region of suppression, where there are no saddle cycles any longer.
- But then, what separates their basins of attraction in the suppression region?
  Are we missing any important information about the structure of the phase space?

In [44] the reactively coupled van der Pol oscillators were considered with non-linear coupling. It was found out that non-linear reactive coupling has lead to disappearance of a stable torus in the system via some bifurcation, which looked mysterious at the first glance. This has lead the authors to pose the above question about the true structure of the phase space in the vicinity of synchronization region.

In [44] it was hypothesized how various objects like fixed points, periodic orbits and tori should be packed in the phase space in order to allow for all the bifurcational transitions observed numerically. In particular, this configuration had to explain the coexistence of two stable cycles in the absence of any saddle cycles around. One of the complications involved in the study of coupled oscillators is the dimension of their phase space. Indeed, the minimal dimension of a system that can demonstrate self-sustained oscillations is two. In mathematical terms, a limit cycle needs at least a two-dimensional phase space (phase plane) to exist, and it cannot arise in systems described by only one first-order differential equation with the one-dimensional phase space. Forced synchronization can be studied by applying the forcing signal to a system with a limit cycle and with dimension two. Since harmonic forcing introduces another dimension into the phase space, the total minimal

---

[4] For the properties of manifolds see Sect. 5.1.

dimension of a forced system is three. A three-dimensional phase space can be easily imagined and visualized, and the number of objects that can be embedded into it is limited. In particular, the largest dimension of a stable torus that can exist in a three-dimensional system is two.

When one considers mutually coupled self-sustained oscillators, the simplest model of their interaction would include two two-dimensional systems, and the total dimension of the phase space will be four. A human brain is not trained to imagine or visualize objects in a four-dimensional space, so we can only consider projections of phase trajectories onto three- or two-dimensional spaces, loosing information while doing so. Or we can consider three-dimensional Poincaré sections of this phase space—this is a better option, since if the section is chosen properly, we will not loose information about the objects apart from fixed points. Thus cycles will then turn into points, tori into closed curves, and three-dimensional manifolds (hypersurfaces) into 2D surfaces.

When considering synchronization of two mutually coupled periodic oscillators, one inevitably encounters a stable two-dimensional (2D) torus. Regardless of the dimension of the phase space of the whole system, its surface is a two-dimensional manifold. While in 3D this two-dimensional surface separate the inner volume of the torus from the outer space, in 4D this surface is no longer a separatrix, and the notion of the "inside" and "outside" the torus makes no sense. With this, 4D allows for the existence of objects more complicated than a stable 2D torus.

4.5.1 Hypothesized Structure of the Phase Space

- The phase space of reactively coupled oscillators is arranged differently to that of a forced two-dimensional system, or of dissipatively coupled systems, the phase space of the latter cases holding only one torus that is stable and can be either ergodic or resonant. Namely, with reactive coupling, inside the central part of locking region of Fig. 4.5(a) containing point 1, there are *two* different tori in the phase space.
- Of these, the first is an *attracting resonant* torus whose dimension is two and whose surface is formed by the two-dimensional unstable manifolds of the saddle cycles $S^*_{1,2}$ that close on stable cycles $S_{1,2}$. A sketch of its Poincaré section is shown in Fig. 4.8(a) by the full black closed curve; circles mark the positions of cycles. This resonant torus is similar to that which exists in a forced system or in dissipatively coupled systems, but now, instead of one pair of cycles, it has two pairs of cycles lying on it.
- The second torus is a *saddle resonant* torus, also of dimension two. In Poincaré section a two-dimensional saddle torus looks like a saddle cycle, and its manifolds will look like those of a saddle cycle, see Fig. 4.8(b). In the full four-dimensional phase space, a saddle torus is the intersection of two three-dimensional manifolds. Note that a saddle torus cannot live in (be embedded into) a three-dimensional space, which is not "spacious" enough for that, and requires the space of dimension four at least.

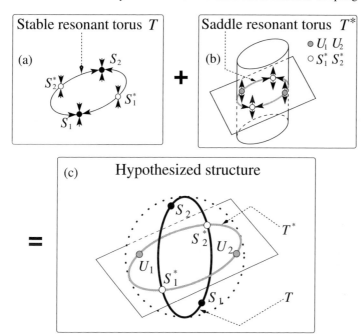

Fig. 4.8. (Color online) Poincaré sections of **a** a stable resonant torus T with two pairs of cycles on it: two stable $S_{1,2}$ and two saddle $S^*_{1,2}$; **b** a saddle resonant torus T^* with two pairs of cycles on it: two saddle $S^*_{1,2}$ and two twice saddle $U_{1,2}$. **c** Hypothesized structure of the phase space: a stable and a saddle resonant tori intersect at $S^*_{1,2}$

- The two tori intersect at the saddle cycles $S^*_{1,2}$ as shown in Fig. 4.8(c). In this figure, circles show cycles: black—$S_{1,2}$, white—$S^*_{1,2}$, grey—$U_{1,2}$. Full closed curves show tori: black—stable T, grey—saddle T^*.
- Two tori lie on the same closed hypersurface ("sphere") sketched in Fig. 4.8(c) by a dotted line. The latter is the unstable manifold of the saddle torus depicted as a cylinder in Fig. 4.8(b) which does not go to infinity, but is closed from below and from above. Horizontal plane is a stable manifold of the saddle torus.

The hypothesized structure of the phase space was verified and visualized by various numerical methods described in [44]. In Fig. 4.9(a) the numerically revealed structure is shown which corresponds to point **1** in Fig. 4.5(a) at which $p = 1.002$ and $B_R = 0.15$. Note that this is not exactly the center of the tongue, since p is slightly larger than 1. Compare this structure with the hypothesized one in Fig. 4.8(c) in order to make sure that it is qualitatively the same.

4.6 Generality of Bifurcational Transitions at Reactive Coupling

The bifurcations in reactively coupled oscillators were mostly revealed by numerical analysis. A question naturally arises: how general is the reported structure of

4 1:1 Mutual Synchronization of Periodic Oscillations

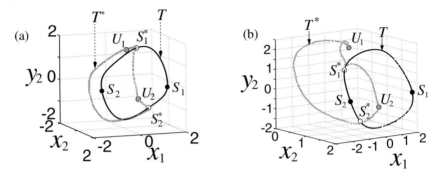

Fig. 4.9. (Color online) **a** The structure of the phase space inside 1:1 locking region for reactively coupled oscillators: **a** Identical van der Pol system (4.3) at $B_D = 0, \lambda_1 = \lambda_2 = 0.5$, $\omega_1 = 1$. Parameters p and B_R correspond to point **1** in Fig. 4.5(a). **b** Non-identical FitzHugh–Nagumo systems (4.22) with $\epsilon_1 = \epsilon_2 = 2$ and $a_1 = a_2 = 0.1$. Parameters p and B_R correspond to point **1** in Fig. 4.10(b). Compare with Fig. 4.8(c)

the bifurcation diagram around 1:1 synchronization tongue? Could it be that the observed bifurcational transitions were partly due to some degeneracy in the system, e.g., due to the fact that the van der Pol oscillators considered were identical apart from their natural frequencies? What happens is we consider non-identical oscillators, or oscillators described by different equations?

In order to demonstrate how general the structure of the 1:1 synchronization region is in two mutually coupled oscillators, we show the similar bifurcation diagram for the case of two slightly non-identical van der Pol oscillators (4.1)–(4.2), with $\lambda_1 = 0.5$ and $\lambda_2 = 0.51$ (Fig. 4.10(a)). This diagram is qualitatively the same as the one for identical van der Pol oscillators shown in Fig. 4.5(a).

In addition, we present the bifurcation diagram around 1:1 synchronization region for two reactively coupled oscillators of a very different type: FitzHugh–Nagumo oscillators. A FitzHugh–Nagumo oscillator is a rough caricature of the famous Hodgkin–Huxley biologically accurate model of a neuron (see [59, 128, 188] for the simplified descriptions of the two models). This system is able to demonstrate periodic oscillations, which would describe repetitive firing of a neuron. The equations for the two reactively coupled systems read

$$\begin{aligned}
\epsilon_1 \dot{x}_1 &= \frac{1}{p}\left(x_1 - \frac{x_1^3}{3} - y_1\right), \\
\dot{y}_1 &= \frac{1}{p}(x_1 + a_1) + B_R(x_2 - x_1), \\
\epsilon_2 \dot{x}_2 &= \left(x_2 - \frac{x_2^3}{3} - y_2\right), \\
\dot{y}_2 &= x_2 + a_2 + B_R(x_1 - x_2).
\end{aligned} \qquad (4.22)$$

Here $\epsilon_1 = \epsilon_2 = 2$ are the time scale separation parameters: they determine how much faster the x-variables are in the equations above as compared to y-variables.

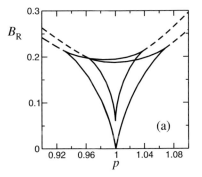 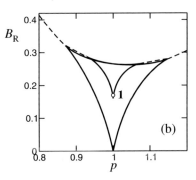

Fig. 4.10. 1 : 1 synchronization tongues on the plane "detuning p"–"coupling strength B_R" for the reactively coupled systems other than identical van der Pol oscillators. **a** Non-identical van der Pol oscillators (4.3) with $\lambda_1 = 0.5$, $\lambda_2 = 0.51$, and $\omega_1 = 1$. **b** Non-identical FitzHugh–Nagumo systems (4.22) with $\epsilon_1 = \epsilon_2 = 2$ and $a_1 = a_2 = 0.1$. *Solid lines* mark saddle-node bifurcations, *dashed lines* mark torus birth (Neimark–Sacker) bifurcations. Compare with Fig. 4.5(a)

$a_1 = a_2 = 0.1$ are bifurcation parameters: while they are within the range $(-1; 1)$, a stable limit cycle exists within each partial FitzHugh–Nagumo system when uncoupled. The frequency detuning between the two oscillators is governed by the parameter p by analogy with the van der Pol systems (4.3), and B_R is the strength of reactive coupling.

The synchronization tongue and its surroundings are depicted in Fig. 4.10(b), which has the same structure as in the van der Pol oscillators, both identical and slightly non-identical (compare with Figs. 4.5(a) and 4.10(a)). For this system the objects in the phase space were calculated using the same algorithm as the one applied to identical van der Pol oscillators in Sect. 4.5, and the structure of the phase space inside the tongue appeared to be qualitatively the same, see Fig. 4.9(b).

4.7 Experiment

As with forced oscillations, in order to convince the reader in the validity of theoretical predictions and of the numerically observed phenomena related to mutual synchronization of periodic oscillations, we present the results of full-scale experiments with electronic circuits. The experimental scheme is given in Fig. 4.11. Now, each of the interacting oscillators is represented by a classical Wien bridge oscillator based on an RC-circuit. The circuit parameters are given in the caption to Fig. 4.11, and the coupling strength is controlled by the value of the variable resistance of the resistor R_c: the larger the R_c, the smaller the coupling is.

It can be shown by deriving the evolution equations for electric currents and voltages in the circuit that the coupling introduced via resistor, as shown in the scheme, results in both a dissipative and a reactive coupling terms. Infinitely large value of R_c is equivalent to the disconnection at the given point of the circuit and means no mutual influence between the subsystems at all. As one can see from

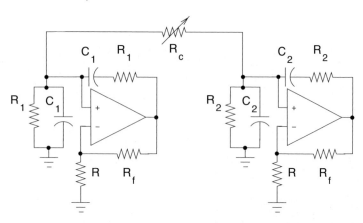

Fig. 4.11. Scheme of an experimental setup used to illustrate the phenomena of mutual synchronization. The Wien bridge oscillators are used with the circuit parameters as follows:
Locking: $R_1 = R_2 = 1.5\,\text{kOhm}$, $R_f = 20\,\text{kOhm}$, $R = 9\,\text{kOhm}$. $C_1 = 33\,\text{nF}$, $C_2 = 35\,\text{nF}$,
Suppression: $R_1 = 1.5\,\text{kOhm}$, $R_2 = 1.65\,\text{kOhm}$, $R_f = 20\,\text{kOhm}$, $R = 9\,\text{kOhm}$, $C_1 = C_2 = 33\,\text{nF}$

the scheme, the parameters R_f and R are identical in two subsystems. The difference between the subsystems can be introduced through the non-equal capacitances C_1 and C_2 or resistances R_1 and R_2 which define the natural frequencies $f^0_{1,2}$ of their oscillations. Note that the given electronic circuits are not weakly non-linear, and oscillations in them are not quasiharmonic: this is well visible in the phase portrait at $R_c = 35.0\,\text{kOhm}$ in Fig. 4.12 (third row), where the shape of the limit cycle tends to a square rather than to a circle.

4.7.1 Phase Locking

First, demonstrate the phenomenon of locking (Fig. 4.12). For that, the parameter set $R_1 = R_2 = 1.5\,\text{kOhm}$, $C_1 = 33\,\text{nF}$, $C_2 = 35\,\text{nF}$ was used. At some large value of $R_c = 48.6\,\text{kOhm}$ when the coupling is weak, the two oscillators behave almost independently of each other, which is especially well visible on the phase portrait in the projection on (x_1, x_2) (compare with Fig. 4.2 at $B_D = 0.015$). The main frequencies of the two subsystems are slightly different, as one can see from the spectra: the highest peak of the first oscillator (second column) is slightly to the left of the middle vertical line on the screen of the oscilloscope, which is highlighted by a white dashed line, while the highest peak of the second oscillator is slightly to the right of it. However, since R_c is not infinitely large, the oscillators do "feel" each others' presence, and each of them demonstrates quasiperiodic oscillations. This is evidenced by the combination frequencies in the spectra, that is somewhat similar to the case of $B_D = 0.015$ and is contrary to $B_D = 0$ in Fig. 4.2.

As R_c decreases to take the value of 35.3 kOhm, the coupling between the subsystems grows, and their spectra are enriched with more combination frequencies. With this, the main oscillation frequencies move towards each other: now they are

4.7 Experiment   101

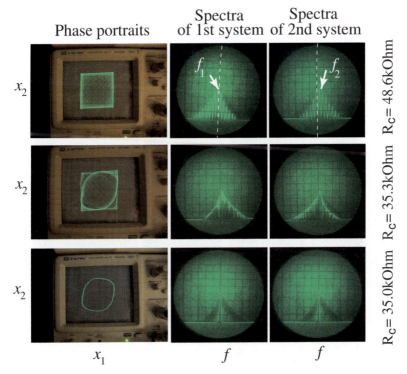

Fig. 4.12. Illustration of 1 : 1 locking in an experiment with the mutually coupled periodic oscillators whose scheme is given in Fig. 4.11. This figure can be compared with Fig. 4.2. All pictures are photographs of the screens of oscilloscopes in which phase portraits on the plane (x_1, x_2) (**first column**), the spectrum of the first subsystem (**second column**), and the spectrum of the second subsystem (**third column**) are shown. *White dashed line* on the spectra emphasize the central axis on the oscilloscope

closer to the vertical line on the oscilloscope screen than in case of $R_c = 48.6$ kOhm, almost coinciding with each other. The phase portrait (first column) reveals that the subsystems are very close to a synchronous regime: the phase trajectory spends a lot of time near a certain closed orbit which is highlighted on the screen of the oscilloscope.

Finally, when R_c is decreased very slightly and achieves the value of 35.0 kOhm, mutual synchronization takes place: the phase portrait is now a limit cycle whose precursor was highlighted in the row above, and the spectra of both systems contain only one peak each at the main frequency of oscillations, which is the same for two systems.

4.7.2 Suppression

Next, demonstrate the occurrence of suppression in mutually coupled oscillators. The parameters of the scheme have slightly changed as compared to the experiment

102   4 1 : 1 Mutual Synchronization of Periodic Oscillations

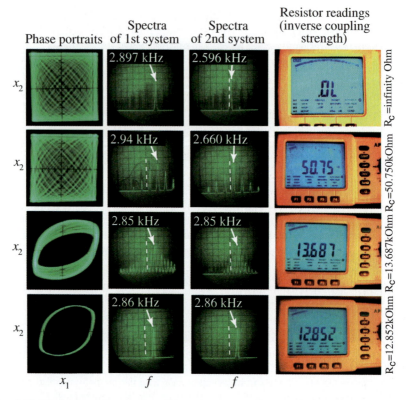

Fig. 4.13. Illustration of 1 : 1 suppression of natural dynamics in an experiment with the mutually coupled periodic oscillators whose scheme is given in Fig. 4.11, to be compared with Fig. 4.4. All pictures are photographs of the screens of oscilloscopes on which phase portraits on the plane (x_1, x_2) and spectra are shown. In the first row *OL* corresponds to an infinite resistance. *Vertical dashed lines* in the spectra indicate the frequency $f_2^0 = 2.596$ kHz of the second system when uncoupled

on locking. Now $C_1 = C_2 = 33$ nF, while the detuning is introduced by setting the non-equal R_1 and R_2 (see caption to Fig. 4.11). The experiment described below does not demonstrate oscillation death, but rather a conventional suppression of oscillations due to coupling, presumably because the coupling appears to be not exclusively dissipative but with the addition of the reactive term. In Fig. 4.13 the results illustrating suppression between the mutually coupled oscillators are summarized. In order to make the illustration more convincing, we provide the snapshots of the resistor readings to the right of each row. Also, the values of the main frequencies of oscillations in each of the coupled subsystems are indicated in the fields of the respective spectra.

At $R_c = \infty$, which is equivalent to a disconnection between the subsystems, the oscillations in them occur at different frequencies (Fig. 4.13 first row, second and third columns) and independently of each other (see phase portrait in the first row, first column). Note that the frequency of the first oscillator is higher than that of the second one.

When R_c takes some large, but finite value of 50.75 kOhm, this is equivalent to a small value of coupling strength. Now, oscillations have started to feel each other, which is reflected in the appearance of numerous peaks at the combination frequencies in their spectra (second row). However, the main spectral peaks stay almost at their initial positions. Also, the oscillations remain pretty much independent which is testified by the phase portrait.

A significant decrease of R_c up to the value of 13.687 kOhm has provided a substantial increase of coupling strength. Now the oscillators behave with a fair account of each other which is reflected in the phase portrait that has a well-defined structure of a two-dimensional torus slightly stretched along the diagonal. The spectra are well enriched with combination frequencies, and a remarkable event has occurred: while the height of the highest peak of the first oscillator remains almost the same as without coupling, in the second oscillator the spectrum peak at the original (smaller) frequency has shrunk, and the dominating peak has grown at the frequency of the first oscillator! Now the main frequencies of the two coupled subsystems are the same. However, this is not synchronization yet, since the oscillations in both subsystems remain quasiperiodic.

Finally at $R_c = 12.852$ kOhm (fourth row) oscillations in both subsystems become strictly periodic with the same frequency $f = 2.86$ kHz, and the phase portrait is a closed curve. Synchronization via suppression of mutually coupled oscillators has taken place.

4.8 Comparison of Synchronization Transitions in Forced and in Mutually Coupled Oscillators

In this section we summarize the similarities and differences between the bifurcational mechanisms of synchronization of periodic oscillations with different forms of coupling that were considered above: unidirectionally coupled, or forced, oscillators; mutually coupled oscillators with dissipative coupling; and mutually coupled oscillators with reactive coupling. Figure 4.14 contains typical two-parameter and one-parameter bifurcation diagrams for all these cases using as an example of the prototype of a self-oscillating system that was considered in this and the previous chapters, namely, the van der Pol oscillator with weak non-linearity.

With certain caution, it would be reasonable to speak about hierarchy of the complexity in the synchronization phenomena in periodic oscillators. A formal criterion of complexity could be the number of periodic orbits involved. With this in mind, bifurcational transitions become more complicated as one goes from forced synchronization with three orbits, through mutual with dissipative coupling with four orbits, to mutual with reactive coupling with six orbits. An important fact to understand and to remember is that if the mutually coupled oscillators are weakly non-linear, the dissipative form of coupling alone cannot induce the complication in the form of phase multistability.

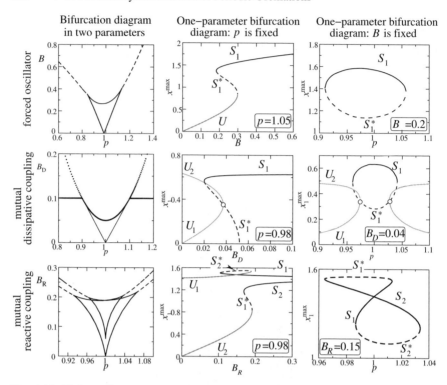

Fig. 4.14. (Color online) Summarized comparison of bifurcation diagrams in the vicinity of 1 : 1 synchronization regions for three cases: **First row**: Forced synchronization in (3.3) with $\lambda = 0.5$, $p = \Omega/\omega_0$ and $\omega_0 = 1$. **Second row**: Mutual synchronization in (4.3) with dissipative coupling, i.e., $B_R = 0$ and $B_D > 0$, with $\lambda_1 = \lambda_2 = 0.1$, $\omega_1 = 1$, $\omega_2 = p\omega_1$. **Third row**: Mutual synchronization in (4.3) with reactive coupling, i.e., $B_D = 0$ and $B_R > 0$, with $\lambda_1 = \lambda_2 = 0.5$, $\omega_1 = 1$, $\omega_2 = p\omega_1$

The behavior of both forced and dissipatively coupled weakly non-linear oscillators is relatively simple and quite similar, apart from oscillator death that cannot occur in a forced system. Phase multistability can only be caused by the introduction of reactive coupling. However, in Chap. 11 we will demonstrate that this is not the case for the oscillators that cannot be regarded as weakly non-linear.

Another important fact to have in mind with regard to the two different forms of coupling, uni- and bidirectional, is that with mutual coupling all spectral peaks move, regardless of the mechanism of synchronization involved. Even when suppression of natural dynamics is being realized, both the peak of the dominating dynamics, and the peak of the slaving dynamics move in the same direction.

5 Homoclinic Mechanism of Synchronization of Periodic Oscillations

In Chap. 3 we considered the two most generic and long known mechanisms of synchronization of periodic oscillations: phase (frequency) locking and suppression of natural dynamics. We have shown that these mechanisms can be associated with local bifurcations of periodic solutions. In this section, we introduce a new synchronization mechanism called the "homoclinic mechanism of synchronization," which involves a non-local bifurcation and is considered in literature to a much less extent.

Let us consider a periodic oscillator driven by a periodic excitation. In the examples considered earlier in Chap. 3 we established that if a weakly non-linear oscillator is forced periodically with *very* small amplitude, and if the forcing frequency is close to the natural frequency of the unperturbed oscillator, then one can expect the phenomenon of 1 : 1 synchronization to occur through the mechanism known as

5 Homoclinic Mechanism of Synchronization of Periodic Oscillations

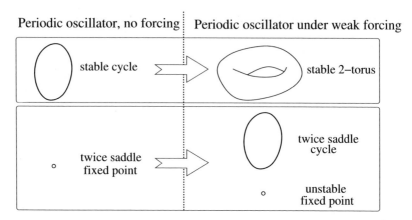

Fig. 5.1. Comparison of the objects that must exist in the phase space of an unperturbed periodic oscillator, and of the same oscillator forced periodically with very small amplitude. There might be more objects in the phase space, but these are compulsory

phase (frequency) locking. Let us briefly recall what the latter means in terms of bifurcations of various objects in the phase space. When there is no forcing, there are at least two objects in the phase space of the system: a stable limit cycle and a fixed point with two unstable directions which will be further referred to as "twice saddle" fixed point (see Fig. 5.1, left panel). Periodic perturbation adds one dimension to the phase space, and at very small forcing amplitudes the forced system generally has: a stable torus produced from the stable cycle (Fig. 5.1, upper panel), plus an unstable fixed point and a twice saddle periodic orbit, both produced from the formerly twice saddle fixed point in the unforced system (Fig. 5.1, lower panel). In the absence of synchronization the torus is ergodic, which means that there are no periodic orbits on its surface.

The state of phase locking in terms of the phase space implies that there is a resonant torus in the system, namely, a two-dimensional toroidal surface on which two cycles are placed: a saddle and a stable one. Importantly, the resonant torus arises from an ergodic (non-resonant) one: a pair of cycles are born *on the torus surface*. For a wide class of forced periodic systems of various origins that have been studied experimentally throughout the last century, the above picture was confirmed and firmly believed in.

Now, let us switch on our mathematical imagination and start thinking as if we do not have this last bit of information. Let us look at the problem of a periodically forced periodic system from the merely geometrical viewpoint. An external forcing can be viewed as an increase of the dimension of the phase space by one. In addition, at very weak strength, the forcing does lead to the birth of a stable 2D torus in the phase space. Moreover, at certain parameters of forcing a pair of cycles can be born via the saddle-node bifurcation. However, *who says these cycles must be born on*

5 Homoclinic Mechanism of Synchronization of Periodic Oscillations 107

the surface of the existing torus?[1] This initial restriction was well justified by the experimental results available by the beginning and the middle of the 20th century. However, if we think of it from the viewpoint of the phase space geometry, there is the whole 3D phase space available for the two cycles to choose the location for their birth.

For an instant, let us discard the assumption that we must expect to find the stable periodic solution exactly on the torus surface, and see what implications this simple thought can lead to. Suppose a tiny forcing is applied to a system with limit cycle with frequency ω_0, and the forcing frequency Ω is slightly different from it. Then, inevitably, a stable torus exists in the phase space. Suppose we increase the forcing amplitude B further, and at some value of it a stable and a saddle cycles are born in some vicinity of the torus. Note that the torus does not undergo any bifurcations, so above the saddle-node bifurcation point two stable regimes coexist! This means the occurrence of *multistability* which was already mentioned in Chap. 4. Depending on the initial conditions, the system can find itself oscillating either quasiperiodically with any main frequency, or periodically with the frequency of forcing—already a complication!

Next, we need to draw our attention to the structure of the synchronization region and stretch the limits of our imagination even further. In Fig. 5.2(a) a sketch of a typical bifurcation diagram of a forced system is shown which is equivalent to the ones we have considered earlier. But here we emphasize a detail which was not given any special attention before: the saddle-node line, which is marked by a solid line on the graph, in fact consists of two portions: the black one marks the saddle-node bifurcation between the saddle and *stable* cycles, while the grey one marks the same bifurcation between the saddle and *twice saddle* cycles. Torus birth (Neimark–Sacker) bifurcation curves hit the saddle-node line exactly at the junctions between its two portions. These special points are bifurcation points of co-dimension two, and they have special names "Takens–Bogdanov points" after the people who discovered and studied them [53, 282]. In this section we will refer to them as to TB-points.

Here, we would like to attract the attention to one more feature of the classical bifurcation diagram in Fig. 5.2(a): small dotted line that emerges from TB point, which was again not shown in the diagrams of Chap. 3. What is this mysterious line? Running a few steps ahead, we can say that this is a homoclinic line whose existence was deduced from the analysis of the truncated equations (3.21)–(3.22) already in [61] in 1948, and then proved rigorously in [88, 117].

However, in 1991 Knudsen et al. [146] have proved the existence of the homoclinic line for a general n-dimensional periodic system under periodic forcing, without reference to the truncated equations and operating in the full phase space of the forced system. Because this result is more general, we would like to comment on it here.

---

[1] Well, there is a good reason for the pair of cycles to be born on the torus at least at small forcing: the torus attracts the trajectories from the surrounding phase space quite strongly.

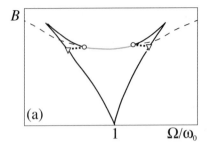 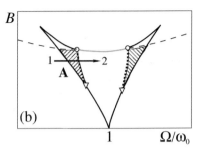

Fig. 5.2. (Color online) Sketches of the possible configurations of bifurcation lines around synchronization region of a periodically forced periodic oscillator. *Solid black lines*: saddle-node bifurcations of a stable and a saddle cycle; *solid grey line*: saddle-node bifurcation of a saddle and a twice saddle cycle; *dashed line*: torus birth bifurcation; *empty circles*: Takens–Bogdanov (TB) points; *dotted line*: homoclinic bifurcation; *triangles*: co-dimension two bifurcation points; *hatched areas*: coexistence of a stable ergodic torus and a stable cycle. **a** "Classical" structure of synchronization tongue that was found analytically in Chap. 3. **b** Possible modification of the classical structure

5.1 Global Bifurcation

Before proceeding further, we need to explain what the global bifurcation is. From the very basics of differential or difference equations, we know that various objects in the phase space, like fixed points or periodic orbits, can undergo bifurcations, the reason behind them being the change in the linear stability of these objects. The term "linear stability" means that we detect the properties of the immediate vicinity of the object only, and do not care about what happens to its larger neighborhood. When the local properties of the object undergo a qualitative change, a local bifurcation occurs. But because local bifurcations are more common than any others, they are normally referred to as simply "bifurcations." However, local bifurcations do not exhaust all possible changes that an object can experience. We remind you that bifurcation is defined as a drastic change in the properties of the object due to a small change in the parameters of the system. Below we will show that there can be even more drastic changes that those caused by local bifurcations.

Consider for simplicity a two-dimensional system with a stable limit cycle, in whose vicinity a saddle fixed point lies (Fig. 5.3(a)). A saddle fixed point lies on the intersection of two manifolds: one stable going towards the fixed point, and one unstable going away from it. Below we remind the reader about the main properties of manifolds.

1. If we place the initial condition exactly on a manifold, we will stay on it *forever*. That is why manifolds are called invariant.
2. If we put the initial condition on the *stable* manifold of a saddle point, with the course of time we will be moving towards the fixed point: the closer to the point, the slower.
3. If we put the initial condition on the *unstable* manifold of a saddle point, with the course of time we will be moving away from the fixed point. There are two

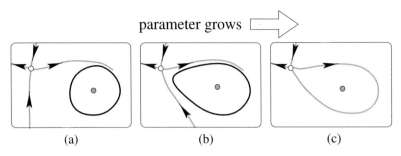

Fig. 5.3. (Color online) Global bifurcation occurring to a stable limit cycle (*black line*) and involving a saddle fixed point (*empty circle*). *Grey circle* shows the unstable fixed point, *grey lines* show the manifolds of the fixed point, and *arrows* show the directions along the manifolds. As we proceed from **a** to **c** a control parameter of the system grows gradually. **c** shows the end of the limit cycle

possibilities: if there is an attractor in the same part of the phase space as the unstable manifold (like the stable cycle in Fig. 5.3(a)), this manifold will be tending towards the attractor. If there is no attractor in this part (see the area to the left of the stable manifold in Fig. 5.3(a)), then the manifold goes to infinity or to some other saddle object.

4. Manifolds whose dimension is equal to the dimension of the full phase space minus one, play the role of separatrices ("walls"): they separate different regions in the phase space, and trajectories cannot cross them.
5. Trajectories that come close to the manifolds roughly follow their direction before they depart from the manifold sufficiently.

In Fig. 5.3(a) the dimension of the phase space is 2, while the manifolds of the fixed point are lines and hence their dimension is one. Because $1 = (2 - 1)$, these manifolds are indeed separatrices. Suppose we fix all parameters of the system except for one, and we gradually increase the latter starting from the value at which the phase space has configuration shown in Fig. 5.3(a). Assume that as the parameter grows, the size of the limit cycle grows, too. This means, that part of it comes closer to the saddle fixed point. From item 5 above it follows that if a part of the cycle comes too close to the saddle fixed point, and therefore to its manifolds, it starts to stretch along the manifolds. With this, its shape is inevitably distorted by adjusting to the shape of the "corner" formed by the manifolds near the saddle point (Fig. 5.3(b)). As our parameter grows further, the limit cycle hits the saddle point (Fig. 5.3(c)). At this instant, the manifolds of the point embrace the cycle and close on it. A homoclinic loop is thus formed. This is called a homoclinic bifurcation.

Note that nothing has happened to the fixed point in this situation, at least on a qualitative level. Most importantly, it did not vanish. Moreover, it did not even change its stability properties: it was a saddle before the bifurcation, and it remains a saddle after it. But what happens to the cycle? The cycle disappears altogether! What is especially interesting here, this catastrophe has happened to cycle through no fault of its own: its *local* properties remained unchanged[2] at the moment of this tragic event, and its size remained large. The only reason for this misfortune is the presence of another object in the phase space that has by mere accident fallen within the vicinity of the cycle. Since this bifurcation is caused by the changes in quite a large neighborhood of the cycle, and is not explained by its local properties, the bifurcation is called non-local, or *global.*

It is worth mentioning in this respect that the stable manifolds of the fixed point form the boundary of the basin of attraction of the limit cycle: only from the initial conditions to the right of the manifold one could arrive at the cycle. By hitting the fixed point, the cycle has in fact touched the boundary of its own basin of attraction. For this reason, this and similar bifurcations are also referred to as *boundary crises.*

5.1.1 Features of a Homoclinic Bifurcation of a Cycle

Let us emphasize the characteristic features of a homoclinic bifurcation of a limit cycle:

- At the moment of disappearance, the cycle has a finite size.
- Before the bifurcation, the cycle becomes distorted with one part stretched and becoming sharper.
- Closely before the bifurcation, the motion on the part that is closer to the saddle point slows down, while away from the saddle point it occurs with normal velocity. Hence, the period of oscillations on the cycle grows substantially and tends to infinity at the moment of bifurcation—which is understandable, since at the fixed point there is no motion at all by definition.
- For the reason above, the motion on the cycle becomes spatially inhomogeneous: during one fraction of the oscillatory cycle the motion is considerably slower than during the other fraction. The realizations become spiky.

All the features above mean that in fact homoclinic bifurcation does not befall on an unsuspecting cycle as a complete surprise, while a parameter of the system is increased. An observer who might be actually changing this parameter will be able to predict the crisis by registering the period and the shape of the cycle. The tendency of the cycle period to infinity is generally a very good criterion of a global bifurcation.

One might ask "What other objects can undergo global bifurcations?" The answer is: any attractor for sure. As an example, take a stable two-dimensional torus whose Poincaré section is a closed curve that looks exactly like the limit cycle in the full phase space. Suppose there is a saddle periodic orbit somewhere near the torus, and its section again looks like the fixed point. Then in the Poincaré section the transitions can be the same as in Fig. 5.3. The most prominent precursor of global

---

[2] To be precise, at the moment of homoclinic bifurcation the Floquet multiplier of the cycle turns to zero, which makes the cycle super-stable. However, this is not regarded as a local bifurcation.

bifurcation will then be the tendency to infinity by the period of amplitude modulation of the realizations of the oscillating system, i.e., the tendency to zero by the beat frequency. At the same time, the section of the torus will be distorted and stretched towards the section of the saddle cycle.

5.2 Homoclinics Inside Synchronization Tongue?

Reverting to Figs. 5.2(a),(b), why would the homoclinic line emerge from the Takens–Bogdanov point at all? The theorem stated in [146] says that this should be the case for the system of any dimension. In this section we follow their argument for a two-dimensional periodic system forced periodically, which is the simplest example of a system that would fall into the required category. Two conditions must be satisfied to make homoclinic bifurcation possible. First, outside the stable torus there must exist a saddle cycle, i.e., an object which the torus could bump into. Well, a saddle cycle does exist within the whole curvy triangle bounded by saddle-node bifurcation lines, so this condition is met near Takens–Bogdanov point as well. The second condition is formally expressed as follows: the torus birth line should point inside the tongue as shown in Fig. 5.2. Suppose this condition is also satisfied.

A proof by contradiction is used. Let us first *make a statement* that there is no homoclinic line in the region of our interest, i.e., bifurcation diagram around Takens–Bogdanov point looks like shown in Fig. 5.4 (this is an enlargement of the respective region of Fig. 5.2(b)). From this it would follow that the torus exists at least everywhere in the shaded area. Let us choose a route shown by a dotted line with an arrow, that leads us from some point just below the torus birth line but as close to it as we wish, to some point on the saddle-node line for non-stable cycles (i.e., for a saddle and a twice saddle ones). The key stages of this route are marked by triangles and numbered as **1, 2, 2*** and **3**. In panels surrounding the bifurcation diagram Poincaré sections of all objects involved are shown whose designations are given in figure caption. Since point **1** is just below the torus birth line, the torus is of small diameter. Importantly, at $B = 0$ this torus was born from, and *around* the now twice saddle cycle, which is denoted by grey circle. The key point in our speculations is that this cycle lies *inside* the torus. Crucially, the torus surface is a manifold, and hence by property 4 of a manifold (see Sect. 5.1) no trajectories, including of course those belonging to any cycles, can cross them. Outside the torus, there are two more cycles, one of which is saddle (empty circle) and the other is stable (black circle). Point **2** is closer to saddle-node line than point **1**, which means that the two cycles (a saddle and a twice saddle) have prepared themselves for a saddle-node bifurcation by approaching each other as shown in the respective panel. Also, point **2** is at a larger distance from the torus birth line than point **1**, so the diameter of the torus is larger as well—but this is not relevant to the problem considered, since we could have chosen a route very close to the torus birth line, along which the torus diameter would not change at all.

At point **3** exactly on the saddle-node line, the saddle and the twice saddle cycles must touch each other. But at point **2** they were separated from each other by a

112   5 Homoclinic Mechanism of Synchronization of Periodic Oscillations

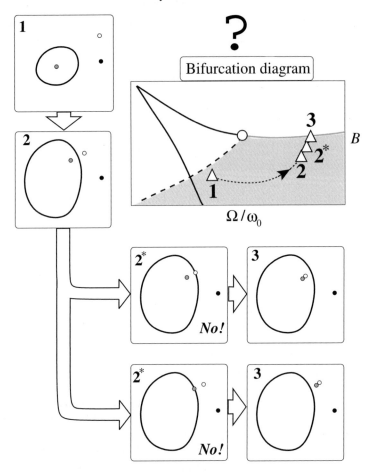

Fig. 5.4. (Color online) Illustration of the argument by Cartwright and Knudsen et al. on the inevitability of the homoclinic bifurcation line emerging from Takens–Bogdanov point, provided that the synchronization tongue looks like in Fig. 5.2(b). Proof by contradiction is employed, for which we first make a statement that there is no homoclinic line inside the synchronization tongue. **Bifurcation diagram**: enlarged part of Fig. 5.2(b) around the Takens–Bogdanov point (*empty circle*), assuming that there is *no homoclinic line*. *Lines* are denoted as in Fig. 5.2. **Panels around bifurcation diagram**: Poincaré sections at points **1**, **2**, **2**[*] and **3** of upper panel. *Grey*, *empty* and *black circles*: twice saddle, saddle and stable cycles, respectively; *black line*: stable torus. This picture is *wrong*, see text

barrier in the form of the torus surface. So, the question is how the two cycles can appear on the same side of this barrier? A naive answer would be: at a certain point **2**[*] one of the cycles should cross the barrier as shown in panels labelled with "**2**[*]." But bear in mind that the barrier is a manifold which is impenetrable by definition! Hence, it is impossible for the two cycles to touch each other if the torus exists. This contradicts the implication of the initial statement, that the torus exists everywhere

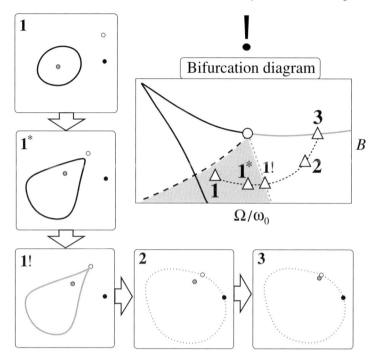

Fig. 5.5. (Color online) Continued illustration of the argument by Cartwright and Knudsen et al. on the inevitability of the homoclinic bifurcation line emerging from the Takens–Bogdanov point, provided that the synchronization tongue looks like in Fig. 5.2(b). **Bifurcation diagram**: enlarged part of Fig. 5.2(b) around the Takens–Bogdanov point. **Panels around bifurcation diagram**: Poincaré sections at points **1**, **1***, **1!**, **2** and **3**. Designations are as in Fig. 5.4. In **2** and **3** the *dotted line* shows a resonant torus that is discussed later in text, but is irrelevant to the argument of the proof. This picture is correct

in the shaded area. Therefore, the statement about the absence of a homoclinic line and about the validity of bifurcation diagram in Fig. 5.4 was wrong!

Now, try to understand how the bifurcation diagram needs to look in order to allow the saddle and twice saddle cycles to merge in a saddle-node bifurcation, and what transitions should occur in the phase space on the chosen route. In Fig. 5.5 the correct bifurcation diagram is shown that now includes the homoclinic line. In order for the non-stable cycles to merge, the torus should disappear. And the only obvious way to do this is via the homoclinic bifurcation as a result of the torus bumping into the saddle cycle. The respective transitions are illustrated in Fig. 5.5 in the panels surrounding the bifurcation diagram.

The simple argument we have given here was first suggested by Cartwright in [61] as applied to truncated equations. It is valid for two-dimensional periodically forced systems only, since their phase space is three-dimensional. This crucially means that the surface of the torus forms a barrier for trajectories, and in order for the two cycles to merge, this barrier has to be destroyed. In systems of higher dimension this argument would not be quite valid, since the torus surface would remain two-dimensional but will no longer be a separatrix. The result by Knudsen et al. will still be valid, but one would need to use a somewhat more sophisticated mathematical concept to demonstrate that: in particular, that of a "central manifold of a bifurcation." For a more rigorous proof we refer the interested reader to the original papers [61, 117, 146].

5.3 How Homoclinics Leads to Synchronization

In the previous section we explained why the homoclinic line must exist inside the synchronization tongue under the specified conditions. However, our argument did not require the knowledge of the details of the structural reforms in the phase space. In this section we will consider the changes in the phase space structure that are due to the homoclinic bifurcation, and how they are related to synchronization.

Imagine that we are following route **A** in Fig. 5.2(b), increasing forcing frequency Ω. In the region containing point **1**, the only stable object in the phase space is the ergodic (non-resonant) torus. The phase space structure inside the hatched area is schematically illustrated in Fig. 5.6(a): in addition to the torus (black line) there is also a stable cycle (black circle), hence multistability occurs. At the same time, there is another important object in the phase space: a saddle cycle (empty circle) born on the black solid line of saddle-node bifurcation. Note that the stable and saddle cycles are inevitably connected by an unstable manifold of the latter, thus forming a heteroclinic structure. Also, note that now the stable manifold of the saddle cycle separates the basins of attraction of the stable cycle and the ergodic torus. As we approach the line of homoclinic bifurcation marked by the grey dotted line in Fig. 5.2(b), the torus approaches the saddle cycle and stretches towards

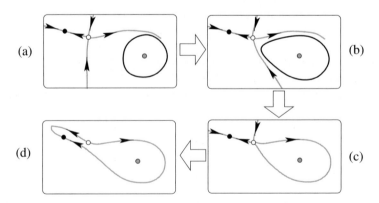

Fig. 5.6. (Color online) Global bifurcation occurring to a stable torus (*black line*) and involving a saddle cycle (*empty circle*) and a stable cycle (*black circle*). Poincaré sections are shown schematically. *Grey circle* shows the twice saddle cycle, *grey lines* show the manifolds of the fixed point, and *arrows* show the directions along the manifolds. As we proceed from **a** to **d** a control parameter of the system grows gradually. **d** shows a newly formed resonant torus

5.3 How Homoclinics Leads to Synchronization   115

it (Fig. 5.6(b)). At the critical value of Ω, the torus is stuck in the saddle cycle (Fig. 5.6(c)), and before this point we can expect nothing new as compared with Fig. 5.3.

But a natural question to ask is: "What happens to this structure next? In particular, what happens to the homoclinic loop as we move deeper into the tongue and closer to point **2** in Fig. 5.2(b)?" One possibility is illustrated in Fig. 5.6: the manifolds of the saddle point, that were glued at the fixed point at the instant of bifurcation (c), can split (d), and the structure can appear that is completely equivalent to a resonant torus. Indeed, there will be a stable cycle and a saddle cycle lying on the same manifold: compare this with the numerical Poincaré sections of the "classical" forced system (Fig. 3.11, fourth raw) and make sure that the structure is the same! Thus, inside the region bounded by homoclinic bifurcation lines a classical resonant torus should exist!

Continue to exploit your imagination: suppose we are performing an experiment with the system that has a synchronization tongue with a structure like in Fig. 5.2(b). Suppose we find the ergodic torus at point **1** and increase the forcing frequency Ω gradually by moving towards the tongue. When we pass by the saddle-node line, we will not notice this, since we will stay in the basin of attraction of the torus and away from the basin of the cycle. However, on the homoclinic bifurcation line, the ergodic torus will abruptly disappear, and the observed trajectory will be bound to jump on the existing stable cycle, whose frequency, by the way, will be equal to the frequency of forcing. An illustration of this phenomenon for a chaotic system is given in Fig. 8.33. What would this transition mean to us? Correct: synchronization. Therefore, by performing this experiment in our imagination, we have derived the principal possibility for a new mechanism of synchronization: via homoclinic bifurcation. Congratulations!

Finally, a question remains: where should the line of homoclinic bifurcation go from the Takens–Bogdanov point? And again, we are able to answer this question by using only the power of our minds. Naturally, in order for the homoclinic bifurcation to be possible, we need at least two objects in the phase space: a torus and a saddle cycle. With this, the saddle cycle required exists only within the area bounded by the lines of saddle-node bifurcations (black and grey in Fig. 5.2). Hence, the line of homoclinic bifurcation must not leave this area. But it cannot end in the middle of the tongue either. Therefore, it should end at some point on the saddle-node bifurcation line for stable and saddle cycles, as shown in Fig. 5.2.

Of course, a skeptic reader would ask: "But how valid are the experiments in our heads? Do they have anything to do with reality?" The experimental evidence of synchronization via the homoclinic bifurcation seems to be obtained by Ueda. In pp. 59–60 of [3] an experiment with a forced van der Pol oscillator is described as follows: "When one chose an intermediate value of external amplitude B... both periodic and beat oscillations would coexist... causing hysteresis and jump phenomena. The parameter range within which such phenomena could occur was... narrow... Stroboscopic sampling (of beat oscillations) filled out a smooth curve." This looks like an accurate description of a homoclinic transition between synchro-

116   5 Homoclinic Mechanism of Synchronization of Periodic Oscillations

nized and non-synchronized (beats) regimes. And this occurred in the simplest van der Pol oscillator already!

How general is this transition to synchronization? In [146] a two-dimensional system describing the behavior of a simple chemical oscillator Brusselator was considered, whose equations read

$$\begin{aligned}\dot{x} &= a + x^2 y - bx - x + B\cos(\Omega t), \\ \dot{y} &= bx - x^2 y.\end{aligned} \qquad (5.1)$$

At $B = 0$ these equations describe a five-component oscillating chemical reaction, where x and y reflect the changes in the concentrations of two species involved in the reaction, while the concentrations of three others species are the control parameters a and b which are assumed to be constant, see [90, 236] for details. Provided that $b > 1 + a^2$, the unforced system ($B = 0$) can demonstrate periodic self-sustained oscillations that are born as a result of Andronov–Hopf bifurcation. In [146] the parameters were fixed as $a = 0.4$ and $b = 1.2$, i.e., slightly above this bifurcation, and the frequency of oscillations at these parameters is $\omega_0 = 0.3750375$. The region of 1 : 1 synchronization was revealed, and here we reproduce their results by plotting the lines of saddle-node and torus birth bifurcations with the help of AUTO [73].

In Fig. 5.7 the bifurcation diagram around the 1 : 1 synchronization tongue for this system is given. For the convenience of comparison with other results in this chapter, we plot this diagram on the plane of parameters "detuning p"–"forcing amplitude B," where $p = \Omega/\omega_0$. One can immediately notice that while the right-hand part of this diagram looks "normal" like in Fig. 5.2(a), the left-hand part has the shape as in Fig. 5.2(b), although it is stretched towards the upper-left corner of the

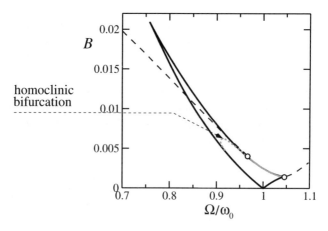

Fig. 5.7. (Color online) Vicinity of 1 : 1 synchronization region in a forced Brusselator (5.1) on the plane of parameters $p = \Omega/\omega_0$ and B. The left-hand border is like in Fig. 5.2(b). *Solid black line*: saddle-node bifurcation of a stable and a saddle cycle; *solid grey line*: saddle-node bifurcation of a saddle and a twice saddle cycles; *dashed lines*: torus birth bifurcations; and *dotted grey line*: homoclinic bifurcation. *Empty circles*: Takens–Bogdanov points

5.4 Synchronization in a Bacteria–Viruses Model   117

parameter plane. The homoclinic bifurcation line emerges from Bogdanov–Takens point and ends on the saddle-node line, as expected.

However, the bifurcation lines of this particular tongue are inclined in such a way, that synchronization transition from an ergodic to resonant torus would occur only if we choose a rather intricate path on the plane (p, B), and the probability to follow this path in a traditional experiment on synchronization does not seem large enough. But at least the principal possibility for the non-classical synchronization mechanism is confirmed in a numerical simulation.

5.4 Synchronization in a Bacteria–Viruses Model

In [218, 220] a system was studied numerically that demonstrated the proper homoclinic synchronization. Namely, a biologically motivated system describing the populations of bacteria and viruses in a pool was considered, whose equations read

$$
\begin{aligned}
\dot{B} &= \frac{\nu BS}{S+K} - B(\rho + \alpha\omega P), \\
\dot{I} &= \alpha\omega BP - \rho I - I/\tau, \\
\dot{P} &= \phi - P\big(\rho + \alpha(B+I)\big) + \beta I/\tau, \\
\dot{S} &= \rho\big(\sigma(t) - S\big) - \frac{\gamma\nu BS}{S+K}, \\
\sigma(t) &= \sigma_0\left(1 - \frac{m}{2}[1 + \sin(\Omega t)]\right).
\end{aligned}
\tag{5.2}
$$

All variables in these equations are dimensional, unlike all previous examples considered in this part. Here, B is the concentration of the bacteria population, P is the concentration of the viruses (phages) population, I is the concentration of the infected bacteria cells, all concentrations being measured in $10^6/\text{ml}$. S is the concentration of resources (food for bacteria) in the pool. $\sigma(t)$ denotes the concentration of nutrients supplied to the pool from outside. The meanings of this model parameters can be found in [218, 220], but here we would like to concentrate on its dynamical properties. At the set of parameter values

$$
\begin{array}{lll}
\nu = 0.024/\text{min}, & & \omega = 0.8, \\
K = 10\,\mu\text{g/ml}, & & \gamma = 0.01\,\text{ng}, \\
\tau = 30\,\text{min}, & \text{and} & \beta = 100, \\
\alpha = 10^{-3}\,\text{ml/min}, & & \phi = 10^{-6},
\end{array}
$$

and without the periodic modulation of σ, the system demonstrates periodic self-oscillations with frequency $\omega_0 = 0.0042$, and its phase space contains a limit cycle and a twice saddle fixed point. When a weak periodic modulation of the supplied nutrients is switched on, the phase space contains the predicted set of objects as

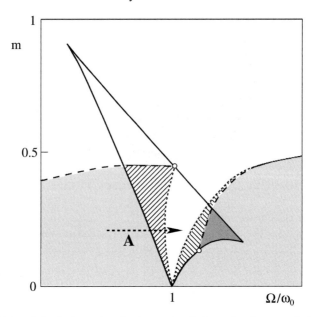

Fig. 5.8. Vicinity of the 1 : 1 synchronization region in bacteria–virus population with periodic modulation of nutrients supply (5.2) on the plane of parameters $p = \Omega/\omega_0$ and modulation depth m. The left-hand side of the tongue border is like in Fig. 5.2(b). *Solid black line*: saddle-node bifurcation of a stable and a saddle cycle; *solid grey line*: saddle-node bifurcation of a saddle and a twice saddle cycles; *dashed lines*: torus birth bifurcations; *empty circles*: Takens–Bogdanov points; and *dotted grey lines*: homoclinic bifurcation. In the *hatched area* a stable torus and a stable cycle coexist

schematically illustrated in Fig. 5.1, namely, a generally ergodic torus, a twice saddle cycle and an unstable fixed point.

A bifurcation diagram around 1 : 1 synchronization tongue was obtained numerically by a number of methods, including AUTO [73] and specially developed software. The diagram is given in Fig. 5.8. Note that the left-hand side of this diagram is like in Fig. 5.8(b).

Consider the left-hand side first and make a note that it looks just like the left-hand side of the one of the forced Brusselator in Fig. 5.7. In agreement with theoretical predictions of [146], from a Takens–Bogdanov point a homoclinic bifurcation line grows in the direction towards the saddle-node line for stable and saddle cycles, and it stops on this line. However, the right-hand side of the synchronization tongue does not look like the one in Fig. 5.7. Note a homoclinic bifurcation line inside the tongue, that connects with the torus birth line above and with the saddle-node line below. The homoclinic line here joins not the saddle-node line, but rather the torus birth line, at the co-dimension-two bifurcation point. Such bifurcations were discussed in [102]. Multistability occurs inside the hatched area where an ergodic torus and a stable cycle coexist (Fig. 5.9).

5.4 Synchronization in a Bacteria–Viruses Model

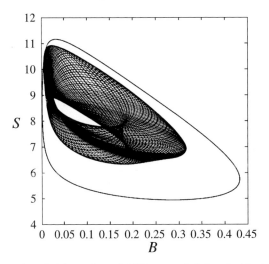

Fig. 5.9. Phase portraits of (5.2) at $\Omega = 0.0039$, $m = 0.3$, i.e., inside the *hatched area* in Fig. 5.8

Unlike in Fig. 5.7, homoclinic bifurcation lines are almost vertical. This means that as one enters the synchronization region along the route like the one marked as **A** in Fig. 5.8, one inevitably crosses it. Let us follow the changes occurring in the phase space on this route by observing the Poincaré sections of the available objects. Fix $m = 0.14$ and change Ω from 0.003 to 0.0041. At $\Omega = 0.003$ ($p = \Omega/\omega_0 \approx 0.714$), there is a stable torus in the system shown in Fig. 5.10(a). A twice saddle cycle is not shown, since it is irrelevant to the particular scenario. At $\Omega = 0.0039$ ($p \approx 0.928$) we are inside the hatched area where the torus coexists with stable and saddle cycles (Fig. 5.10(b)). In this figure we also see a numerically revealed trace of both sides of unstable manifolds of the saddle cycle—see a sequence of dots that go from the empty circle to the torus on one side, and to the stable cycle on the other side. Compare this with the sketch in Fig. 5.6(a) and make sure that the visualized picture repeats the picture we imagined earlier. On the line of homoclinic bifurcation the stable torus has touched the saddle and disappeared, and there is a homoclinic loop formed by the manifolds of the saddle point. The dots in Fig. 5.10(c) belong to these manifolds and were numerically revealed in [220] using an extension of the method proposed in [137]. Note that at these parameters the stable cycle has complex-conjugate Floquet multipliers, and its Poincaré section looks like a stable focus, rather than a node unlike in Fig. 5.6. In Fig. 5.10(c) this is especially well visible, since the trajectory goes to the cycle along a spiral. However, this does not change anything, since the node is not much different from a focus from the viewpoint of topology. Also, at the moment of bifurcation, the manifolds still intersect, and one can notice that they go through the saddle at some angle and not along smooth lines, exactly like in Fig. 5.6(c).

With a further increase of Ω, we are inside the tongue, and the Poincaré section (Fig. 5.10(d)) shows how the manifolds have split near the saddle cycle and the new

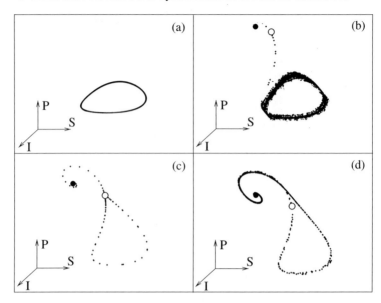

Fig. 5.10. Poincaré sections of the system describing a bacteria–virus population with periodic modulation of nutrients supply (5.2) obtained by numerical methods. Detuning p is being increased following route **A** in Fig. 5.8. The figure illustrates a homoclinic bifurcation occurring to a stable torus (*black line*) and involving a saddle cycle (*empty circle*) and a stable cycle (*black circle*). As we proceed from **a** to **d** the detuning p (see Fig. 5.8) grows gradually. In **b** the trace of the manifold going from saddle cycle (*empty point*) to the torus (*black line*) is shown. **d** shows a newly formed resonant torus

structure in the form of a resonant torus has born, in full analogy with Fig. 5.6(d), except that the unstable manifolds of the saddle close on the stable cycle after making a few loops around it.

5.5 Summary

Let us discuss the meaning of the homoclinic bifurcation line inside the synchronization tongue. Coming back to the description of what 1 : 1 phase locking is (see second paragraph of Chap. 5), in physics terms, 1 : 1 synchronization implies that the frequency of forced system coincides with frequency of forcing at appropriate values of forcing parameters from inside a certain range. In mathematical terms synchronization is most often associated with going from an ergodic torus to a resonant one. What happens in the system (5.2) is exactly the same, except that the details of this transition are different from the classical phase locking. Hence, we can state that a synchronization transition occurs in this system, but via a mechanism that is neither phase locking, nor suppression. We can call this *synchronization via homoclinic bifurcation*, or via crisis.

6 $n:m$ Synchronization of Periodic Oscillations

In Chap. 3 we considered synchronization phenomena in periodically forced periodic oscillators, when the forcing frequency was close to the natural frequency of unforced oscillations. A natural question arises: "What happens if we are perturbing the system with a frequency which is substantially different from its natural frequency? Are there any synchronization phenomena in that case?" The answer is "yes," and below we will illustrate this. Before we proceed with considering the effects involved, the next section introduces a few fundamental concepts—but a reader with the appropriate mathematical background can skip it.

6.1 Important Definitions Relevant to $n:m$ Synchronization

As promised above, in this section we give definitions of several mathematical concepts that are immediately relevant to $n:m$ synchronization.

6.1.1 Poincaré Return Time

Poincaré return times, or simply "return times" for brevity, $T_i, i = 1, 2, \ldots$, are the time intervals between two successive crossings by the phase trajectory of the

122   6 $n:m$ Synchronization of Periodic Oscillations

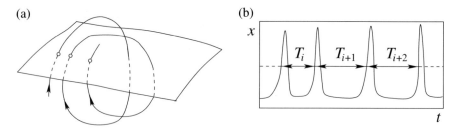

Fig. 6.1. Illustration of the concept of **a** Poincaré section and the return times; and **b** threshold-crossing interspike intervals

Poincaré secant surface in one direction, e.g., from below to above like illustrated in Fig. 6.1(a). If the system oscillates periodically, the values of T_i are the same for all i and are equal to the period T of oscillations.[1] Then $2\pi/T_i = 2\pi/T$ is the main frequency of periodic oscillations at which the highest spectrum peak is located. If the system behaves quasiperiodically, all T_i's are generally different. Then their average value is the mean period of oscillations, whose inverse is the mean frequency of oscillations. Note, that the latter might not be associated with any particular spectral peak. Another name for T_i is (threshold-crossing) *interspike interval*, that comes from neurobiology where the return time is defined as a time interval T_i between two successive intersections by the realization x of a certain threshold (Fig. 6.1(b)).

6.1.2 Phase of Oscillations

In Sect. 3.1 the phase of quasiharmonic oscillations was defined. However, most real-life oscillations are not quasiharmonic, like the ones in the systems that are *not* weakly non-linear. For them one needs to define phase as well. It needs to be emphasized that there is no unique way to introduce a phase for any oscillations of any complexity. Moreover, phase is a subsidiary concept that only serves as a helpful means to quantify certain effects. Therefore, for the same process, one can introduce several phases in different ways, depending on the needs of the particular problem. However, we haste to say, if the phases are introduced correctly, they should all be in agreement with each other. Two methods for the phase introduction are described in Sect. 8.3, but below we will consider the method which is most relevant to this chapter.

6.1.3 Phase of Oscillations via Poincaré Section

One particular approach for the phase definition is related to the Poincaré section, according to which the phase φ is introduced as a function of time that increases

---

[1] This is true, provided that the limit cycle has only one loop. If it has n loops, then the secant surface might be defined in such a way, that the cycle crosses it up to n times during one period. In that case, the subsequent T_i's are different, but they behave periodically.

exactly by 2π between two successive returns to the secant surface in the same direction (e.g., from above to below). Moreover, in between the two returns, the phase grows linearly and hence the full information about it is contained in the values of time moments of return. To summarize, one return to the secant surface corresponds to one full oscillation and to the change of phase by 2π. The respective formula is given by (8.10).

6.1.4 Poincaré Winding (Rotation) Number

Poincaré winding number ρ is introduced as

$$\rho = \lim_{i \to \infty} \frac{\varphi_i - \varphi_0}{2\pi i}, \tag{6.1}$$

where i is the number of return to the given secant surface, and φ_i is the phase of the forced system at the instant of ith return. It is known that the winding number does not depend on the initial phase φ_0 [133]. Using the definition of phase above, the denominator of ρ in (6.1) is the increment of the phase of the *external forcing* between the launch of the process at $i = 0$, and the end of ith full oscillation of the force. The numerator is the increment of the phase φ of the *forced system*, that has occurred while the forcing has made i full oscillations. If the forcing is periodic, this is equivalent to saying that ρ is the ratio of the period of the forcing to the mean period of the forced system.

What is the relevance of the winding number ρ to synchronization problem? If the forcing system makes m oscillations while the forced system makes n oscillations, the winding number ρ is equal to $n : m$, i.e., is a rational number,[2] and this implies the occurrence of $n : m$ synchronization. If ρ is irrational, this means the absence of synchronization.

6.1.5 Synchronization Order $n : m$

The order of synchronization is equal to $n : m$ if m periods of external forcing correspond to n periods of response oscillations; i.e. it is the same as rotation number, only sometimes this term becomes more convenient to use.

van der Pol seems to be the first to investigate the response of the circuit to periodic forcing in a wide range of its frequencies [292]. $1 : m$ and $m : 1$ synchronization was studied analytically by Hayashi [110] and Landa [160] for the special cases of a harmonically forced oscillator for the full range of the values of forcing amplitude B.

6.2 $1 : m$ and $m : 1$ Forced Synchronization in Weakly Non-linear Oscillators

Consider a system of two van der Pol oscillators, assuming that one subsystem is forcing another as follows:

---

[2] A rational number is a number that can be represented as the ratio of two integer numbers.

$$\dot{x}_1 = y_1,$$
$$\dot{y}_1 = (\lambda - x_1^2)y_1 - x_1 + Bx_2, \quad (6.2)$$
$$\dot{x}_2 = y_2,$$
$$\dot{y}_2 = (\lambda - x_2^2)y_2 - p^2 x_2. \quad (6.3)$$

Here, (6.3) represent a single autonomous van der Pol system, while (6.2) describe the van der Pol oscillator that is forced by a signal from (6.3). One can also say that the two subsystems are coupled unidirectionally.

Note, that in Sect. 3 we considered a forced van der Pol oscillator (3.3), assuming that the frequency of forcing was close to its natural frequency, and the bifurcation diagram in the vicinity of a 1 : 1 synchronization region was shown, e.g., in Fig. 3.10(a) for $\lambda = 0.5$. The difference between an oscillator to which a harmonically oscillating force is applied, and an oscillator forced by another self-oscillating system is that in the latter case the applied signal Bx_2 is crucially *not harmonic*.[3] This means that its spectrum contains peaks not only at the main frequency, but also at the multiples of it. The use of a non-harmonic forcing will help us to illustrate the phenomena that were difficult to detect with harmonic forcing, because they occurred only within very narrow ranges of control parameters.

Here, we set a relatively small value of the non-linearity parameter $\lambda = 0.5$, which is the same as the one used for numerical illustrations in Sect. 3.9. The phase portrait, realization and spectrum of the autonomous van der Pol system at $\lambda = 0.5$ are shown in Fig. 6.2.

In order to illustrate the effects induced by the external non-harmonic forcing applied to a van der Pol oscillator (6.2), we calculate the interspike intervals (return

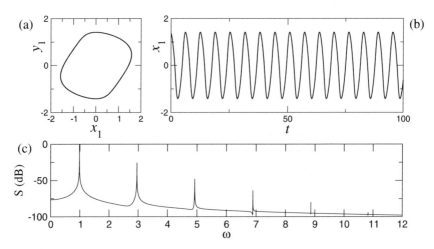

Fig. 6.2. a Phase portrait, **b** realization x_1 and **c** spectrum of x_1 of (6.2) at $\lambda = 0.5$ and $B = 0$

---
[3] Because self-sustained oscillations can occur only in non-linear systems.

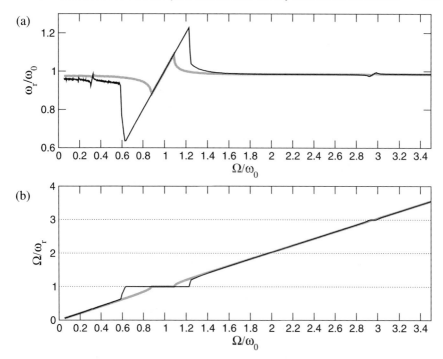

Fig. 6.3. $1:m$ and $m:1$ synchronization in a weakly non-linear van der Pol oscillator forced by another van der Pol oscillator (6.2)–(6.3) at $\lambda = 0.5$. Forcing amplitudes are $B = 0.2$ (*grey line*) and $B = 0.4$ (*black line*), and p changes in a wide range. **a** The response frequency ω_r against the forcing frequency Ω, both normalized by the natural frequency $\omega_0 = 0.984721005$. *Linear segments* with different slopes are visible. **b** Inverse rotation number $1/\rho = \Omega/\omega_r$ against detuning Ω/ω_0. Plateaus at $1/\rho = 1$ and 3 are clearly visible. This is a trace of "devil's staircase," for more detail see Fig. 6.4

times) both for the forcing and for the forced systems. We define them as the time intervals between the successive crossings of zero in one direction (from below to above) by the variables x_1 and x_2, respectively. Because the forcing is periodic, the inverse of any of its interspike intervals gives the true value of the forcing frequency Ω. The inverse of the *mean* interspike interval of the forced system provides the value of the mean response frequency ω_r. In the table below all designations are given that will be used in this section.

Set the strength of forcing at two constant values $B = 0.2$ and $B = 0.4$, and let p change between 0.01 and 3.5. Figure 6.3 summarizes the frequency relationships in the two unidirectionally coupled van der Pol oscillators (6.2)–(6.3) at $\lambda = 0.5$. The abscissa shows the classical parameter used to explain synchronization phenomena: frequency detuning between the systems, which is the ratio of the forcing frequency Ω to the natural frequency of unforced oscillations ω_0, rather than p.

126   6 $n:m$ Synchronization of Periodic Oscillations

The reason is that p is not equal to Ω/ω_0, which will also be illustrated later.[4] Figure 6.3(a) shows normalized response frequency ω_r/ω_0.

One can make a few observations. First, ω_r/ω_0 changes with detuning Ω/ω_0 in the whole range of the values of the latter. Second, ω_r/ω_0 changes in a non-monotonic way. Moreover, there are segments when ω_r/ω_0 grows linearly with Ω/ω_0. Third, at a larger value of B the linear segments are wider, and also more of such segments become visible. Note, that the widest linear segment can be found around $\Omega/\omega_0 = 1$, and the next most pronounced linear segments are located around the values $\Omega/\omega_0 = 1/\rho = 1:3$ and $3:1$, i.e., at the inverses of the respective rotation numbers ρ.

Consider Fig. 6.3(b) where the same information as in (a) is presented, only at a slightly different angle. Here, we do not plot the (normalized) response frequency ω_r alone, but rather the ratio Ω/ω_r, i.e., an inverse of rotation number ρ. This representation allows one to reveal the horizontal plateaus of $1/\rho$ at $1:m$ or $m:1$, where m's are integer (odd) numbers. Each such plateau implies that in the given range of Ω/ω_0, the forcing entrains the subsystem being forced, so that its frequency remains equal to a multiple or a fraction of the one of forcing. Note, that this is a robust effect that occurs not just in one point, but in a whole range of forcing frequency values.

In Fig. 6.3 only the general view of the frequency dependencies is given, but for the finer structure of this graph let us turn to Fig. 6.4 where the smaller values of Ω/ω_0 are illustrated in more detail. In (a) one can see that in fact there are a lot of linear segments in the graph considered, which were not detectable from Fig. 6.3 just because they were much smaller than the one around $\Omega/\omega_0 = 1$. Figure 6.4(b) presents the same information as Fig. 6.3(b), but here the vertical axis is inverted to show the rotation number $\rho = \omega_r/\Omega$. This is done in order to allow the reader to see for sure that the plateaus occur at the levels of $\rho = m:1$ corresponding to odd m. Namely, plateaus at m equal to 3, 5, 7, 9, 11, 13 and 15 can be found in this figure. The occurrence of a plateau with rotation number $m:1$ implies that the response frequency is exactly m times larger than the forcing one. In other words, while forcing makes one oscillation, the forced system makes m oscillations, and hence synchronization of the order $m:1$ occurs. Note, that plateaus of synchronization with rotation number $\rho = m:1$ appear at the values of Ω/ω_0 which are close to the respective values of $1/\rho$, i.e., to $1:11$, $1:13$, etc. This means that the $m:1$ synchronization can occur when the forcing frequency is initially sufficiently close to m times the natural frequency.

To summarize, Figs. 6.3 and 6.4 have illustrated synchronization of two kinds: $1:m$ and $m:1$, where $m > 1$. These phenomena are also called synchronization on

---

[4] When (6.2)–(6.3) are uncoupled ($B = 0$), the natural frequencies of their self-oscillations at Andronov–Hopf bifurcation ($\lambda = 0$) are equal to the square roots of the coefficients before x_1 or x_2, i.e., to 1 and to p, respectively. However, with the non-linearity λ being not vanishingly small in both subsystems, their natural frequencies are just a tiny bit smaller than 1 and p; e.g. in (6.2) the natural frequency $\omega_0 = 0.984721005 < 1$. Parameter p does determine the frequency Ω of the forcing signal Bx_2, and hence the detuning between the two subsystems, but is not equal to the latter.

6.2 1 : m and m : 1 Forced Synchronization in Weakly Non-linear Oscillators

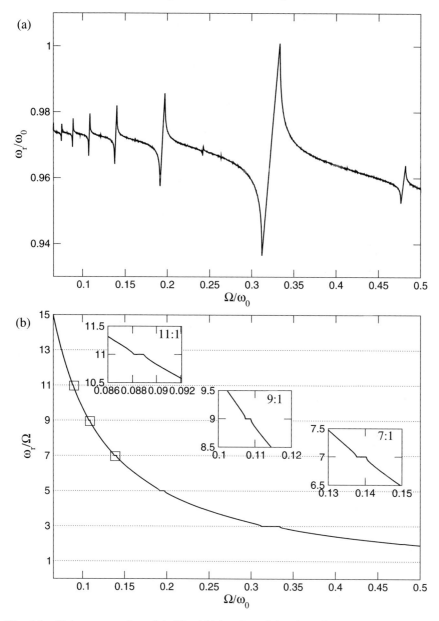

Fig. 6.4. a Enlargement of graph in Fig. 6.3(a) at $B = 0.4$ and small Ω/ω_0. A lot of *linear segments* are visible. **b** The same data as in Fig. 6.3(b), but the ordinate is inverted for convenience to show the rotation number $\rho = \omega_r/\Omega$: *plateaus* at odd numbers are visible. *Insets* show the enlargements of this graph around the plateaus at ρ equal to 11, 9 and 7

overtones and on undertones, respectively [160]. Panels (b) of these figures show the structure reminding the devil's staircase which we will describe in Sect. 6.3.

128   6 $n:m$ Synchronization of Periodic Oscillations

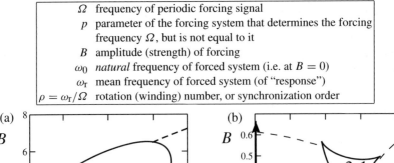

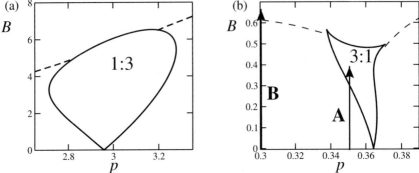

Fig. 6.5. Bifurcation diagrams in the vicinities of **a** 1 : 3 and **b** 3 : 1 synchronization regions in a van der Pol oscillator forced by another van der Pol oscillator (6.2)–(6.3) at $\lambda = 0.5$. *Solid black lines*: saddle-node bifurcations; *dashed lines*: torus birth bifurcations

What we have seen so far, were the analogs of one-parameter bifurcation diagrams: in (6.2)–(6.3) we changed only one parameter p that defined the frequency detuning between the two subsystems (although was not equal to it). However, the transitions to synchronous regimes are better understood on the plane of two classical synchronization parameters: detuning and forcing strength B. We choose two particular synchronization regions where non-1 : 1 synchronization occurs: 1 : 3 and 3 : 1 synchronization. The vicinities of the respective synchronization tongues are shown in Fig. 6.5(a), (b) on the planes (p–B). Detailed analysis of 1 : m and n : 1 synchronization tongues can be found in [110, 160].

These figures are reminiscent of that for the 1 : 1 synchronization (compare with Fig. 3.9). Namely, synchronization occurs within the region whose lower part is a curvy triangle with the tip at $\Omega/\omega_0 = 3$ (although the respective value of p is slightly less than 3). The borders of the synchronization regions are formed by the lines of saddle-node bifurcation in their lower parts, and by torus birth bifurcations in their upper parts. We illustrate the transitions in the vicinity of 3 : 1 synchronization with the same characteristics as were used to illustrate 1 : 1 synchronization.

6.2.1 3 : 1 Phase (Frequency) Locking

Let us fix $p = 0.35$ and go along route **A** in Fig. 6.5(b) to enter synchronization region via the line of saddle-node bifurcation, i.e., consider phase (frequency) lock-

6.2 1 : m and m : 1 Forced Synchronization in Weakly Non-linear Oscillators

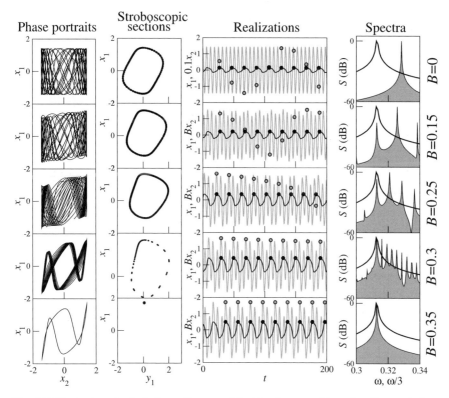

Fig. 6.6. Illustration of a 3 : 1 phase (frequency) locking in two unidirectionally coupled van der Pol oscillators (6.2)–(6.3) at $\lambda = 0.5$. In Fig. 6.5(b) we move along route **A** corresponding to $p = 0.35$, as we enter the synchronization tongue via saddle-node bifurcation line. Phase portraits, stroboscopic sections, realizations and spectra are shown for each value of forcing strength B given on the right of each row. **First column**: Phase portraits on the plane (x_2, x_1). **Second column**: Stroboscopic sections on the plane $(\dot{x}_1, x_1)$, with $\dot{x}_1 = y_1$. **Third column**: *Black line*—forcing $Bx_2(t)$ (except at $B = 0$ where it is $x_2/10$), *grey line*—response $x_1(t)$; *black circles*—maxima of forcing Bx_2, *grey circles*—value of x_1 when x_2 is at its maximum. **Fourth column**: Spectra: *black line*—forcing Bx_2, *shaded*—response x_1. The scales of the frequency axis are *different* for two spectra: the spectrum of forcing is shown against ω, while the one of response against $\omega/3$

ing. In Fig. 6.6 the transitions are illustrated with phase portraits, stroboscopic sections, realizations and spectra. For the stroboscopic section, we collect all values of x_1 and y_1 of (6.2)–(6.3) at which the forcing x_2 takes its maximal values. At 3 : 1 synchronization, i.e., with the response making three oscillations while the forcing makes just one, there will be only one point in the stroboscopic section.

First, consider the evolution of spectra (fourth column of Fig. 6.6) with the increase of forcing strength B. In order to demonstrate the effect most clearly, the frequency axis is chosen to be different for the spectrum of the forcing and of the response: the frequencies of response are squeezed by the factor of 3, so that at

6 $n:m$ Synchronization of Periodic Oscillations

the state of locking ($B = 0.35$) the main peaks of the forcing and of the forced system coincide. With this in mind, let us consider how the distance between the main spectral peaks changes. At zero strength $B = 0$, the forcing frequency is at some distance from one-third of the natural frequency of the system, and even with the rescaled frequencies the spectral peaks are apart. As forcing is switched on and becomes stronger ($B = 0.15$), the peaks at combination frequencies appear in the spectrum of the response, and one of them necessarily coincides with three times the forcing frequency, which in the rescaled graph is visible as the coincidence of the first peak from the left and the one of forcing. At the same time, the main peak of response has shifted towards one-third of the frequency of the forcing peak. At even larger $B = 0.25$, the situation is the same as with $B = 0.15$, only enhanced. At $B = 0.3$ the main peak of response signal is now located at exactly three times the forcing frequency! However, the motion is still not periodic, and there is no synchronization yet. Finally, at $B = 0.35$ the $3:1$ synchronization is achieved: forced oscillations become periodic and their frequency is three times larger than the forcing one. To summarize, the main spectrum peak moves gradually towards the third of forcing frequency to become locked with it.

In terms of the stroboscopic section (second column of Fig. 6.6), on the route to locking we go through the invariant closed curve that is densely filled with points ($B < 0.3$), to the closed curve on which points are placed non-homogeneously ($B = 0.3$). In the upper part of the closed curve one can see condensation of the points, which is a precursor of locking. And indeed, at $B = 0.35$ a stable point appears exactly at the place of condensation. This means that the locked regime is born on the surface of the torus from the "condensation of phase trajectories," as they sometimes say.

Realizations of both forcing Bx_2 and response x_1 (third column of Fig. 6.6) provide a good illustration of the phase relationships between the two subsystems. At small forcing strengths $B < 0.3$, each time the forcing takes its maximal value, the response is at a different stage of its oscillations. Closer to the locking boundary ($B = 0.3$), one can see how the phase of the response starts to adjust to the phase of forcing: the grey circles are almost in one horizontal line, but not quite. Finally, when locking takes place at $B = 0.35$, the response is always at the same phase when the forcing takes its maximal values: phase locking has been achieved. Note, that while the forcing makes one full oscillation (black line), the forced system makes exactly three oscillations (grey line).

The phase portraits on the planes "forcing"–"response" (first column of Fig. 6.6) are in line with other observations. When the system was quite away from the state of locking ($B < 0.3$), the mutual phase portrait is quite disordered: the two systems pay little attention to each other and oscillate almost independently. However, the stronger the forcing, the more the forced system feels the forcing, and the more structure appears in the phase portrait. At $B = 0.3$ the phase trajectory spends most of the time around the future stable limit cycle, which reveals itself inside the locking region at $B = 0.35$. This cycle has three loops to signify that one of the systems is three times faster than the other.

6.2 1 : m and m : 1 Forced Synchronization in Weakly Non-linear Oscillators   131

The transition to phase $3 : 1$ locking appears very similar to the transition to $1 : 1$ locking (compare with Fig. 3.11).

6.2.2 3 : 1 Suppression of Natural Dynamics

Next, let us enter the synchronization region via route **B** in Fig. 6.5(b) that crosses the dashed line of torus birth bifurcation, by fixing $p = 0.3$ and increasing B. By analogy with $1 : 1$ synchronization, we suggest that suppression of natural dynamics should occur, the transition to which is illustrated in Fig. 6.7 by phase portraits, stroboscopic sections, realizations and spectra. We will start with considering the evolution of spectra with the increase of B (fourth column of Fig. 6.7), and this will immediately reveal the crucial distinction of this route from route **A** illustrated above. First, the frequency axis is *not* rescaled for the spectrum of the response. From the very beginning ($B = 0$) the frequency of forcing was approximately three times less than the frequency of natural oscillations in the system to be forced. In order to show this without rescaling the abscissas, we embrace a larger range of frequencies. As a result, one can see that the spectrum of forcing, besides the component at Ω, does contain the components at 3Ω and 5Ω (and more, but they are not visible in the given range of frequencies), so it is indeed not harmonic.

At $B = 0$ the main peaks of forcing and of the response are well separated. At $B = 0.15$ the spectrum of response is enriched with combination frequencies, and also contains a component at Ω. With too many spectrum components, it might be difficult to locate the main peaks, so we mark them by $\triangle$ (forcing at Ω) and $\triangledown$ (response). The main peak of the response almost does not move with the growth of B, however, its height is gradually decreasing with the simultaneous growth of the component at Ω: the power is being redistributed between these two frequencies. At $B = 0.55$ the peak at Ω becomes higher than the one that originates from natural dynamics (which is now slightly shifted from ω_0), but the motion is quasi-periodic and there is no synchronization yet. Finally, at $B = 0.65$ the oscillations in the system become periodic with the frequency of forcing. Note, that the transition from oscillations at about ω_0 to the ones at Ω has taken place abruptly. First, the main frequency of oscillations has jumped from ω_0 to Ω, and then the component at ω_0 was gradually suppressed to complete extinction. Note, that unlike in case of $3 : 1$ locking, the $3 : 1$ suppression has resulted in oscillations with the *frequency of forcing* Ω itself, rather than 3Ω. This means that most of the system's own dynamics is extinct. Exactly how much of it is left, will be clearer from the realizations which will be commented on below.

In parallel to spectra, follow the route by observing stroboscopic sections (second column of Fig. 6.7). In full analogy with $1 : 1$ suppression (Fig. 3.12), with the increase of B the invariant closed curve shrinks, until it becomes a point when synchronization occurs.

Realizations (third column) carry valuable information about both phase and amplitude relationships between the interacting systems. When the forcing is disconnected from the main system ($B = 0$), the latter oscillates approximately three

132 6 *n* : *m* Synchronization of Periodic Oscillations

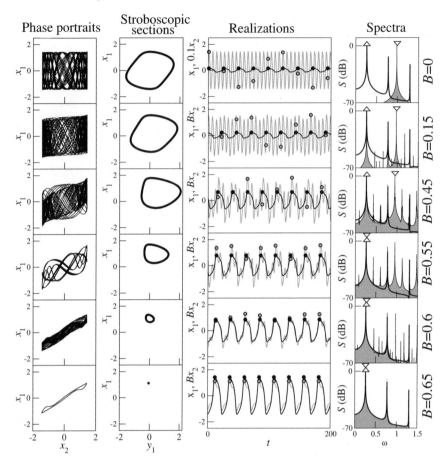

Fig. 6.7. Illustration of a 3 : 1 suppression in two unidirectionally coupled van der Pol oscillators (6.2)–(6.3) at $\lambda = 0.5$. In Fig. 6.5(b) we move along route **B** corresponding to $p = 0.3$, as we enter the synchronization tongue via torus birth bifurcation line. Phase portraits, stroboscopic sections, realizations and power spectral densities (spectra) are shown for each value of forcing strength B given on the right of each row. **First column**: Phase portraits on the plane (x_2, x_1). **Second column**: Stroboscopic sections on the plane $(\dot{x}_1, x_1)$. **Third column**: *Black line*—forcing $Bx_2(t)$ (except at $B = 0$ where it is $x_2/10$), *grey line*—response $x_1(t)$; *black circles*—maxima of forcing Bx_2, *grey circles*—values of x_1 when x_2 is at its maximum. **Fourth column**: *Black line*—spectrum of forcing x_2, *shaded*—spectrum of response x_1. △ marks the main peak of forcing at Ω, and ▽ the main peak of response. Note, that the scales of the frequency axis are *the same* for two spectra (unlike in Fig. 6.6)

times faster than the driving system: during one full forcing cycle there are approximately three full-amplitude oscillations in the forced system. The non-zero forcing ($B > 0$), modulates the mean value of response realizations which is in phase with the driving system. But around the mean value there are fast oscillations that originate from the system's own dynamics, and whose amplitude decreases as forcing

6.3 $n : m$ Synchronization in Strongly Non-linear Oscillators with Spiky Forcing 133

grows stronger. The larger the external driving is, the stronger the modulation of the mean value and the smaller the fast oscillations are. Thus, the forcing imposes onto the system the lower-frequency oscillations and at the same time suppresses the fast ones. At $B = 0.65$ the realization of the forced system almost coincides with the one of forcing. Natural oscillations are now completely suppressed, although their trace is visible in the form of the small bumps.

The phase portraits on the plane "forcing"–"response" (first column of Fig. 6.7) illustrate the transition from oscillations independent of each other to the ones that are synchronized. This is a classical transition to synchronization via the suppression mechanism.

In the example that was considered in this section, we were able to reveal $1 : m$ and $m : 1$ synchronization for many values of m, but only for odd ones. For even values of m, synchronization tongues are very narrow and are difficult to detect. The fundamental reason for that is that the spectrum of the autonomous van der Pol oscillator does not contain components at even multiples of ω_0, i.e., at $2\omega_0$, $4\omega_0$, etc. (Fig. 6.2(c)), because of the special form of non-linearity in its equations which is of the cubic type. The even harmonics would have enhanced the respective synchronization regions, but in their absence it becomes quite a challenge to find them.

6.3 $n : m$ Synchronization in Strongly Non-linear Oscillators with Spiky Forcing

In Sect. 6.2 it was demonstrated that if the forcing frequency is close to an integer multiple of the natural frequency of oscillations, $1 : m$ synchronization can occur. If, on the other hand, the forcing frequency is about m times smaller than the natural frequency, the system can be synchronized with the rotation number $m : 1$. In this section we introduce an even more general phenomenon, namely, $n : m$ synchronization, where neither m nor n are equal to 1.

Such synchronization is likely to occur when both the forcing and the system being forced contain many frequency components in the spectra of their oscillations, that could interact and enhance synchronization regions of higher orders n and m. Quite a good model of such oscillations can again be the well-familiar van der Pol system with large non-linearity λ, which we set to 2. The phase portrait of the respective unforced oscillations is given in Fig. 6.8(a) (compare with Fig. 6.2(a)). Further, we need to introduce a periodic forcing whose spectrum contains a lot of harmonics of the main frequency. A most suitable realization of forcing would thus be a function of time that contains sharp spikes. As a model capable of generating such a signal, we can again use van der Pol oscillator with large non-linearity $\lambda = 2$. The full system modelling the required oscillator under external forcing now reads as follows:

$$
\begin{aligned}
\dot{x}_1 &= y_1, \\
\dot{y}_1 &= \left(\lambda_1 - x_1^2\right)y_1 - x_1 + B\frac{|y_2| + y_2}{2},
\end{aligned}
\tag{6.4}
$$

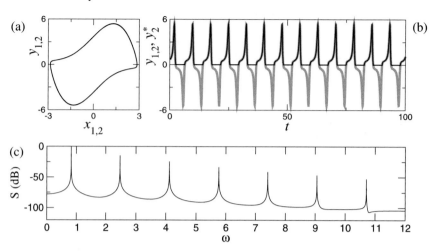

Fig. 6.8. **a** Phase portrait, **b** realization $y_{1,2}$ (*grey*) and $y_2^* = (|y_2|+y_2)/2$ (*black*), **c** spectrum of $x_{1,2}$ of (6.2)–(6.3) at $\lambda = 2$ and $B = 0$. Only odd harmonics of the main frequency are present in the spectrum

$$\begin{aligned}\dot{x}_2 &= y_2, \\ \dot{y}_2 &= (\lambda_2 - x_2^2)y_2 - p^2 x_2.\end{aligned} \qquad (6.5)$$

The difference from (6.2)–(6.3) is in the way the external forcing is applied to the first system. In order to construct the spiky forcing signal, we take the y_2-coordinate (shown by grey line in Fig. 6.8(b)) and ignore its negative values (black line in Fig. 6.8(b)), which mathematically can be described as $(|y_2| + y_2)/2$, as shown in the second line of (6.4).

By analogy with the case of the weakly non-linear oscillator under weakly non-linear forcing considered in Sect. 6.2, let us fix the strength of forcing at $B = 0.4$ and change p in a range $[0.1; 3.5]$ to see how the response frequency ω_r and rotation number ρ evolve. The summary of frequency relationships is given in Fig. 6.9, where the same quantities as in Figs. 6.3 and 6.4 are shown against Ω/ω_0. At first glance, the same kinds of transitions are observed. However, there are some features here that were not present in Figs. 6.3 and 6.4. Namely, there are considerably more linear segments in (a) and considerably more plateaus in (b). Here, plateaus of rotation number ρ occur not only at the values $1 : m$ or $m : 1$, but also at $n : m$, where neither n nor m are equal to one (although the plateau at $1 : 1$ is also present!). How do we interpret the occurrence of a plateau at $\rho = n : m$? Well, quite simply, when the external force makes m oscillations, the forced system makes exactly n oscillations. This is called $n : m$ synchronization.

Let us illustrate the mechanisms of $n : m$ synchronization. For this purpose single out some particular rotation number, for example $2 : 3$, and consider its vicinity in more detail. The bifurcation diagram around the $2 : 3$ synchronization region on the plane (p, B) is given in Fig. 6.10. Note, that with λ being substantially larger than zero, the frequency of unforced oscillations in the van der Pol system is noticeably

6.3 $n:m$ Synchronization in Strongly Non-linear Oscillators with Spiky Forcing 135

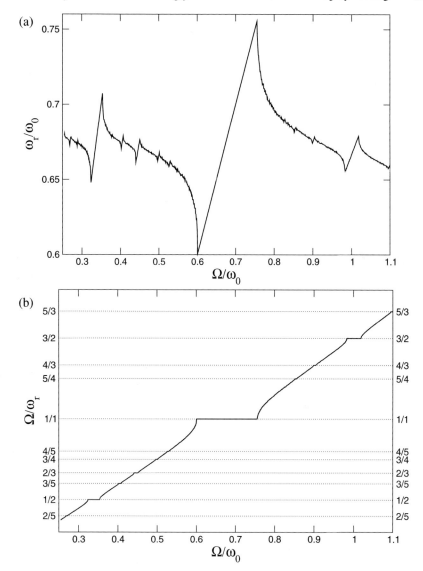

Fig. 6.9. $n:m$ synchronization in a strongly non-linear van der Pol oscillator forced by a spiky signal from another van der Pol oscillator (6.4)–(6.5) at $\lambda = 2$ and $B = 0.4$. **a** The response frequency ω_r against the forcing frequency Ω, both normalized by the natural frequency $\omega_0 = 1.209$. *Linear segments* are visible. **b** "Devil's staircase": inverse of rotation number $\rho = \Omega/\omega_r$ against Ω/ω_0. Plateaus at rational values of $\rho = n:m$ are clearly visible

less than one, see the location of the highest peak in the spectrum of its oscillations (Fig. 6.8(c)) and the value of Ω at $p = 1$ in Fig. 6.11. Similarly, the frequency of the forcing system (6.5) is noticeably less than p in the range considered, as demonstrated in Fig. 6.11. Moreover, the dependence of Ω on p is not linear. For

136   6 $n:m$ Synchronization of Periodic Oscillations

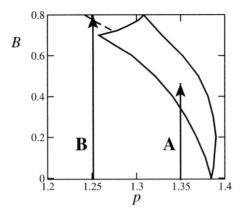

Fig. 6.10. Bifurcation diagram in the vicinity of 3:2 synchronization region in two strongly non-linear unidirectionally coupled van der Pol oscillators (6.4)–(6.5) at $\lambda = 2$. *Solid black lines*: saddle-node bifurcations; *dashed line*: torus birth bifurcations

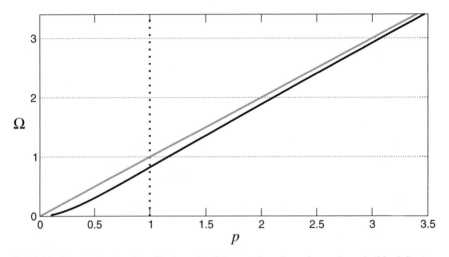

Fig. 6.11. Frequency Ω of oscillations in (6.5) as a function of p at $\lambda = 2$ (*black line*) as compared to the diagonal (*grey line*). *Vertical line* marks $p = 1$ to highlight the fact that Ω is noticeably smaller than 1 at this point

this reason, the tip of the 2:3 tongue in Fig. 6.10 hits the abscissa at $p = 1.384$ rather than 3:2. The synchronization region is formed by the same typical lines of saddle-node and torus birth bifurcations. Also, it is inclined considerably, stretching towards the lower values of p that are closer to lower rotation numbers.

6.3.1 2:3 Phase (Frequency) Locking

Consider route **A** that leads inside the synchronization tongue through the line of saddle-node bifurcation, and illustrate the changes that occur to the system dynamics

6.3 $n:m$ Synchronization in Strongly Non-linear Oscillators with Spiky Forcing

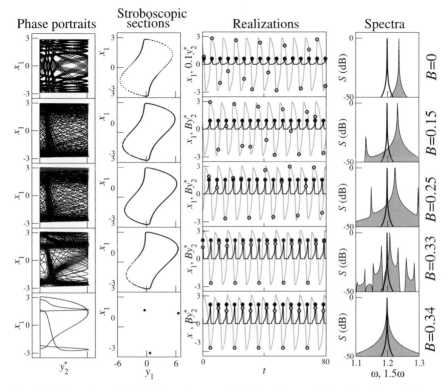

Fig. 6.12. Illustration of a $2:3$ phase (frequency) locking in two unidirectionally coupled van der Pol oscillators (6.4), (6.5) at $\lambda = 2$. In Fig. 6.10 we move along route **A** corresponding to $p = 1.35$, as we enter the synchronization tongue via the saddle-node bifurcation line. Phase portraits, stroboscopic sections, realizations and spectra are shown for each value of forcing strength B given on the right of each row. **First column**: Phase portraits on the plane (y_2^*, x_1), where $y_2^* = (|y_2| + y_2)/2$ is forcing. **Second column**: Stroboscopic sections on the plane (y_1, x_1). **Third column**: *Black line*—forcing $Bx_2(t)$ (except at $B = 0$ where it is $y_2^*/10$), *grey line*—response $x_1(t)$; *black circles*—maxima of forcing By_2^*, *grey circles*—values of x_1 when y_2^* is at its maximum. **Fourth column**: Spectra: *black line*—forcing By_2^*, *shaded*—response x_1. The scales of the frequency axis are *different* for two spectra: the spectrum of forcing is shown against ω, while the one of response against $\omega \times 3/2$

on that way. As before, use phase portraits, stroboscopic sections, realizations and spectra as the helpful aids (Fig. 6.12). Start with spectra in the fourth column. The frequency axis is different for the forcing signal (unchanged) and for the response signal (divided by rotation number $2:3$). With forcing strength set to zero $B = 0$, the forcing frequency is slightly smaller than $1.5\omega_0$. When B is being gradually increased from zero ($B = 0.15, 0.25, 0.33$ in Fig. 6.12), the main spectrum peak of the response signal (shaded area) is shifted towards $2\Omega/3$, which in the figure can be seen as the two highest spectral peaks approaching each other. At $B = 0.34$ the two peaks coincide on the picture, which means that the response frequency

138   6 $n : m$ Synchronization of Periodic Oscillations

became equal to two thirds of the forcing frequency. The combination frequencies have disappeared, and $2 : 3$ locking has taken place.

The same transition is accompanied by the changes in the stroboscopic section (second column in Fig. 6.12). We go from a closed invariant curve representing the surface of an ergodic torus, to three points that represent a stable limit cycle that is born on the torus surface. Why do we see three points instead of one? Just because of the way the stroboscopic section is constructed: we collect a point (x_1, y_1) each time the forcing system is at the same phase. Look at realizations at $B = 0.34$, observe the maxima of y_2^* marked by black circles, and look up the respective value of x_1 is (grey circles). It is clear that each three maxima of y_2^* correspond to one full period of the grey line. Hence, we will have three different points in the stroboscopic section.

Phase portraits (first column) on the plane "forcing y_2^*"–"response x_1" illustrate the transition from the independent behavior of the two systems at $B = 0$, to phase-locked oscillations at $B = 0.34$. Just before the locking has occurred, at $B = 0.33$ the trajectories start to concentrate around the future limit cycle that is born at $B = 0.34$.

This is a manifestation of the classical phase (frequency) locking for the most general type of synchronization with the order $n : m$, which was equal $2 : 3$ here.

6.3.2 The Route to $2 : 3$ Suppression

The upper part of the $2 : 3$ synchronization region in Fig. 6.10 is bounded by the line of torus birth bifurcation, which is associated with the suppression of natural oscillations. However, if by analogy with the traditional studies of the suppression mechanism, we choose a route like **B**, we would not be able to observe the classical behavior that normally occurs on the way to suppression. The reason for that is that the synchronization tongue is inclined so much, that many synchronization regions with various rotation numbers lie in between its border and $B = 0$. The evolution of the dynamical regimes along route **B** is illustrated in Fig. 6.13 where the x_1 coordinate of the stroboscopic section is shown against the value of forcing strength B. The region of $2 : 3$ suppression is at the right end of the diagram and is represented by three points. On the way to suppression, the transitions between quasiperiodic behavior, and the synchronized regimes with various rotation numbers are evident.

6.4 Circle Map: Derivation

Earlier in this chapter we illustrated some particular cases of $n : m$ synchronization, n and m being positive integer numbers. In Sect. 6.3 we considered a strongly nonlinear oscillator subject to spiky forcing. We have shown that unlike in the case of a weakly non-linear oscillator under nearly harmonic forcing, it can demonstrate synchronization of $n : m$ orders which can be quite easily detected numerically. Let us consider a limiting case of such an oscillator.

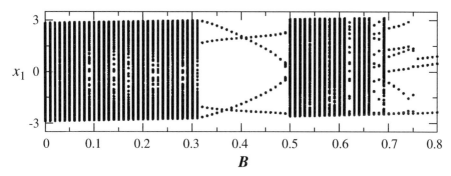

Fig. 6.13. A zoo of oscillation regimes as one moves along route **B** in Fig. 6.10: one-parameter diagram showing coordinate x_1 in (6.4)–(6.5) against the value of coupling strength B. Changes between quasiperiodic motion and regimes of $n:m$ synchronization are illustrated

Start with forcing. Among the spiky functions of time t, the most spiky of all is Dirac delta function $\delta(t)$ that is defined as

$$\delta(t) = \begin{cases} \infty, & t = 0, \\ 0, & t \neq 0, \end{cases} \quad \int_{-\infty}^{\infty} \delta(t)\,dt = 1. \tag{6.6}$$

The most spiky periodic forcing $F(t)$ with the most pronounced harmonics in the spectrum is a periodic sequence of delta-spikes that can be formally written as

$$F(t) = \sum_{j=-\infty}^{\infty} \delta(t - t_j), \quad t_j = jT, \tag{6.7}$$

where T is the period of forcing, and j is the spike number.

Next, we need to choose an oscillator to be forced. We need a very non-linear oscillator, which implies that the shape of its realizations is far from being harmonic. A classical example of such a system is a relaxation oscillator, whose characteristic feature is a very non-circular shape of its limit cycle. Another important feature of a strongly non-linear oscillator is temporal inhomogeneity of its oscillations: depending on the stage of oscillations, and in fact on the current position on the limit cycle, the motion either slows down or speeds up. It is convenient to describe an oscillator in terms of amplitude and phase rather than in terms of the original variables.

6.4.1 Amplitude and Phase of Oscillations

Let us introduce the concepts of amplitude and phase of oscillations. In Sect. 3 we dealt with quasiharmonic oscillations and used this concept semi-intuitively. When the oscillations are not quasiharmonic, the definition of amplitude becomes a problem with several possible solutions, as well as the definition of a phase. One of the popular ways to define both phase and amplitude is by considering a projection of the system trajectory onto a suitable plane. If the oscillator is two-dimensional, this

6 $n:m$ Synchronization of Periodic Oscillations

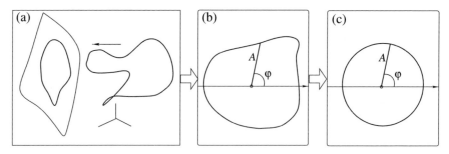

Fig. 6.14. a, b Illustration for the introduction of amplitude A and phase φ on a limit cycle of complex shape. **c** Transition to a cycle with constant amplitude

would simply be its phase plane. If the dimension of the oscillator is larger than 2, then one can choose a convenient surface in the phase space to project the phase trajectories onto it. In Fig. 6.14(a) a limit cycle of a complex shape is shown in 3D (right) together with its projection on a suitable plane (left). For a periodic oscillator the projection will be a closed curve.[5] Place the origin somewhere inside the projection of the limit cycle (Fig. 6.14(b)). As time passes by, the phase point travels along the closed curve, and it does so periodically. At each time moment, connect the phase point with the origin by a straight line and call the length of this line instantaneous amplitude $A(t)$. Call the angle between this line and the abscissa the phase $\varphi(t)$. After one full rotation the phase increases exactly by 2π. It is obvious, that if the projection of the limit cycle on the chosen plane is not a circle, the amplitude defined in such a way will oscillate with the period of the cycle. At the same time, phase grows in magnitude monotonously and unboundedly. The location of the point on the limit cycle can be completely defined by the phase φ. Then the respective value of the amplitude A is described by some periodic function of φ, $\tilde{A}(\varphi)$. Coming back to the strongly non-linear oscillator, its limit cycle is usually not a circle, and its phase usually grows with different velocity at different parts of this cycle. But because the motion on the cycle is periodic, the intervals of fast and of slow phase growth repeat themselves periodically. The velocity $\dot{\varphi}(t)$ of phase growth at each time moment t will depend essentially on the current position of the phase point within the limit cycle, i.e., in fact on the current value of φ. The equation for the evolution of phase can thus be written as

$$\dot{\varphi} = \Lambda + f(\varphi), \tag{6.8}$$

where Λ is the constant average velocity of the phase growth, and $f(\varphi)$ is some non-linear function of φ, which is periodic in its argument and has zero mean. The time derivative $\dot{\varphi}$ is called instantaneous frequency and oscillates periodically. Let us consider perturbation of the limit cycle, and assume that the amplitude component of this perturbation decays in time almost instantly. At the same time, perturbation of

---

[5] Sometimes the limit cycle can have loops, and its projections might have loops, too. A convenient surface would be such that the cycle projection onto it *does not contain loops*.

6.4 Circle Map: Derivation   141

the phase component does not decay at all. This implies that if at the state (A_1, φ_1), where $A_1 = \tilde{A}(\varphi_1)$, we kick the system in such a way that both its amplitude and phase change to some new values (A_2, φ_2) that would correspond to a point *outside* the limit cycle, the amplitude almost instantly relaxes to the value of $\tilde{A}(\varphi_2)$, i.e., to its value corresponding to the position *on* the limit cycle at the perturbed phase φ_2 [92]. At the same time, perturbations of the phase will not decay or vanish, and φ will just continue to evolve starting from the value φ_2.

Remember, that in the problem of synchronization the most essential bit of information comes from the behavior of phases rather than amplitudes of oscillations. Normally, amplitudes match the behavior of phases, but nevertheless, phases come first. In fact, perhaps, the most popular name for synchronization in general is "phase locking," although we know already that this is only one of its possible mechanisms. Let us concentrate only on phases and throw the oscillations of amplitudes out of the problem. Mathematically this would imply that the amplitude is fixed at a certain value, i.e., oscillations occur on a perfect circle (Fig. 6.14(c)). However, we emphasize that the speed of phase growth $\dot{\varphi}$ continues to depend essentially on the position within the cycle.

6.4.2 From Differential to Discrete Equation for Phase

After we formulate a simplified model for a phase of a strongly non-linear autonomous (unforced) oscillator, let us incorporate forcing into the problem. Remember that we have earlier decided to represent forcing as a sequence of delta-spikes (6.7). In strongly non-linear oscillators is does matter a lot at what stage (phase) of the oscillation the next kick occurs: there are normally regions on the cycle when the system is less sensitive to perturbation, and there are regions where the response is profound. This can be modelled for example by making the forcing amplitude depend on the current phase φ as follows:

$$\dot{\varphi} = \Lambda + f(\varphi) \sum_{j=-\infty}^{\infty} \delta(t - t_j). \tag{6.9}$$

Let us try to simplify the problem further. We are not really interested in how exactly φ behaves in between the delta-kicks. What matters to us is the direct consequence of ith kick, that presumably results in the change of the value of φ at which the next kick number $(i + 1)$ would arrive. Let us therefore introduce the difference between the phase at the moment t_i and the phase at t_{i+1}:

$$\varphi_{i+1} - \varphi_i = \int_{t_i}^{t_{i+1}} \frac{d\varphi}{dt} dt$$

$$= \Lambda(t_{i+1} - t_i) + \int_{t_i}^{t_{i+1}} f(\varphi(t)) \sum_{j=-\infty}^{\infty} \delta(t - t_j) dt$$

$$= \Lambda(t_{i+1} - t_i) + \sum_{j=-\infty}^{\infty} \int_{t_i}^{t_{i+1}} f(\varphi(t))\delta(t - t_j) dt.$$

142   6 $n:m$ Synchronization of Periodic Oscillations

Note, that in the equations above we are taking the integral from t_i to t_{i+1}. Because for any i, $(t_{i+1} - t_i) = T$ is the same, within any such interval there will necessarily be a single delta-spike number i, i.e., $j = i$. All other delta-spikes with $j \neq i$ lie outside this interval and therefore are not taken into account. Use the property (3.96) of the delta-function to calculate the integral

$$\varphi_{i+1} - \varphi_i = \Lambda(t_{i+1} - t_i) + f(\varphi(t_i)).$$

Recall that $\varphi(t_i) = \varphi_i$, and rearrange the expression above as

$$\varphi_{i+1} = \varphi_i + \Lambda T + f(\varphi_i). \tag{6.10}$$

This is the most general form of the circle map. It describes the dynamics of a phase on the limit cycle under external forcing.

In order to perform theoretical analysis of the properties of the circle map, we need to define the function $f(\varphi_i)$, which should be periodic in φ_i. The simplest periodic function is sine with some amplitude $B > 0$, which for this map is usually taken with a negative sign. Denote $\Lambda T = \Delta$ and obtain

$$\varphi_{i+1} = \varphi_i + \Delta - B \sin(\varphi_i) = F(\varphi_i), \tag{6.11}$$

which is called sine circle map, or just circle map for brevity.

6.5 Circle Map: Properties

Sine circle map is a paradigmatic toy model which has been extensively studied with regard to synchronization problem (see [92] for properties, history and references). Here, we will briefly describe the essential features of this map that make it so useful for the understanding of synchronization. Parameter Δ is an analog of frequency detuning, and parameter B is an analog of the strength of forcing. As any other one-dimensional map, circle map can be described in terms of the phase plane $(\varphi_i, \varphi_{i+1})$. Note, that the phase φ_i grows unboundedly with i. But for convenience of the analysis, it is customary to illustrate the dynamics of this map by taking function $F(\varphi_i)$ in (6.11) by modulus of 2π,[6] so that all points φ_i and φ_{i+i} lie within the limits $[0; 2\pi]$. Note, that if we set $\Delta > 2\pi$, the use of the modulus will effectively reduce the value of Δ by $2\pi n$, $n = 1, 2, \ldots$ so it makes no sense to consider Δ outside the interval $[0; 2\pi]$.

Some typical phase portraits for four different sets of parameters Δ and B are shown in Fig. 6.15. In (a)–(c) the cases of $|B| < 1$ are illustrated. In (a) the value of Δ is larger than B, and the function $F(\varphi_i)$ does not cross the diagonal. Hence, there are no fixed points in the map, and the phase point jumps along the segments of $F(\varphi_i)$. This behavior corresponds to the absence of $1:1$ phase synchronization,

---

[6] This means that if the current value of φ_i or of φ_{i+i} is larger than 2π, we subtract from it 2π consecutively until it falls within $[0; 2\pi]$. Similarly, if φ_i or φ_{i+i} are smaller than 0, we add 2π repeatedly until it falls inside $[0; 2\pi]$.

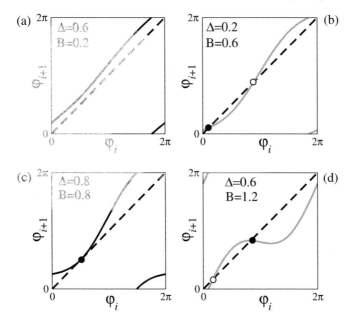

Fig. 6.15. Typical phase portraits of sine circle map (6.11) at different sets of control parameters Δ and B, whose values are given in the respective panels

but we need additional analysis to find out if synchronization of some higher orders $n:m$ occurs. Drawing analogy with a periodic oscillator forced periodically, the detuning Δ is too big, and the forcing strength B is not enough to synchronize the system.

In (b) Δ is less than B, and the graph of $F(\varphi_i)$ crosses the diagonal. This means the occurrence of two fixed points: one stable (black circle), and one unstable (white circle). The function $F(\varphi_i)$ now describes the surface of a resonant torus, on whose surface the pair of cycles was born. This is $1:1$ phase locking.

In (c) Δ is equal to B and the graph of $F(\varphi_i)$ touches the diagonal. This is the instant of saddle-node bifurcation. The equation $\Delta = B$ thus defines the borderline of $1:1$ phase locking region. Note, that $1 - \Delta = B$ is also the condition for saddle-node bifurcation, and so this is another borderline for the respective region.

In (d) an interesting case is illustrated: the forcing strength is larger than 1. At such values, the map becomes non-invertible, and can demonstrate dynamical chaos.

One can introduce rotation (winding) number for the circle map as

$$\rho = \lim_{i \to \infty} \frac{\varphi_i - \varphi_0}{2\pi i}. \tag{6.12}$$

If ρ is a rational number $n:m$, then phase locking of the respective order occurs, just like in the continuous-time forced periodic oscillator. The circle map demonstrates all kinds of resonances $n:m$ and this is why it remains the classical model used for the understanding of synchronization.

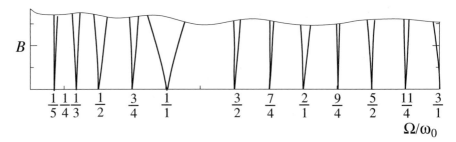

Fig. 6.16. General picture of Arnold tongues at small amplitudes B of external forcing. Ω is the forcing frequency, and ω_0 is the natural frequency of unforced oscillations

6.6 Arnold Tongues

To complete the chapter on $n:m$ synchronization of periodic oscillations by periodic forcing, it needs to be noted that the most general structure of the bifurcation diagram on the plane of parameters "detuning Ω/ω_0"–"forcing strength B" at small B would look like in Fig. 6.16. This figure shows only the regions where phase locking occurs, while suppression or homoclinic synchronization are not illustrated. This structure was revealed by V. Arnold [30], after whom the synchronization tongues were named "Arnold tongues." In particular, Arnold has proved the following important statement. Suppose we have integer numbers n, m, n^* and m^*, and there are synchronization tongues with the rotation numbers $n:m$ and $n^*:m^*$. Then in between them on the plane $(\Omega/\omega_0, B)$ there is a tongue with rotation number $(n+n^*):(m+m^*)$. This is called Farey order.

One might note that the picture in Fig. 6.16 is shown for small forcing B only. What happens at larger B? Without going into detail, it has to be mentioned that at large B different synchronization tongues can overlap resulting in multistability and hysteresis [93]. Moreover, chaos can occur and synchronization can be destroyed.

6.7 $n:m$ Synchronization: Experiment

There have been quite a few experimental results on $n:m$ synchronization. One of the first observations of this phenomenon in a biological system was reported in [94], where forcing in the form of pulses of an electric current was applied to a spontaneously beating aggregate of cardiac cells from embryonic chick heart. Synchronization of several orders was detected and the traces of Arnold tongues were revealed from the experimental data. However, one might wonder: "Although the system studied here is a biological one, it is a bit artificial because the result quoted required a special, presumably expensive experimental setup and a piece of biological tissue isolated from the living organism. But what about natural living systems? Is synchronization of the order $n:m$ a sufficiently robust phenomenon that it can be detected without an expensive experimental setup in an almost every-day situa-

6.7 $n:m$ Synchronization: Experiment   145

tion?" Below we will describe how this phenomenon can be observed inexpensively in a human being.

In Chap. 2 we mentioned human cardiovascular system within which several rhythmic processes interact and might synchronize. Two of such processes are heartbeats and breathing. In a healthy human both of these processes are not strictly periodic, but one can argue that they can be approximately viewed as periodic self-oscillatory processes under the influence of a large number of randomly fluctuating factors which smear the observed behavior.

It has been known for a while that heart beats can in principle be synchronous with the breathing [152, 259] under certain conditions. However, it seems that in freely breathing humans spontaneous cardiorespiratory synchronization is quite rare. In [257] a set of experiments were described, in the course of which six healthy volunteers underwent controlled breathing with the prescribed frequencies from a wide range between 3 and 30 breaths per minute. Realizations describing the heart beating were the electrocardiograms (ECG) that reflected the electrical activity of a heart (Fig. 6.17(b)). Respiration was measured by means of wrapping an elastic band around the chest and measuring the change in extension of the band while the subject breathed. This way, both the amplitude (depth) of breathing and its frequency were taken into account.

In Fig. 6.17 typical plots for (a) respiration and (b) ECG are given as functions of time, both in dimensionless units. For both ECG and the forcing in the form of breathing one can introduce phases using (8.10) and then calculate the so-called generalized phase difference $\varphi_{n,m}(t)$ as follows:

$$\varphi_{n,m}(t) = n\psi_1(t) - m\psi_2(t), \tag{6.13}$$

where $\psi_1(t)$ and $\psi_2(t)$ are the phases of the interacting systems, and $n:m$ is synchronization order. The condition of phase synchronization is then the existence of a sufficiently long plateau of $\varphi_{n,m}(t)$. An example of a phase difference between respiration and ECG for the $2:7$ synchronization is given in Fig. 6.18(a). One can observe a long and noisy plateau, which is indeed an evidence of synchronization. In order to better illustrate this phenomenon, consider the stroboscopic section of respiration signal x_{resp} that consists of the values x_{resp}^i taken at the time instants when R-peaks of ECG, which are the highest and sharpest peaks, have crossed the threshold as shown in Fig. 6.17(b). In Fig. 6.18(b) numbers are assigned to the successive values of x_{resp}^i in order to single out their cyclic behavior: two large oscillations of the function include exactly seven points. This means that while the subject inhales and exhales twice, his heart beats seven times.

The full picture illustrating the response of the given subject to paced breathing is given in Fig. 6.19 where the plane of parameters "frequency detuning f_{resp}/f_0"– "strength of forcing A" is shown for the same subject that was illustrated above. Here, each point $(f_{\text{resp}}/f_0, A)$ represents a single experiment with the same frequency of breathing f_{resp}. With this, the amplitude A of breathing was automatically adjusted by the subject himself: the faster the subject was breathing, the shallower (A is small), and the slower, the deeper (A is large). A point was marked as empty

146   6 $n:m$ Synchronization of Periodic Oscillations

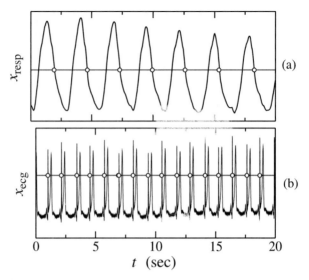

Fig. 6.17. **a** Respiration and **b** electrocardiogram (ECG) of a healthy volunteer undergoing paced breathing. Both quantities are given in dimensionless units. *Horizontal lines* define the threshold which is crossed by the functions in one direction: from above to below. *Empty circles* are the points of intersection with the threshold

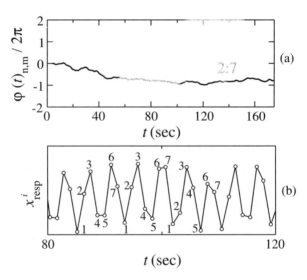

Fig. 6.18. **a** Generalized phase difference φ between breathing and electrocardiogram (ECG) that correspond to synchronization order $2:7$. **b** A fragment of the realization of a stroboscopic section of respiration signal: the values x^i_{resp} of respiration are taken at the time moments when ECG crosses the threshold level. Numbers from *1* to *7* are attached to successive points of this graph: during two large oscillations, there are exactly seven points of the stroboscopic section. This is another evidence of a $2:7$ synchronization

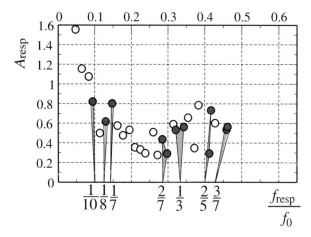

Fig. 6.19. The cutoff of the plane of parameters "frequency detuning f_{resp}/f_0"–"strength of forcing A" for a subject undergoing paced respiration, where f_0 is his average heart rate at rest and f_{resp} is respiration frequency. *Empty circles* denote the points at which no $n:m$ synchronization was detected for n and m less than 10. *Filled circles* mark the points where some synchronization was detected. The start and the end points at which synchronization of each particular order $n:m$ was observed, were connected with the tip of the supposed synchronization tongue in order to roughly outline its borderlines. The area in between the borderlines of the tongue is *shaded*. The structure resembling the one of Arnold tongues is revealed (compare with Fig. 6.16)

circle if no $n:m$ synchronization was observed with n and m less than 10. A point was filled with color or pattern if such synchronization was observed during sufficiently long time intervals. The start and end points of an interval of breathing frequencies, at which synchronization with a certain order $n:m$ was observed, are connected with the points $(n:m, 0)$, i.e., with the supposed tips of the respective tongues. The shaded areas roughly outline synchronization tongues. The picture in Fig. 6.19 is only a cutoff of the full plane of parameters. The structure of synchronization regions in the given parameter plane resembles the one that would be obtained by crossing Fig. 6.16 along the same route. One can say that Arnold tongues were revealed in a full-scale biological experiment.

$n:m$ frequency synchronization between another pair of rhythms in the human cardiovascular system was systematically studied recently in [237], where the rhythm with the basic frequency around 0.1 Hz was synchronized by means of paced breathing in a range of frequencies with various synchronization orders.

6.8 Summary

In this chapter we have defined and illustrated the $n:m$ synchronization in forced, or unidirectionally coupled, systems. We hope to have convinced the reader that with different synchronization orders, the mechanisms of synchronization remain the same as with the simplest 1 : 1 synchronization. Namely, both locking and suppression are observed, although it was more difficult to observe the latter because the synchronization tongues were bended strongly on the plane of the forcing parameters "detuning"–"forcing strength."

The same phenomenon of $n : m$ synchronization can occur if two or more oscillators are coupled mutually, as demonstrated in [260] and in Chap. 11. Moreover, not only periodic, but also chaotic and noise-induced oscillations can become synchronized with the order $n : m$ as will be shown in Sects. 8.6 and 13.2, respectively.

We would like to emphasize that synchronization of any order, i.e., with any rotation number, is a *robust* effect. This means that it does not occur only at a single carefully selected set of values of control parameters, but rather within a finite range over each of them. A slight variation of, say, detuning between the interacting subsystems does not lead to the disappearance of the effect. An important consequence of this is that synchronization is not destroyed by a small random perturbation—while it remains small! This is of extreme importance from the viewpoint of applications and experiments with real-life and man-made devices, where random perturbations are inevitable. The case when random perturbation is not always small is considered in Chap. 7.

7 1 : 1 Forced Synchronization of Periodic Oscillations in the Presence of Noise

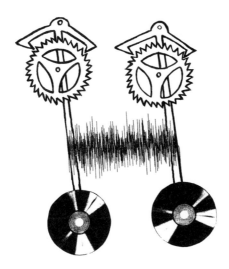

So far we have been considering forced synchronization of periodic oscillations in the almost unrealistic situation where there was no other influence on the system beyond the external periodic forcing. However, real-life objects are normally influenced by random fluctuations, or noise, of various origins. Noise can originate from random motion of molecules and atoms inside and outside the object, from fluctuations of external parameters like humidity, temperature, illumination, concentration of chemical elements, etc., influencing the values of the physical parameters of the system. For example, random motion of electrons and ions in the elements of electric circuits leads to fluctuations of instantaneous values of conductance or capacitance.

The question naturally arises if addition of random input to the system will influence its response to external periodic perturbation and generally the phenomenon of synchronization.

Consider again the well-familiar van der Pol model of a periodic oscillator under harmonic forcing, which is in addition subjected to the influence of noise $\xi(t)$

$$\ddot{x} - (\lambda - x^2)\dot{x} + \omega_0^2 x = B\cos(\Omega t) + \xi(t). \tag{7.1}$$

150   7 1:1 Forced Synchronization of Periodic Oscillations in the Presence of Noise

The problem of synchronization of periodic oscillations in the van der Pol system in the presence of noise was solved by Stratonovich and co-authors [153, 277].

7.1 Introductory Comments on Random Processes

The readers who are familiar with the main ideas of the theory of random processes are suggested to skip this section.

In the previous chapters we considered purely deterministic processes. The main feature of such processes is that they are completely predictable: starting from exactly the same initial conditions, one can run the process many times, and its realizations will be identical.

A random process is very different: one can launch the random process several times from exactly the same initial conditions (perform a random experiment), and realizations from different runs will generally be different. Moreover, one cannot predict the outcome of a random experiment for sure, and any predictions can be made only in probabilistic sense. They say that a random process is a random function of time.

In view of the above, one needs a special mathematical approach to characterize a random process. The most general idea behind it is averaging over the ensemble of realizations. Suppose we can run the same random process $X(t)$ with the same initial conditions as many times as we like: ideally, infinitely many times. With each run, we record its realization $a_i(t)$, $i = 1, 2, \ldots$. A random process $X(t)$ can be characterized by its average value $\langle X(t) \rangle$ (another term is "mean") estimated by averaging over the ensemble of its realizations $a_i(t)$ as follows:

$$\langle X(t) \rangle = \lim_{N \to \infty} \frac{1}{N} \sum_{i=-N}^{N} a_i(t). \tag{7.2}$$

$\langle X(t) \rangle$ can change in time. However, although this approach might be convenient for an experimentalist, it is not convenient for analytical estimates. A very helpful function that allows one to perform analytical calculations related to random processes is a probability density distribution (PDD).

7.1.1 One-Dimensional Probability Density, Mean and Variance

One-dimensional probability density distribution (PDD) $p^X(x, t)$ is introduced as

$$p^X(x, t) = \lim_{\Delta x \to 0} \frac{P\{X(t) \in [x, x + \Delta x)\}}{\Delta x} \tag{7.3}$$

and means the probability P with which the random process $X(t)$ at the time moment t takes the value that falls within the interval $[x, x + \Delta x)$, normalized by the size of the interval Δx. If $p^X(x, t)$ is known, one does not need to repeat a random experiment infinitely many times in order to find the average, since $p^X(x, t)$ al-

7.1 Introductory Comments on Random Processes

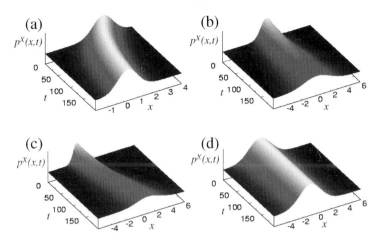

Fig. 7.1. Schematic illustrations of the one-dimensional probability density distributions $p^X(x, t)$ of random processes with different averages $\langle X(t)\rangle$ and variances $\sigma_X^2(t)$. **a** $\langle X(t)\rangle$ is changing in time, $\sigma_X^2(t)$ is constant. **b** $\langle X(t)\rangle$ is constant, $\sigma_X^2(t)$ is changing. **c** Both $\langle X(t)\rangle$ and $\sigma_X^2(t)$ are changing. **d** Both $\langle X(t)\rangle$ and $\sigma_X^2(t)$ are constant. This $p^X(x, t)$ corresponds to the first-order stationary process

ready contains all necessary information. Several kinds of behavior of $p^X(x, t)$ are schematically illustrated in Fig. 7.1. In particular, in (a) $p^X(x, t)$ is shown with average value changing in time, which is visually accompanied by the moving position of the maximum of $p^X(x, t)$.

With the help of $p^X(x, t)$, $\langle X(t)\rangle$ can be found as

$$\langle X(t)\rangle = \int_{-\infty}^{\infty} x p^X(x, t)\, dx, \qquad (7.4)$$

which is an equivalent of averaging over the ensemble of realizations. In what follows, angular brackets $\langle\rangle$ will denote average over the ensemble of realizations. In the integral above, the value x that the process $X(t)$ can take, enters with the order one, and average $\langle X(t)\rangle$ is called the moment of the first order. Obviously, if $p^X(x, t) = p^X(x)$, i.e., does not depend on time, $\langle X(t)\rangle$ does not depend on time, too (see Fig. 7.1(d)).

One can introduce other characteristics of the random process, e.g., mean square $\langle X^2(t)\rangle$ as

$$\langle X^2(t)\rangle = \int_{-\infty}^{\infty} x^2 p^X(x, t)\, dx, \qquad (7.5)$$

which has the meaning of the ensemble average value of the square of the process. However, sometimes it is more convenient to use variance $\sigma_X^2(t)$

$$\sigma_X^2(t) = \langle X^2(t)\rangle - \langle X(t)\rangle^2. \qquad (7.6)$$

The broader the $p^X(x, t)$, the larger the variance is. It is worth noting that while average $\langle X(t)\rangle$ might be constant, $\sigma_X^2(t)$ does not have to be constant. In Fig. 7.1(b)

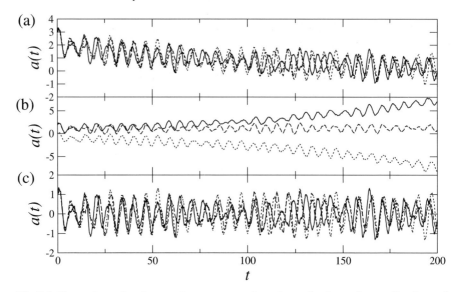

Fig. 7.2. Illustrations of various random processes. In each panel **a**, **b** or **c** three realizations of the same random processes are shown, launched from the same initial conditions. **a** Random process whose average value changes in time. **b** Random process whose variance changes in time. **c** Stationary random process

the $p^X(x,t)$ is given with variance $\sigma_X^2(t)$ growing with time, which is reflected in $p^X(x,t)$ becoming broader. With this, the average $\langle X(t) \rangle$ does not change in time. In Fig. 7.1(c) $\langle X(t) \rangle$ moves from negative to positive values, and $\sigma_X^2(t)$ grows with time, making $p^X(x,t)$ broader and shifting its maximum towards positive values of x. In Fig. 7.1(a)–(c) the $p^X(x,t)$ are changing with time in this or that way, which is an indication of the non-stationary processes which are often the characteristics of some transient, not established behavior.

In Fig. 7.2 each panel shows three different realizations of the same random process: (a) illustrates the process whose average value changes in time which can be seen as a trend in the realizations, (b) illustrates the process whose average is constant in time, but the variance grows, (c) illustrates the stationary process.

7.1.2 Two-Dimensional Probability Density, Correlation and Covariance

One-dimensional PDD carries a limited amount of information about the random process and does not describe if and how the values of the process at different time moments are related to each other. To take account of the latter, one can introduce a two-dimensional PDD $p_2^{XX}(x,t,x_\tau,t+\tau)$ of the random process $X(t)$ as follows:

$$p_2^{XX}(x,t,x_\tau,t+\tau) = \lim_{\substack{\Delta x \to 0, \\ \Delta x_\tau \to 0}} \frac{P\{X(t) \in [x, x+\Delta x] \ \& \ X(t+\tau) \in [x_\tau, x_\tau + \Delta x_\tau]\}}{\Delta x \, \Delta x_\tau}. \quad (7.7)$$

7.1 Introductory Comments on Random Processes 153

It has the meaning of the probability with which two events happen simultaneously: at the time moment t the process $X(t)$ takes the values from $[x, x + \Delta x)$, and at the time moment $t + \tau$ the values from $[x_\tau, x_\tau + \Delta x_\tau)$. The superscript ^{XX} is used in order to signify that the two events correspond to the same process X. It is difficult to visualize p_2^{XX}, since it generally depends on four different arguments and hence is rarely used on its own to characterize the process.

However, p_2^{XX} is used for characterization of the statistical relations between different values of random processes at different time moments by means of correlation function which is denoted here as $K[X, X_\tau]$ in order to comply with the designations of [276, 277]. The letter K stands for correlation, square brackets $[X, X_\tau]$ refer to the processes between which correlation is considered—in our case between the process $X(t)$ (symbol X in square brackets) and its delayed version $X(t + \tau)$ (symbol X_τ in square brackets). Correlation function $K[X, X_\tau]$ is defined as

$$K[X, X_\tau] = \langle X(t)X(t + \tau) \rangle$$
$$= \int_{-\infty}^{\infty} \int_{-\infty}^{\infty} xx_\tau \, p_2^{XX}(x, t, x_\tau, t + \tau) \, dx \, dx_\tau. \tag{7.8}$$

When calculating $K[X, X_\tau]$, an average has been made over all values the process can take, and therefore the resulting function $K[X, X_\tau]$ does not depend on them, being a function of only two arguments: the current time moment t and the temporal distance τ from t. Correlation function has the meaning of the average product of the values of the process at two different time moments. It is obvious, that the largest value of $K[X, X_\tau]$ occurs at $\tau = 0$, since the largest statistical dependence is between the values of the process at the same time moment. The argument τ defines the temporal separation of the considered values x and x_τ of the random process. It is natural to assume that generally for real processes, the larger the time separation τ between the moments is, the smaller the statistical dependence is between the respective values of the random process. However, this dependence is not necessarily monotonous.

Perhaps, a more convenient function is covariance $\Psi[X, X_\tau]$ defined as

$$\Psi[X, X_\tau] = \langle \left(X(t) - \langle X(t) \rangle\right) \times \left(X(t + \tau) - \langle X(t + \tau) \rangle\right) \rangle$$
$$= \int_{-\infty}^{\infty} \left(x(t) - \langle X(t) \rangle\right)\left(x(t + \tau) - \langle X(t + \tau) \rangle\right)$$
$$\times p_2^{XX}(x, t, x_\tau, t + \tau) \, dx \, dx_\tau. \tag{7.9}$$

The value of $\Psi[X, X_\tau]$ at $\tau = 0$ is in fact variance σ_X^2 introduced above

$$\sigma_X^2(t) = \Psi[X, X_\tau]|_{\tau=0}. \tag{7.10}$$

The meaning of $\Psi[X, X_\tau]$ for some process $X(t)$ is exactly the same as the meaning of $K[\tilde{X}, \tilde{X}_\tau]$ for the centered process $\tilde{X}(t)$, that is constructed by removing the average value $\langle X(t) \rangle$ from the process $X(t)$, i.e., $\tilde{X}(t) = X(t) - \langle X(t) \rangle$.

154   7 1:1 Forced Synchronization of Periodic Oscillations in the Presence of Noise

The correlation and covariance are related as follows:

$$
\begin{aligned}
\Psi[X, X_\tau] &= \langle \big(X(t) - \langle X(t) \rangle \big) \big(X(t+\tau) - \langle X(t+\tau) \rangle \big) \rangle \\
&= \langle X(t)X(t+\tau) - X(t)\langle X(t+\tau) \rangle \\
&\quad - \langle X(t) \rangle X(t+\tau) + \langle X(t) \rangle \langle X(t+\tau) \rangle \rangle \\
&= \langle X(t)X(t+\tau) \rangle - \langle X(t) \rangle \langle X(t+\tau) \rangle \\
&\quad - \langle X(t) \rangle \langle X(t+\tau) \rangle + \langle X(t) \rangle \langle X(t+\tau) \rangle \\
&= K[X, X_\tau] - \langle X(t) \rangle \langle X(t+\tau) \rangle.
\end{aligned}
\tag{7.11}
$$

7.1.3 Stationary Process

In various applications, however, one is often interested in the processes that are established after all the transients have died out. Such processes are referred to as stationary. There are many degrees of stationarity, but in practical applications only a couple of them appear most useful: first-order stationarity and wide sense stationarity. If the process is first-order stationary, its $p^X(x, t)$ does not change in time (Fig. 7.1(d)). This does mean that the average $\langle X \rangle$ and the variance σ_X^2 are constants, but does not say anything about the relationship between the values at different time moments.

A wide-sense stationary process is a process whose average $\langle X \rangle$ is constant, power P_X is finite, and covariance $\Psi[X, X_\tau]$ depends only on the temporal distance τ between any two time moments considered, but does not depend on the current time t. The function $\Psi[X, X_\tau]$ of a wide-sense stationary process is even, and the variance $\sigma_X^2 = \Psi[X, X_\tau]|_{\tau=0}$ is the largest value the covariance can take. Note that the power P_X of the wide-sense stationary centered process is equal to its variance σ_X^2, as will be shown later (see (7.20)). If a random process has some well defined time scale, which is often visible in its oscillating realizations (like in Fig. 7.2), its covariance oscillates as well. Moreover, for most real stationary processes the envelope of covariance decays with τ, and the faster it decays, the more random (more disordered, less coherent, less correlated) the process is. A typical covariance of a wide-sense stationary random process is shown schematically in Fig. 7.3(a).

7.1.4 Correlation Time

Note that often it is not convenient to characterize the random process by the whole function $\Psi[X, X_\tau]$, especially as one needs to compare the properties of different processes. It is much more convenient to have just one number to characterize the degree of randomness of the process, and one can introduce correlation time t_{cor} as, e.g., in [276]

$$
t_{\text{cor}} = \frac{1}{\sigma_X^2} \int_0^\infty \big| \Psi[X, X_\tau](\tau) \big| \, d\tau.
\tag{7.12}
$$

The faster the envelope of $\Psi[X, X_\tau](\tau)$ decays, the smaller the t_{cor} is. Note that in the definition (7.12) the integral is normalized by the value of variance, i.e., in fact

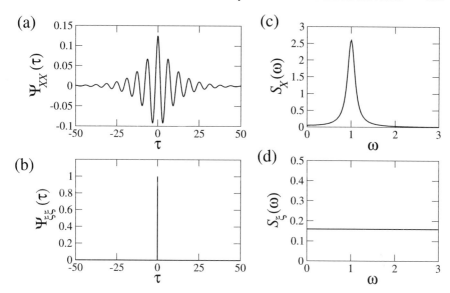

Fig. 7.3. **a** A typical covariance of a wide-sense stationary random process with a well-defined time scale. **b** Covariance of white noise shown schematically, since delta-function has infinite value at $\tau = 0$. **c, d** Spectra of the processes whose covariances are shown in **a**, **b**, respectively

by the power of the process, in order to make the quantity t_{cor} to be independent of it. Processes with different powers can have the same degree of order, and likewise the processes with the same power can have different degrees of order.

7.1.5 Correlation Between Two Different Processes

In some applications one needs to assess the statistical relationships between different random processes $X(t)$ and $Y(t)$. One can introduce a joint two-dimensional probability density distribution $p_2^{XY}(x, t, y_\tau, t + \tau)$ for them as

$$p_2^{XY}(x, t, y_\tau, t + \tau)$$
$$= \lim_{\substack{\Delta x \to 0, \\ \Delta y_\tau \to 0}} \frac{P\{X(t) \in [x, x + \Delta x) \,\&\, Y(t + \tau) \in [y_\tau, y_\tau + \Delta y_\tau)\}}{\Delta x \, \Delta y_\tau}. \quad (7.13)$$

By analogy, we can introduce the cross-correlation $K[X, Y_\tau]$ between the two processes

$$K[X, Y_\tau] = \langle X(t) Y(t + \tau) \rangle$$
$$= \int_{-\infty}^{\infty} \int_{-\infty}^{\infty} x x_\tau \, p_2^{XY}(x, t, y_\tau, t + \tau) \, dx \, dy_\tau, \quad (7.14)$$

156   7 1:1 Forced Synchronization of Periodic Oscillations in the Presence of Noise

and cross-covariance $\Psi[X, Y_\tau]$

$$\Psi[X, Y_\tau] = \langle (X(t) - \langle X(t) \rangle) \times (Y(t + \tau) - \langle Y(t + \tau) \rangle) \rangle$$
$$= \int_{-\infty}^{\infty} (x(t) - \langle X(t) \rangle)(y(t + \tau) - \langle Y(t + \tau) \rangle)$$
$$\times p_2^{XY}(x, t, y_\tau, t + \tau) \, dx \, dy_\tau. \tag{7.15}$$

The relationship between $K[X, Y_\tau]$ and $\Psi[X, Y_\tau]$ is, by analogy with (7.11),

$$\Psi[X, Y_\tau] = K[X, Y_\tau] - \langle X(t) \rangle \langle Y(t + \tau) \rangle. \tag{7.16}$$

7.1.6 Spectrum of a Wide-Sense Stationary Process

The last characteristic of a wide-sense stationary random process $X(t)$ which we will be using in this chapter is the power spectral density $S_X(\omega)$, which we will call "spectrum" for brevity. Spectrum is introduced via Wiener–Khintchine theorem as a Fourier transform of the covariance.[1] Hence, covariance is an inverse Fourier transform of the spectrum, namely,

$$S_X(\omega) = \int_{-\infty}^{\infty} \Psi[X, X_\tau](\tau) e^{-i\omega\tau} \, d\tau, \tag{7.17}$$

$$\Psi[X, X_\tau](\tau) = \frac{1}{2\pi} \int_{-\infty}^{\infty} S_X(\omega) e^{i\omega\tau} \, d\omega, \tag{7.18}$$

where ω is the radial frequency. From the property of the Fourier transform it follows that if the covariance oscillates with a certain time scale, this time scale will be visible in the spectrum $S_X(\omega)$ as a peak around some central frequency. If there are two or more time scales involved, they will be visible in the spectrum as two or more peaks. The spectrum of a process whose covariance is shown in Fig. 7.3(a) is given in Fig. 7.3(c).

Meaning of the Spectrum

The power spectral density (spectrum) has the meaning of the distribution of the power of the process over frequencies. If the spectrum has one peak, it means that the power of the process is concentrated around the central frequency of this peak. This property is illustrated in Fig. 7.3(a),(c): the covariance in (a) makes about eight full oscillations within 50 time units, which corresponds to the average frequency $\omega \approx 8/50 \times 2\pi \approx 1$. This frequency is the central frequency of the spectral peak visible in (c).

---

[1] Strictly speaking, Wiener–Khintchine theorem introduces the spectrum as a inverse Fourier transform of correlation function $K[X, X_\tau]$. But in literature it is often assumed that the mean value of the process $X(t)$ is zero, and therefore correlation turns into covariance according to (7.11).

7.1 Introductory Comments on Random Processes 157

Power of the Process

The total power P_X of a wide-sense stationary process $X(t)$ is the integral of the spectrum over all frequencies, divided by 2π:

$$P_X = \frac{1}{2\pi} \int_{-\infty}^{\infty} S_X(\omega)\, d\omega. \tag{7.19}$$

Division by 2π has to be done for the following reason. In real physical experiments, people normally measure spectra as a power versus *plain* frequency f, rather than radial frequency ω. The power is then calculated as $P_X = \int_{-\infty}^{\infty} S(f)\, df$. In (7.19) integration is made over radial frequencies ω which are related to f as $\omega = 2\pi f$, and coefficient 2π is introduced in order to comply with the physically motivated definition of the power as above. The relationship between the variance and the power of the centered process can be explained by considering $\Psi[X, X_\tau]$ at $\tau = 0$ in (7.18) and remembering that it is equal to variance σ_X^2 according to (7.10)

$$\sigma_X^2 = \Psi[X, X_\tau]|_{\tau=0} = \frac{1}{2\pi} \int_{-\infty}^{\infty} S_X(\omega) e^{\mathrm{i}\omega \times 0}\, d\omega$$

$$= \frac{1}{2\pi} \int_{-\infty}^{\infty} S_X(\omega)\, d\omega = P_X. \tag{7.20}$$

Therefore, variance of a wide-sense stationary centered process is equal to its power.

Calculation of the Spectrum

Since the covariance of a wide-sense stationary process is an even function of τ, its Fourier transform can be calculated as

$$S_X(\omega) = \frac{1}{2\pi} \int_{-\infty}^{\infty} \Psi[X, X_\tau](\tau) \cos(\omega\tau)\, d\tau$$

$$- \mathrm{i}\frac{1}{2\pi} \underbrace{\int_{-\infty}^{\infty} \Psi[X, X_\tau](\tau) \sin(\omega\tau)\, d\tau}_{=0}, \tag{7.21}$$

and finally

$$S_X(\omega) = \frac{1}{2\pi} \int_{-\infty}^{\infty} \Psi[X, X_\tau](\tau) \cos(\omega\tau)\, d\tau. \tag{7.22}$$

White Noise

When one needs to consider a process which is most random, it is convenient to introduce an idea of "white noise." Mathematically, white noise $\xi(t)$ is a process with zero mean $\langle \xi(t) \rangle = 0$, whose covariance is a delta-function, i.e., $\Psi[\xi, \xi_\tau] = \langle \xi(t)\xi(t+\tau) \rangle = \delta(\tau)$, see Fig. 7.3(b). They say that this process is "delta-correlated". Note that, strictly speaking, white noise is not a wide-sense stationary process,

158   7 1:1 Forced Synchronization of Periodic Oscillations in the Presence of Noise

since its power P_ξ, which is the value of delta-function at $\tau = 0$, is infinite. The spectrum of white noise can be introduced by Wiener–Khintchine theorem (7.17) and as was shown by (3.97) is equal to a constant,

$$S_\xi(\omega) = \frac{1}{2\pi}, \tag{7.23}$$

which is illustrated by Fig. 7.3(d).

7.2 Truncated Equations

Our aim is to write down the truncated equations for the amplitude and phase of forced oscillations with noise. We introduce new state variables A and φ such that the solution of (7.1) has the form (3.5). We follow exactly the same approach as without noise, and arrive at the equation similar to (3.17), but its right-hand part contains one more term

$$\frac{\xi}{i\Omega} e^{-i\Omega t}.$$

We proceed by analogy with (3.17), and then get rid of the deterministic fast terms by averaging them over the period T of external forcing using (3.18), and arrive at the following equation for the complex amplitude a:

$$\dot{a} + \frac{(\omega_0^2 - \Omega^2)}{2i\Omega} a - \frac{\lambda}{2} a + \frac{1}{8} a|a|^2 = -i\frac{B}{2\Omega} - i\frac{\xi}{\Omega} e^{-i\Omega t}.$$

Note that $\xi e^{-i\Omega t}$ is a random process, i.e., not a deterministic term, and we cannot apply time averaging to it like to deterministic terms. Simplification of this term will be considered in the next section. After we go from complex amplitude a to the real amplitude A and phase φ, we arrive at the following set of truncated equations with noise:

$$\dot{A} = \frac{A}{2}\left(\lambda - \frac{1}{4}\right) - \frac{B}{2\Omega}\sin\varphi - \frac{\xi}{\Omega}\sin(\Omega t + \varphi) = F_A,$$

$$\dot{\varphi} = \Delta - \frac{B}{2\Omega A}\cos\varphi - \frac{\xi}{\Omega A}\cos(\Omega t + \varphi) = F_\varphi. \tag{7.24}$$

7.3 Simplification of the Fluctuational Terms in Truncated Equations

In (7.24) in the right-hand parts F_A and F_φ one can single out the terms that depend only on A and φ, and the terms that in addition depend on noise ξ

$$F_A = G_A(A, \varphi) + H_A(A, \varphi, \xi),$$
$$F_\varphi = G_\varphi(A, \varphi) + H_\varphi(A, \varphi, \xi),$$

7.3 Simplification of the Fluctuational Terms in Truncated Equations   159

where

$$G_A = \frac{A}{2}\left(\lambda - \frac{1}{4}\right) - \frac{B}{2\Omega}\sin\varphi, \qquad H_A = -\frac{\xi}{\Omega}\sin(\Omega t + \varphi), \qquad (7.25)$$

$$G_\varphi = \Delta - \frac{B}{2\Omega A}\cos\varphi, \qquad H_\varphi = \frac{\xi}{\Omega A}\cos(\Omega t + \varphi). \qquad (7.26)$$

The terms depending on ξ were called fluctuational terms by Stratonovich [277], and their form as defined by (7.25)–(7.26) presents some difficulties for further analysis. In [277] it was proposed to simplify (7.24) in order to make the fluctuational terms more convenient. Of course, any simplification will be possible after one makes additional assumptions on the properties of noise.

The simplification algorithm proposed by Stratonovich in [277] involves:

- Going from a stochastic differential equation to the Fokker–Planck (FP) equation that describes the evolution in time of the joint probability density distribution (PDD) p of the variables A and φ (for brevity, we omit subscript $_2$ and superscripts $^{A,\varphi}$ in the designation for the PDD).
- Simplification of the FP equation.
- Reconstructing the stochastic differential equations that correspond to the simplified FP equation.

In this book we will not describe the general procedure of the derivation of a FP equation from the stochastic differential equation, and we refer the reader to Chap. 4, Sect. 9 of [276]. Here, we only quote the result we need: for a system of two stochastic differential equations of the form

$$\dot{A} = G_A(A, \varphi) + H_A(A, \varphi, \xi) = F_A,$$
$$\dot{\varphi} = G_\varphi(A, \varphi) + H_\varphi(A, \varphi, \xi) = F_\varphi$$

one can write a FP equation describing the evolution of the joint probability density distribution $p(A, \varphi, t)$ according to the following prescription:

$$\frac{\partial p}{\partial t} = -\frac{\partial}{\partial A}\left\{\left(\langle F_A \rangle + \int_{t_0-t}^{0}\Psi\left[\frac{\partial F_A}{\partial A}, F_{A\tau}\right]d\tau + \int_{t_0-t}^{0}\Psi\left[\frac{\partial F_A}{\partial \varphi}, F_{\varphi\tau}\right]d\tau\right)p\right\}$$
$$-\frac{\partial}{\partial \varphi}\left\{\left(\langle F_\varphi \rangle + \int_{t_0-t}^{0}\Psi\left[\frac{\partial F_\varphi}{\partial A}, F_{A\tau}\right]d\tau + \int_{t_0-t}^{0}\Psi\left[\frac{\partial F_\varphi}{\partial \varphi}, F_{\varphi\tau}\right]d\tau\right)p\right\}$$
$$+\frac{\partial^2}{\partial A^2}\left\{\left(\int_{t_0-t}^{0}\Psi[F_A, F_{A\tau}]d\tau\right)p\right\} + \frac{\partial^2}{\partial A\,\partial \varphi}\left\{\left(\int_{t_0-t}^{0}\Psi[F_A, F_{\varphi\tau}]d\tau\right)p\right\}$$
$$+\frac{\partial^2}{\partial \varphi\,\partial A}\left\{\left(\int_{t_0-t}^{0}\Psi[F_\varphi, F_{A\tau}]d\tau\right)p\right\} + \frac{\partial^2}{\partial \varphi^2}\left\{\left(\int_{t_0-t}^{0}\Psi[F_\varphi, F_{\varphi\tau}]d\tau\right)p\right\}.$$
$$(7.27)$$

Here, $\Psi[R, Q_\tau]$ is the cross-covariance of the two random processes R and Q defined as (see (7.15), (7.16))

$$\Psi[R, Q_\tau] = \langle RQ_\tau \rangle - \langle R \rangle \langle Q_\tau \rangle.$$

160   7 1:1 Forced Synchronization of Periodic Oscillations in the Presence of Noise

Here $\langle\rangle$ denote averaging over the ensemble of realizations of the random process, R is the first random process at the time moment t, and Q_τ is another random process at the time moment $t + \tau$.

We further proceed by analogy with [277] where the equations of a slightly different form were considered. Obviously, due to the external random influence ξ, the functions F_A and F_φ are random functions of time. The various covariances that appear in (7.27) can be expressed as follows. For the start, consider the first covariance

$$\Psi\left[\frac{\partial F_A}{\partial A}, F_{A\tau}\right] = \left\langle\frac{\partial F_A}{\partial A} \times F_{A\tau}\right\rangle - \left\langle\frac{\partial F_A}{\partial A}\right\rangle\langle F_{A\tau}\rangle$$
$$= \left\langle\frac{\partial(G_A + H_A)}{\partial A} \times (G_{A\tau} + H_{A\tau})\right\rangle$$
$$- \left\langle\frac{\partial(G_A + H_A)}{\partial A}\right\rangle\langle G_{A\tau} + H_{A\tau}\rangle. \tag{7.28}$$

Because G_A and $\partial G_A/\partial A$ are deterministic functions of time, they are going to be the same for any realization of random process ξ. Hence, their ensemble averages $\langle G_A\rangle$ and $\langle\partial G_A/\partial A\rangle$ are going to be the functions G_A and $\partial G_A/\partial A$ themselves, i.e.,

$$\langle G_A\rangle = G_A, \qquad \left\langle\frac{\partial G_A}{\partial A}\right\rangle = \frac{\partial G_A}{\partial A}. \tag{7.29}$$

Also, the average of a product of a deterministic and a random functions, is the product of the deterministic function and the average of the random function, i.e.,

$$\Psi\left[\frac{\partial F_A}{\partial A}, F_{A\tau}\right] = \frac{\partial G_A}{\partial A} \times G_{A\tau} + \frac{\partial G_A}{\partial A} \times \langle H_{A\tau}\rangle + \left\langle\frac{\partial H_A}{\partial A}\right\rangle \times G_{A\tau}$$
$$+ \left\langle\frac{\partial H_A}{\partial A} \times H_{A\tau}\right\rangle - \frac{\partial G_A}{\partial A} \times G_{A\tau} - \left\langle\frac{\partial H_A}{\partial A}\right\rangle \times G_{A\tau}$$
$$- \left\langle\frac{\partial H_A}{\partial A}\right\rangle \times G_{A\tau} - \left\langle\frac{\partial H_A}{\partial A}\right\rangle \times \langle H_{A\tau}\rangle. \tag{7.30}$$

Some terms above cancel each other. Also, remember that the random process ξ has zero average, hence from (7.25) $H_{A\tau}$ has zero average, too. Finally one obtains

$$\Psi\left[\frac{\partial F_A}{\partial A}, F_{A\tau}\right] = \left\langle\frac{\partial H_A}{\partial A} \times H_{A\tau}\right\rangle. \tag{7.31}$$

By analogy, one can calculate averages and covariances of other terms in (7.27)

$$\langle F_A\rangle = G_A, \tag{7.32}$$
$$\langle F_\varphi\rangle = G_\varphi, \tag{7.33}$$

$$\Psi\left[\frac{\partial F_A}{\partial A}, F_{A\tau}\right] = \left\langle\frac{\partial H_A}{\partial A} \times H_{A\tau}\right\rangle = \langle 0 \times H_{A\tau}\rangle = 0, \tag{7.34}$$

7.3 Simplification of the Fluctuational Terms in Truncated Equations   161

$$\Psi\left[\frac{\partial F_A}{\partial \varphi}, F_{\varphi\tau}\right]$$

$$= \left\langle \frac{\partial H_A}{\partial \varphi} \times H_{\varphi\tau}\right\rangle$$

$$= \left\langle -\frac{\xi}{\Omega}\cos(\Omega t + \varphi) \times \left(-\frac{\xi_\tau}{\Omega A_\tau}\cos(\Omega t + \Omega\tau + \varphi_\tau)\right)\right\rangle$$

$$= \langle\xi\xi_\tau\rangle\frac{1}{A_\tau\Omega^2}\cos(\Omega t + \varphi)\cos(\Omega t + \Omega\tau + \varphi_\tau)$$

$$= \langle\xi\xi_\tau\rangle\frac{1}{A_\tau\Omega^2}\frac{1}{2}[\cos(\Omega\tau + \varphi_\tau - \varphi) + \cos(2\Omega t + \varphi + \Omega\tau + \varphi_\tau)].$$

For (7.27) we will need to calculate an integral of Ψ with respect to τ. With that in mind, in the calculations above let us separate the terms which are independent on τ:

$$\Psi\left[\frac{\partial F_A}{\partial \varphi}, F_{\varphi\tau}\right]$$

$$= \langle\xi\xi_\tau\rangle\frac{1}{2A_\tau\Omega^2}\big[\cos(\Omega\tau + \varphi_\tau - \varphi) + \cos(2\Omega t + 2\varphi)\cos(\Omega\tau + \varphi_\tau - \varphi)$$

$$- \sin(2\Omega t + 2\varphi)\sin(\Omega\tau + \varphi_\tau - \varphi)\big]$$

$$= \langle\xi\xi_\tau\rangle\frac{1}{2A_\tau\Omega^2}\big[\cos(\Omega\tau + \varphi_\tau - \varphi)\big(1 + \cos(2\Omega t + 2\varphi)\big)$$

$$- \sin(\Omega\tau + \varphi_\tau - \varphi)\sin(2\Omega t + 2\varphi)\big].$$

Now, we need to take an integral of $\Psi[\partial F_A/\partial\varphi, F_{\varphi\tau}]$ over τ from $(t_0 - t)$ to 0, where t_0 is some initial time moment from which we start to consider the process. Because we are interested in the established processes, we set t_0 to minus infinity

$$\int_{-\infty}^{0}\Psi\left[\frac{\partial F_A}{\partial \varphi}, F_{\varphi\tau}\right]\mathrm{d}\tau$$

$$= \frac{1 + \cos(2\Omega t + 2\varphi)}{2A_\tau\Omega^2}\int_{-\infty}^{0}\langle\xi\xi_\tau\rangle\cos(\Omega\tau + \varphi_\tau - \varphi)\,\mathrm{d}\tau$$

$$- \frac{\sin(2\Omega t + 2\varphi)}{2A_\tau\Omega^2}\int_{-\infty}^{0}\langle\xi\xi_\tau\rangle\sin(\Omega\tau + \varphi_\tau - \varphi)\,\mathrm{d}\tau. \qquad (7.35)$$

Formally, we are going to integrate over an infinite time interval. However, we make an assumption that noise ξ is a fast random process with correlation time t_{cor} much less than the relaxation time of the system (7.1) and (7.24) that is of the order $1/(\varepsilon\omega_0)$, i.e.,

$$t_{\mathrm{cor}} \ll \frac{1}{\varepsilon\omega_0}.$$

In that case, the correlation function $\langle\xi\xi_\tau\rangle$ of ξ decays to zero in such time intervals, during which the slow variables of the system (7.24) A and φ almost do not change.

162   7 1:1 Forced Synchronization of Periodic Oscillations in the Presence of Noise

Hence, in the calculations above we regard A and φ as constants that do not change with τ, i.e.,

$$A = A_\tau, \qquad \varphi = \varphi_\tau. \qquad (7.36)$$

Substitution of (7.36) into (7.35) gives

$$\int_{-\infty}^{0} \Psi \left[\frac{\partial F_A}{\partial \varphi}, F_{\varphi\tau} \right] d\tau$$
$$= \frac{1 + \cos(2\Omega t + 2\varphi)}{2A\Omega^2} \int_{-\infty}^{0} \langle \xi \xi_\tau \rangle \cos(\Omega\tau) \, d\tau$$
$$- \frac{\sin(2\Omega t + 2\varphi)}{2A\Omega^2} \int_{-\infty}^{0} \langle \xi \xi_\tau \rangle \sin(\Omega\tau) \, d\tau. \qquad (7.37)$$

Assume that ξ is a stationary process, then its correlation function $\langle \xi \xi_\tau \rangle$ depends only on τ. In the expression above, the first integral is half of the Fourier transform (FT) of $\langle \xi \xi_\tau \rangle$ at the frequency Ω, i.e., by Wiener–Khintchine theorem is half the value of the power spectral density S_ξ of the random process ξ at the frequency Ω. The second integral is the imaginary part of the FT and is equal to zero. With this, we obtain

$$\int_{-\infty}^{0} \Psi \left[\frac{\partial F_A}{\partial \varphi}, F_{\varphi\tau} \right] d\tau = \frac{1 + \cos(2\Omega t + 2\varphi)}{4A\Omega^2} S_\xi(\Omega). \qquad (7.38)$$

We would like to further simplify the FP equation (7.27) and hence the term defined by (7.38). We can again employ the Bogoliubov–Krylov method of averaging and recall that A and φ are slowly varying functions of time (see (3.6)). Then we can average each term of FP equation on a period of the external force Ω using (3.18). Finally, we obtain

$$\int_{-\infty}^{0} \Psi \left[\frac{\partial F_A}{\partial \varphi}, F_{\varphi\tau} \right] d\tau = \frac{1}{4A\Omega^2} S_\xi(\Omega). \qquad (7.39)$$

By analogy consider other terms of (7.27)

$$\Psi \left[\frac{\partial F_\varphi}{\partial A}, F_{A\tau} \right] = \left\langle \frac{\partial H_\varphi}{\partial A} \times H_{A\tau} \right\rangle = 0, \qquad (7.40)$$

$$\Psi \left[\frac{\partial F_\varphi}{\partial \varphi}, F_{\varphi\tau} \right] = \left\langle \frac{\partial H_\varphi}{\partial \varphi} \times H_{\varphi\tau} \right\rangle = 0, \qquad (7.41)$$

$$\Psi[F_A, F_{A\tau}] = \langle H_A H_{A\tau} \rangle = \frac{1}{4\Omega^2} S_\xi(\Omega), \qquad (7.42)$$

$$\Psi[F_A, F_{\varphi\tau}] = \langle H_A H_{\varphi\tau} \rangle = 0, \qquad (7.43)$$

$$\Psi[F_\varphi, F_{A\tau}] = \langle H_\varphi H_{A\tau} \rangle = 0, \qquad (7.44)$$

$$\Psi[F_\varphi, F_{\varphi\tau}] = \langle H_\varphi H_{\varphi\tau} \rangle = \frac{1}{4\Omega^2 A^2} S_\xi(\Omega). \qquad (7.45)$$

7.3 Simplification of the Fluctuational Terms in Truncated Equations 163

In view of the above, (7.27) can be rewritten as

$$
\begin{aligned}
\frac{\partial p}{\partial t} = & -\frac{\partial}{\partial A}\left\{\left(G_A + \int_{-\infty}^{0}\left\langle\frac{\partial H_A}{\partial A}\times H_{A\tau}\right\rangle d\tau + \int_{-\infty}^{0}\left\langle\frac{\partial H_A}{\partial \varphi}\times H_{\varphi\tau}\right\rangle d\tau\right)p\right\} \\
& -\frac{\partial}{\partial \varphi}\left\{\left(G_\varphi + \int_{-\infty}^{0}\left\langle\frac{\partial H_\varphi}{\partial A}\times H_{A\tau}\right\rangle d\tau + \int_{-\infty}^{0}\left\langle\frac{\partial H_\varphi}{\partial \varphi}\times H_{\varphi\tau}\right\rangle d\tau\right)p\right\} \\
& +\frac{\partial^2}{\partial A^2}\left\{\left(\int_{-\infty}^{0}\langle H_A H_{A\tau}\rangle d\tau\right)p\right\} + \frac{\partial^2}{\partial A\,\partial \varphi}\left\{\left(\int_{-\infty}^{0}\langle H_A H_{\varphi\tau}\rangle d\tau\right)p\right\} \\
& +\frac{\partial^2}{\partial \varphi\,\partial A}\left\{\left(\int_{-\infty}^{0}\langle H_\varphi H_{A\tau}\rangle d\tau\right)p\right\} + \frac{\partial^2}{\partial \varphi^2}\left\{\left(\int_{-\infty}^{0}\langle H_\varphi H_{\varphi\tau}\rangle d\tau\right)p\right\}.
\end{aligned}
$$

(7.46)

Substitute all terms in (7.32)–(7.33) and (7.40)–(7.45) into (7.27) or (7.46) to obtain

$$
\begin{aligned}
\frac{\partial p}{\partial t} = & -\frac{\partial}{\partial A}\left\{\left(G_A + \frac{S_\xi(\Omega)}{4A\Omega^2}\right)p\right\} - \frac{\partial}{\partial \varphi}\{G_\varphi p\} \\
& +\frac{\partial^2}{\partial A^2}\left\{\frac{S_\xi(\Omega)}{4\Omega^2}p\right\} + \frac{\partial^2}{\partial \varphi^2}\left\{\frac{S_\xi(\Omega)}{4\Omega^2 A^2}p\right\},
\end{aligned}
$$

(7.47)

where G_A and G_φ are as in (7.25). We have arrived at the Fokker–Planck equation which is simplified by means of averaging over the period of the external force, and thus of getting rid of fast terms. Now we would like to reconstruct stochastic equations in the form

$$
\dot{A} = \tilde{G}_A(A,\varphi) + \tilde{H}_A(A,\varphi,\boldsymbol{\eta}),
$$

(7.48)

$$
\dot{\varphi} = \tilde{G}_\varphi(A,\varphi) + \tilde{H}_\varphi(A,\varphi,\boldsymbol{\eta}),
$$

(7.49)

that would result in the simplified FP equation (7.47), if one wanted to construct it following the recipe (7.27). Note that in general the stochastic equation can involve more than one source of noise which is emphasized here by writing a noise vector $\boldsymbol{\eta} = (\eta_1, \eta_2, \ldots)$ rather than a scalar. In particular, we need to find the explicit forms of functions $\tilde{G}_A$, $\tilde{H}_A$, $\tilde{G}_\varphi$ and $\tilde{H}_\varphi$ which might be different from G_A, H_A, G_φ and H_φ. In order to do this, compare separate terms of (7.47) with the respective terms of (7.46), remembering that all functions in the latter would be marked by tildes. We observe that

$$
\int_{-\infty}^{0}\langle\tilde{H}_A\tilde{H}_{\varphi\tau}\rangle d\tau = \int_{-\infty}^{0}\langle\tilde{H}_\varphi\tilde{H}_{A\tau}\rangle d\tau = 0,
$$

which can be true if the processes $\tilde{H}_A$ and $\tilde{H}_\varphi$ are not correlated. This can be realized if e.g. $\tilde{H}_A$ depends on the noise η_1, while $\tilde{H}_\varphi$ on η_2, η_1 and η_2 being uncorrelated. If this is so, then the two pairs of processes $\partial\tilde{H}_A/\partial\varphi$ and $\tilde{H}_\varphi$, and $\partial\tilde{H}_\varphi/\partial A$ and $\tilde{H}_A$, are not correlated, too, i.e.,

$$
\int_{-\infty}^{0}\left\langle\frac{\partial\tilde{H}_A}{\partial\varphi}\times\tilde{H}_{\varphi\tau}\right\rangle d\tau = \int_{-\infty}^{0}\left\langle\frac{\partial\tilde{H}_\varphi}{\partial A}\times\tilde{H}_{A\tau}\right\rangle d\tau = 0.
$$

(7.50)

164   7 1:1 Forced Synchronization of Periodic Oscillations in the Presence of Noise

Then

$$\tilde{G}_A + \int_{-\infty}^{0} \left\langle \frac{\partial \tilde{H}_A}{\partial A} \times \tilde{H}_{A\tau} \right\rangle d\tau = G_A + \frac{S_\xi(\Omega)}{4A\Omega^2},\tag{7.51}$$

$$\tilde{G}_\varphi + \int_{-\infty}^{0} \left\langle \frac{\partial \tilde{H}_\varphi}{\partial \varphi} \times \tilde{H}_{\varphi\tau} \right\rangle d\tau = G_\varphi,\tag{7.52}$$

where $\tilde{G}$ and $\tilde{H}$ correspond to (7.48), (7.49), and G and H to (7.25), (7.26). Now consider

$$\int_{-\infty}^{0} \langle \tilde{H}_A \tilde{H}_{A\tau} \rangle d\tau = \frac{S_\xi(\Omega)}{4\Omega^2}.\tag{7.53}$$

If $\tilde{H}_A$ depended on A, the integral above would have depended on A, too. But it does not, so we conclude that $\tilde{H}_A$ is independent of A, and therefore $\partial \tilde{H}_A/\partial A = 0$. This leads to the disappearance of the integral in (7.53), and the final expression for $\tilde{G}_A$ is

$$\tilde{G}_A = G_A + \frac{S_\xi(\Omega)}{4A\Omega^2}.\tag{7.54}$$

Next, consider

$$\int_{-\infty}^{0} \langle \tilde{H}_\varphi \tilde{H}_{\varphi\tau} \rangle d\tau = \frac{S_\xi(\Omega)}{4\Omega^2 A^2}.\tag{7.55}$$

Here, the integral depends on A, but does not depend on φ, which means that $\tilde{H}_\varphi$ explicitly depends on A, but not on φ. Then in (7.52) the term involving $\partial \tilde{H}_\varphi/\partial \varphi$ vanishes, and $\tilde{G}_\varphi$ is

$$\tilde{G}_\varphi = G_\varphi.\tag{7.56}$$

Equation (7.55) can be valid if $\tilde{H}_\varphi$ is expressed as

$$\tilde{H}_\varphi = \frac{\sqrt{S_\xi(\Omega)}}{\sqrt{2}\Omega A}\eta_2,\tag{7.57}$$

where η_2 is delta-correlated noise with zero mean and unity variance, i.e.,

$$\langle \eta_2(t) \rangle = 0, \qquad \langle \eta_2(t)\eta_2(t+\tau) \rangle = \delta(\tau), \qquad \langle \eta_2^2(t) \rangle = 1.$$

One can substitute (7.57) into (7.55) to check that the equality would hold. Finally, we need to find $\tilde{H}_A$. From (7.53) we deduce that $\tilde{H}_A$ is independent both of A and of φ. The following expression for $\tilde{H}_A$ would make (7.53) valid:

$$\tilde{H}_A = \frac{\sqrt{S_\xi(\Omega)}}{\sqrt{2}\Omega}\eta_1,$$

where η_1 is delta-correlated noise with zero mean and unity variance, i.e.,

$$\langle \eta_1(t) \rangle = 0, \qquad \langle \eta_1(t)\eta_1(t+\tau) \rangle = \delta(\tau), \qquad \langle \eta_1^2(t) \rangle = 1.$$

7.4 Probability Density Distribution of the Phase Difference   165

In order to enable $\tilde{H}_A$ and $\tilde{H}_\varphi$ to be uncorrelated, we require that η_1 and η_2 are uncorrelated, too, i.e.,

$$\langle \eta_1(t)\eta_2(t+\tau)\rangle \equiv 0.$$

Finally, the simplified stochastic differential equations that are roughly equivalent to the original (7.24) have the form

$$\dot{A} = \frac{A}{2}\left(\lambda - \frac{1}{4}\right) - \frac{B}{2\Omega}\sin\varphi + \frac{S_\xi(\Omega)}{4A\Omega^2} + \frac{\sqrt{S_\xi(\Omega)}}{\sqrt{2\Omega}}\eta_1, \tag{7.58}$$

$$\dot{\varphi} = \Delta - \frac{B}{2\Omega A}\cos\varphi + \frac{\sqrt{S_\xi(\Omega)}}{\sqrt{2\Omega A}}\eta_2. \tag{7.59}$$

7.4 Probability Density Distribution of the Phase Difference

Consider the equations for the amplitude A and phase difference φ with simplified fluctuational terms (7.58)–(7.59). Analysis of these equations still remains quite a difficult problem, in spite of the simplification performed in the previous section. Let us consider a special case when the amplitude of forcing signal is small

$$B \ll \varepsilon A_0, \tag{7.60}$$

where A_0 is the amplitude of the oscillations without harmonic forcing or noise, i.e., at $B = 0$ and $S_\xi(\Omega) = 0$. By analogy with the deterministic case considered in Sect. 3.4, the instantaneous amplitude A will not be very different from A_0 in average, which can be mathematically expressed as

$$\frac{\langle (A - A_0)^2\rangle}{A_0^2} \ll 1. \tag{7.61}$$

In that case, a good approximation will be to replace A by A_0 in (7.59)

$$\dot{\varphi} = \Delta - \frac{B}{2\Omega A_0}\cos\varphi + \frac{\sqrt{S_\xi(\Omega)}}{\sqrt{2\Omega A_0}}\eta_2, \tag{7.62}$$

and to treat it separately. For convenience denote

$$\frac{B}{2\Omega A_0} = \Delta_s, \qquad \frac{\sqrt{S_\xi(\Omega)}}{\sqrt{2\Omega A_0}} = D, \tag{7.63}$$

and rewrite (7.62) as

$$\dot{\varphi} = \Delta - \Delta_s\cos\varphi + D\eta_2 = F, \tag{7.64}$$

$$F = G + H, \qquad G = \Delta - \Delta_s\cos\varphi, \qquad H = D\eta_2. \tag{7.65}$$

166   7 1:1 Forced Synchronization of Periodic Oscillations in the Presence of Noise

Let us write down a Fokker–Planck equation for the evolution of the one-dimensional probability density $p(\varphi, t)$ corresponding to (7.64) using the one-dimensional version of the recipe (7.27)

$$\frac{\partial p}{\partial t} = -\frac{\partial}{\partial \varphi}\left\{\left(\langle F\rangle + \int_{-\infty}^{0} \Psi\left[\frac{\partial F}{\partial \varphi}, F_\tau\right] d\tau\right) p\right\}$$
$$+ \frac{\partial^2}{\partial \varphi^2}\left\{\left(\int_{-\infty}^{0} \Psi[F, F_\tau] d\tau\right) p\right\}. \tag{7.66}$$

We find the components of the FP equation above

$$\langle F\rangle = G = \Delta - \Delta_s \cos\varphi, \tag{7.67}$$

$$\Psi\left[\frac{\partial F}{\partial \varphi}, F_\tau\right] = \left\langle\frac{\partial H}{\partial \varphi} H_\tau\right\rangle, \qquad \frac{\partial H}{\partial \varphi} = 0 \quad \Rightarrow \quad \Psi\left[\frac{\partial F}{\partial \varphi}, F_\tau\right] = 0, \tag{7.68}$$

$$\Psi[F, F_\tau] = \langle H H_\tau\rangle = D^2\langle\eta_2\eta_{2\tau}\rangle = D^2\delta(\tau). \tag{7.69}$$

Substitution of (7.67)–(7.69) into (7.66) gives the FP equation for $p(\varphi, t)$

$$\frac{\partial p}{\partial t} = -\frac{\partial}{\partial \varphi}\{(\Delta - \Delta_s \cos\varphi)p\} + \frac{D^2}{2}\frac{\partial^2 p}{\partial \varphi^2}. \tag{7.70}$$

Denote

$$J(\varphi) = (\Delta - \Delta_s \cos\varphi)p - \frac{D^2}{2}\frac{\partial p}{\partial \varphi}, \tag{7.71}$$

where $J(\varphi)$ is the probability current. In order to reduce the lengths of the formula below, introduce the following designations:

$$Q = \frac{2}{D^2}\Delta, \qquad Q_s = \frac{2}{D^2}\Delta_s, \tag{7.72}$$

so that $J(\varphi)$ is rewritten as

$$J(\varphi) = \frac{D^2}{2}\left[(Q - Q_s \cos\varphi)p - \frac{\partial p}{\partial \varphi}\right]. \tag{7.73}$$

Then (7.70) can be rewritten as

$$\frac{\partial p}{\partial t} + \frac{\partial J}{\partial \varphi} = 0, \tag{7.74}$$

which is the law of "conservation of probability." We are interested in the stationary PDD, i.e., the one that does not change in time, $\partial p/\partial t = 0$. Then from (7.74) the probability current J does not depend on φ, i.e., $\partial J/\partial \varphi = 0$. Differentiate (7.73) with respect to φ remembering that now J is a function of only φ and not of t, i.e., changing the partial derivatives to the straight ones

$$\frac{d^2 p}{d\varphi^2} - \frac{d}{d\varphi}[(Q - Q_s \cos\varphi)p] = 0. \tag{7.75}$$

7.4 Probability Density Distribution of the Phase Difference   167

Let us solve this equation in order to find the stationary probability density distribution $p(\varphi)$. After integrating (7.75) once, we find

$$\frac{dp}{d\varphi} - (Q - Q_s \cos \varphi)p = C_1. \tag{7.76}$$

First, we find a solution for the homogeneous form of the (7.76), i.e., when $C_1 = 0$

$$\frac{dp}{p} = (Q - Q_s \cos \varphi)\, d\varphi, \quad p \geq 0,$$
$$\ln p = (Q\varphi - Q_s \sin \varphi) + C_2. \tag{7.77}$$

Taking exponents of both parts, we obtain

$$p(\varphi) = Ce^{(Q\varphi - Q_s \sin \varphi)},$$

where $C = e^{C_2}$. The solution of a non-homogeneous equation will be sought for in the form

$$p(\varphi) = C(\varphi)e^{(Q\varphi - Q_s \sin \varphi)}, \tag{7.78}$$

i.e., where C is no longer a constant, but a function of φ. After substitution of (7.78) into (7.76) we obtain

$$\frac{dC(\varphi)}{d\varphi}e^{(Q\varphi - Q_s \sin \varphi)} + C(\varphi)e^{(Q\varphi - Q_s \sin \varphi)}(Q - Q_s \cos \varphi)$$
$$- (Q - Q_s \cos \varphi)C(\varphi)e^{(Q\varphi - Q_s \sin \varphi)} = C_1,$$

where the last two terms in the right-hand side cancel each other. The equation for the unknown function $C(\varphi)$ is therefore

$$\frac{dC(\varphi)}{d\varphi}e^{(Q\varphi - Q_s \sin \varphi)} = C_1.$$

Direct integration gives

$$C(\varphi) = \int_{C_3}^{\varphi} C_1 e^{(-Q\psi + Q_s \sin \psi)}\, d\psi. \tag{7.79}$$

Thus, from (7.78) we find the solution of (7.75), which is

$$p(\varphi) = C_1 e^{(Q\varphi - Q_s \sin \varphi)} \int_{C_3}^{\varphi} e^{(-Q\psi + Q_s \sin \psi)}\, d\psi. \tag{7.80}$$

Two constants need to be determined: C_1 and C_3, and there are two conditions that the function $p(\varphi)$ has to satisfy, from which we can find them. The first is periodicity condition stating that the PDD of some phase difference φ is the same as PDD of $\varphi + 2\pi$, i.e.,

$$p(\varphi) = p(\varphi + 2\pi).$$

168   7 1:1 Forced Synchronization of Periodic Oscillations in the Presence of Noise

Consider $p(\varphi + 2\pi)$

$$p(\varphi + 2\pi) = C_1 e^{Q2\pi} e^{(Q\varphi - Q_s \sin \varphi)} \int_{C_3}^{\varphi+2\pi} e^{(-Q\psi + Q_s \sin \psi)} \, d\psi. \tag{7.81}$$

Under the integral change variables $\tilde{\psi} = \psi - 2\pi$, then the limits of integration will change as well, so that

$$\begin{aligned} p(\varphi + 2\pi) &= C_1 e^{Q2\pi} e^{(Q\varphi - Q_s \sin \varphi)} \int_{C_3 - 2\pi}^{\varphi} e^{(-Q\tilde{\psi} - Q2\pi + Q_s \sin \tilde{\psi})} \, d\tilde{\psi} \\ &= C_1 e^{(Q\varphi - Q_s \sin \varphi)} \int_{C_3 - 2\pi}^{\varphi} e^{(-Q\psi + Q_s \sin \psi)} \, d\psi, \end{aligned} \tag{7.82}$$

where in the last integral tilde over ψ is omitted since it is only a dummy variable. In order to enable the last expression to be equal to $p(\varphi)$ defined by (7.80), one has to set C_3 to plus or minus infinity, $C_3 = \pm\infty$, bearing in mind that $-\infty - 2\pi = -\infty$ and $\infty - 2\pi = \infty$. Let us choose $C_3 = -\infty$ for now, and come back to $p(\varphi + 2\pi)$

$$p(\varphi + 2\pi) = C_1 e^{2\pi Q} e^{(Q\varphi - Q_s \sin \varphi)} \int_{-\infty}^{\varphi+2\pi} e^{(-Q\psi + Q_s \sin \psi)} \, d\psi. \tag{7.83}$$

Split the range of integration $(-\infty; \varphi + 2\pi]$ into an infinite number of intervals of equal size 2π: $\dots, (\varphi - 2\pi n - 2\pi; \varphi - 2\pi n], \dots, (\varphi - 4\pi - 2\pi; \varphi - 4\pi], (\varphi - 2\pi - 2\pi; \varphi - 2\pi], (\varphi - 2\pi; \varphi], (\varphi; \varphi + 2\pi]$. Consider integrals over each of these intervals

$$\begin{aligned} p(\varphi + 2\pi) = C_1 e^{2\pi Q} e^{(Q\varphi - Q_s \sin \varphi)} \\ \times \Bigg(\int_{\varphi}^{\varphi+2\pi} e^{(-Q\psi + Q_s \sin \psi)} \, d\psi + \int_{\varphi-2\pi}^{\varphi} e^{(-Q\psi + Q_s \sin \psi)} \, d\psi \\ + \int_{\varphi-4\pi}^{\varphi-2\pi} e^{(-Q\psi + Q_s \sin \psi)} \, d\psi + \cdots \Bigg). \end{aligned}$$

Consider, e.g., the third integral in the equation above, and introduce the change of variables $\tilde{\psi} = \psi + 4\pi$

$$\begin{aligned} \int_{\varphi-4\pi}^{\varphi-2\pi} e^{(-Q\psi + Q_s \sin \psi)} \, d\psi &= \int_{\varphi}^{\varphi+2\pi} e^{(-Q(\tilde{\psi} - 4\pi) + Q_s \sin \tilde{\psi})} \, d\tilde{\psi} \\ &= e^{4\pi Q} \int_{\varphi}^{\varphi+2\pi} e^{(-Q\psi + Q_s \sin \psi)} \, d\psi. \end{aligned}$$

Similarly, each integral over the interval $(\varphi - 2\pi n - 2\pi; \varphi - 2\pi n]$ can be reduced to the integral over $(\varphi; \varphi + 2\pi]$ that is multiplied by a constant $\exp(2\pi(n+1)Q)$. Finally, with account of periodicity condition we put $p(\varphi)$ instead of $p(\varphi + 2\pi)$ and

7.4 Probability Density Distribution of the Phase Difference 169

obtain

$$p(\varphi) = C_1 e^{2\pi Q}\left(1 + e^{2\pi Q} + e^{4\pi Q} + \cdots\right)e^{(Q\varphi - Q_s \sin\varphi)}$$
$$\times \int_{\varphi}^{\varphi + 2\pi} e^{(-Q\psi + Q_s \sin\psi)}\, d\psi. \tag{7.84}$$

Denote

$$C_1 e^{2\pi Q}\left(1 + e^{2\pi Q} + e^{4\pi Q} + \cdots\right) = \frac{1}{N}, \tag{7.85}$$

so that N is

$$N = \frac{e^{-2\pi Q}}{C_1(1 + e^{2\pi Q} + e^{4\pi Q} + \cdots)}.$$

Note that provided that $Q < 0$, the infinite sum

$$\left(1 + e^{2\pi Q} + e^{4\pi Q} + \cdots\right) = \frac{1}{1 - e^{2\pi Q}}, \tag{7.86}$$

and

$$N = \frac{e^{-2\pi Q} - 1}{C_1}. \tag{7.87}$$

7.4.1 Case of $Q > 0$

One might wonder what happens if $Q > 0$, since then (7.86) is no longer valid. Remember, that in (7.82) one can choose $C_3 = +\infty$ and write $p(\varphi + 2\pi)$ by analogy with the above as

$$p(\varphi + 2\pi)$$

$$= -C_1 e^{2\pi Q} e^{(Q\varphi - Q_s \sin\varphi)} \int_{\varphi + 2\pi}^{\infty} e^{(-Q\psi + Q_s \sin\psi)}\, d\psi$$

$$= -C_1 e^{2\pi Q} e^{(Q\varphi - Q_s \sin\varphi)}$$
$$\times \left(\int_{\varphi + 2\pi}^{\varphi + 4\pi} e^{(-Q\psi + Q_s \sin\psi)}\, d\psi + \int_{\varphi + 4\pi}^{\varphi + 6\pi} e^{(-Q\psi + Q_s \sin\psi)}\, d\psi + \cdots\right)$$

$$= -C_1 e^{2\pi Q} e^{(Q\varphi - Q_s \sin\varphi)}\left(e^{-2\pi Q} + e^{-4\pi Q} + e^{-6\pi Q} + \cdots\right)$$
$$\times \int_{\varphi}^{\varphi + 2\pi} e^{(-Q\psi + Q_s \sin\psi)}\, d\psi$$

$$= -C_1 e^{(Q\varphi - Q_s \sin\varphi)}\left(1 + e^{-2\pi Q} + e^{-4\pi Q} + \cdots\right)$$
$$\times \int_{\varphi}^{\varphi + 2\pi} e^{(-Q\psi + Q_s \sin\psi)}\, d\psi$$

$$= -C_1 e^{(Q\varphi - Q_s \sin\varphi)} \frac{1}{1 - e^{-2\pi Q}} \times \int_{\varphi}^{\varphi + 2\pi} e^{(-Q\psi + Q_s \sin\psi)}\, d\psi$$

$$= C_1 e^{(Q\varphi - Q_s \sin\varphi)} \frac{1}{e^{-2\pi Q} - 1} \times \int_{\varphi}^{\varphi + 2\pi} e^{(-Q\psi + Q_s \sin\psi)}\, d\psi.$$

170   7 1:1 Forced Synchronization of Periodic Oscillations in the Presence of Noise

One can denote N as in (7.87) and arrive at the equation below for $p(\varphi)$. Therefore, both for positive and negative Q we obtain the same expression for $p(\varphi)$

$$p(\varphi) = \frac{1}{N} e^{(Q\varphi - Q_s \sin \varphi)} \int_\varphi^{\varphi + 2\pi} e^{(-Q\psi + Q_s \sin \psi)} \, d\psi. \tag{7.88}$$

We still need to find the constant C_1, which can be done using normalization condition

$$\int_0^{2\pi} p(\varphi) \, d\varphi = 1,$$

$$\frac{1}{N} \int_0^{2\pi} \left(e^{(Q\varphi - Q_s \sin \varphi)} \int_\varphi^{\varphi + 2\pi} e^{(-Q\psi + Q_s \sin \psi)} \, d\psi \right) d\varphi = 1.$$

Introduce the new variable $\chi = \psi - \varphi$. Note that $\psi = \chi + \varphi$ and $d\psi = d\chi$, since φ is regarded as a constant while one considers the integral over ψ. Also, we need to change the limits of integration of the inner integral: when $\psi = \varphi$, $\chi = 0$, and when $\psi = \varphi + 2\pi$, $\chi = 2\pi$. Then

$$\frac{1}{N} \int_0^{2\pi} \left(\int_0^{2\pi} e^{-Q\chi + Q_s(\sin \varphi - \sin \psi)} \, d\chi \right) d\varphi = 1. \tag{7.89}$$

Next, we need to transform the difference $(\sin \varphi - \sin \psi)$ using trigonometric identity

$$\sin \varphi - \sin \psi = 2 \cos\left(\frac{\psi + \varphi}{2}\right) \sin\left(\frac{\psi - \varphi}{2}\right).$$

Substituting into (7.89) gives

$$\frac{1}{N} \int_0^{2\pi} \left(\int_0^{2\pi} e^{-Q\chi + 2Q_s \cos(\chi/2 + \varphi) \sin(\chi/2)} \, d\chi \right) d\varphi = 1 \tag{7.90}$$

and

$$N = \int_0^{2\pi} e^{-Q\chi} \underbrace{\left(\int_0^{2\pi} e^{2Q_s \cos(\chi/2 + \varphi) \sin(\chi/2)} \, d\varphi \right)}_{\hat{I}} \, d\chi. \tag{7.91}$$

In the last expression an integral $\hat{I}$ is involved that cannot be expressed through simpler functions. Such integrals can be, however, expressed through special functions called Bessel functions, which will be described in the next section.

7.5 Bessel Functions

Bessel functions arise, e.g., as one tries to expand in a Fourier series the function $\exp(ix \sin t)$, where i is an imaginary unit, x is a real number and t is time, namely,

$$e^{ix \sin t} = \sum_{n=-\infty}^{\infty} J_n(x) e^{int}.$$

7.5 Bessel Functions

The Fourier coefficients (see, e.g., [134] for the basics of Fourier analysis) denoted here as J_n are equal to

$$J_n(x) = \frac{1}{2\pi} \int_0^{2\pi} e^{-i(nt - x \sin t)} \, dt. \tag{7.92}$$

Functions $J_n(x)$ form a special class of functions that cannot be represented through simpler functions—they are called Bessel functions of the first kind. With this, n is the order of Bessel function, and x is its argument. $J_n(x)$ in the form of (7.92) are also called "integral representation of Bessel functions." We will not go into mathematical detail of the origin and properties of these functions, but this information can be found, e.g., in [4].

Note that the function $J_0(x)$, i.e., the function of the zeroth order, has the form

$$J_0(x) = \frac{1}{2\pi} \int_0^{2\pi} e^{ix \sin t} \, dt.$$

One can also introduce modified, or hyperbolic, Bessel functions $I_n(x)$ which are expressed via $J_n(x)$ as

$$I_n(x) = i^{-n} J_n(ix). \tag{7.93}$$

A modified Bessel function of the zeroth order has the form

$$I_0(x) = \frac{1}{2\pi} \int_0^{2\pi} e^{-x \sin t} \, dt. \tag{7.94}$$

The plot of $I_0(x)$ is given in Fig. 7.4(a). Note that this function is even, real, $I_0(0) = 1$, it increases monotonously as $|x|$ grows, and its asymptotic behavior is as follows [143]:

$$I_0(x) \to \frac{e^{|x|}}{\sqrt{2\pi |x|}}, \quad |x| \to \infty, \tag{7.95}$$

as illustrated in Fig. 7.4(a) by thick grey line, as compared to thin black line showing $I_0(x)$.

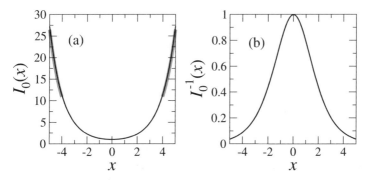

Fig. 7.4. a Graph of modified Bessel function of zeroth order $I_0(x)$ (*black line*) and of the function $e^{|x|}/\sqrt{2\pi |x|}$ (*grey line*). b Graph of $I_0^{-1}(x)$

7.6 Probability Density Distribution of the Phase Difference, Continued

Compare (7.94) with $\widehat{I}$ in (7.91) of Sect. 7.4, and observe that $\widehat{I}$ can be reduced to the form (7.94) by writing

$$\widehat{I} = \int_0^{2\pi} e^{2Q_s \sin(\chi/2) \cos(\chi/2+\varphi)} \, d\varphi = \int_0^{2\pi} e^{-2Q_s \sin(\chi/2) \sin(\varphi+\chi/2-\pi/2)} \, d\varphi.$$

Introduce $\varphi' = \varphi + \chi/2 - \pi/2$, then $d\varphi' = d\varphi$, and the limits of integration are then from $(\chi/2 - \pi/2)$ to $(\chi/2 + 3\pi/2)$

$$\widehat{I} = \int_{(\chi/2-\pi/2)}^{(\chi/2+3\pi/2)} e^{-2Q_s \sin(\chi/2) \sin(\varphi')} \, d\varphi'.$$

Note that $\exp[-2Q_s \sin(\chi/2)]$ does not depend on φ' and does not participate in integration. With this, $\sin \varphi'$ is a periodic function with respect to φ' with period 2π, and so is the function $\exp[\sin(\varphi')]$. The integral of the latter function over any interval over φ of the length 2π will be the same as the integral from 0 to 2π. Hence, we can write

$$\widehat{I} = \int_0^{2\pi} e^{-2Q_s \sin(\chi/2) \sin(\varphi')} \, d\varphi',$$

which by comparison with (7.94) is

$$\widehat{I} = 2\pi I_0 \left(2Q_s \sin\left(\frac{\chi}{2} \right) \right). \tag{7.96}$$

With account of (7.96), (7.91) can be rewritten as

$$N = 2\pi \int_0^{2\pi} e^{-Q\chi} I_0 \left(2Q_s \sin\left(\frac{\chi}{2} \right) \right) d\chi. \tag{7.97}$$

Split the integral into two as follows:

$$N = 2\pi \left[\int_0^{\pi} e^{-Q\chi} I_0 \left(2Q_s \sin\left(\frac{\chi}{2} \right) \right) d\chi \right.$$
$$\left. + \int_{\pi}^{2\pi} e^{-Q\chi} I_0 \left(2Q_s \sin\left(\frac{\chi}{2} \right) \right) d\chi \right] \tag{7.98}$$

and introduce variable substitution which would be different for the first and the second integrals

$$\chi' = \begin{cases} \frac{1}{2}(\pi - \chi), & 0 < \chi < \pi, \\ \frac{1}{2}(\chi - \pi), & \pi < \chi < 2\pi. \end{cases}$$

7.6 Probability Density Distribution of the Phase Difference, Continued   173

The borders of integration limits in terms of χ' are going to be as follows:

1st integral: $\chi = 0 \;\Rightarrow\; \chi' = \dfrac{\pi}{2},$ $\chi = \pi \;\Rightarrow\; \chi' = 0,$

2nd integral: $\chi = \pi \;\Rightarrow\; \chi' = 0,$ $\chi = 2\pi \;\Rightarrow\; \chi' = \dfrac{\pi}{2}.$

Now we have to express χ and $d\chi$ via χ', which will be different for different integration intervals, namely,

1st integral: $\chi = (\pi - 2\chi'),$ $d\chi = -2\,d\chi',$

2nd integral: $\chi = (\pi + 2\chi'),$ $d\chi = 2\,d\chi'.$

Substitute χ, $d\chi$ and new integration limits into (7.98)

$$
\begin{aligned}
N &= 2\pi \int_{\pi/2}^{0} e^{-Q(\pi-2\chi')} I_0\!\left(2Q_s \sin\!\left(\frac{\pi}{2} - \chi'\right)\right) \times (-2)\,d\chi' \\
&\quad + 2\pi \int_{0}^{\pi/2} e^{-Q(\pi+2\chi')} I_0\!\left(2Q_s \sin\!\left(\frac{\pi}{2} + \chi'\right)\right) \times 2\,d\chi' \\
&= 4\pi \int_{0}^{\pi/2} e^{-\pi Q} e^{2\chi' Q} I_0(2Q_s \cos \chi')\,d\chi' \\
&\quad + 4\pi \int_{0}^{\pi/2} e^{-\pi Q} e^{-2\chi' Q} I_0(2Q_s \cos \chi')\,d\chi' \\
&= 4\pi e^{-Q\pi} \int_{0}^{\pi/2} I_0(2Q_s \cos \chi')2 \underbrace{\frac{(e^{2\chi' Q} + e^{-2\chi' Q})}{2}}_{\cosh(2Q\chi')}\,d\chi'.
\end{aligned}
$$

From that,

$$
N = 8\pi e^{-Q\pi} \int_{0}^{\pi/2} \cosh(2Q\chi') I_0(2Q_s \cos \chi')\,d\chi'. \tag{7.99}
$$

In the final step of calculating N let us use the following integral that can be found in [97] on p. 716, Sect. 6.681, formula 3:

$$
\int_{0}^{\pi/2} \cos(2\mu x) I_{2\nu}(2a \cos x)\,dx = \frac{\pi}{2}\, I_{\nu-\mu}(a)\, I_{\nu+\mu}(a). \tag{7.100}
$$

Compare (7.99) with (7.100) and observe that (7.99) can be rewritten in the form of (7.100) if one uses the following identity:

$$
\cosh(x) = \cos(ix) \tag{7.101}
$$

and takes $\nu = 0$. Namely,

$$
N = 8\pi e^{-Q\pi} \int_{0}^{\pi/2} \cos(2iQ\chi') I_0(2Q_s \cos \chi')\,d\chi', \tag{7.102}
$$

and using (7.100) it is equal to

$$
N = 4\pi^2 e^{-Q\pi} I_{-iQ}(Q_s) I_{iQ}(Q_s). \tag{7.103}
$$

174   7 1:1 Forced Synchronization of Periodic Oscillations in the Presence of Noise

This is an expression for N given in terms of Bessel functions, which is an analytic expression.

In order to understand how the product $I_{-iQ}(Q_s)I_{iQ}(Q_s)$ can be calculated numerically, let us also give N in terms of integrals by using (7.92) and (7.93)

$$I_{-iQ}(Q_s) = i^{iQ}\frac{1}{2\pi}\int_0^{2\pi} e^{-i(-iQt - iQ_s \sin t)}\,dt$$

$$= i^{iQ}\frac{1}{2\pi}\int_0^{2\pi} e^{(-Qt - Q_s \sin t)}\,dt,$$

$$I_{iQ}(Q_s) = i^{-iQ}\frac{1}{2\pi}\int_0^{2\pi} e^{-i(iQt - iQ_s \sin t)}\,dt \qquad (7.104)$$

$$= i^{-iQ}\frac{1}{2\pi}\int_0^{2\pi} e^{(Qt - Q_s \sin t)}\,dt,$$

$$I_{-iQ}(Q_s)I_{iQ}(Q_s) = \frac{1}{(2\pi)^2}\int_0^{2\pi} e^{(-Qt - Q_s \sin t)}\,dt\int_0^{2\pi} e^{(Qt' - Q_s \sin t')}\,dt'.$$

Hence, N can be rewritten as

$$N = e^{-Q\pi}\int_0^{2\pi} e^{(-Qt - Q_s \sin t)}\,dt\int_0^{2\pi} e^{(Qt' - Q_s \sin t')}\,dt', \qquad (7.105)$$

where Q and Q_s are defined by (7.72) together with (7.63). Finally, the probability density $p(\varphi)$ is expressed by (7.88) with N defined by (7.103) or (7.105).

7.7 Mean Frequency of Forced Oscillations with Noise

Remember, that the full phase $\psi(t)$ of forced oscillations is defined by (3.7). The derivative $\dot\psi(t)$ of the full phase defines the instantaneous frequency of forced oscillations, and the derivative averaged over the ensemble of realizations (statistical average) $\langle\dot\psi(t)\rangle$ defines the mean frequency of forced oscillations. From (3.7) it follows that

$$\langle\dot\psi(t)\rangle = \Omega + \langle\dot\varphi(t)\rangle. \qquad (7.106)$$

We know forcing frequency Ω, therefore we need to estimate $\langle\dot\varphi(t)\rangle$, i.e., the statistical average of the right-hand side of (7.64)

$$\langle\dot\varphi\rangle = \langle\Delta - \Delta_s \cos\varphi + D\eta_2\rangle \quad \text{where } \langle\eta_2\rangle = 0. \qquad (7.107)$$

In order to do this, we need stationary probability density distribution $p(\varphi)$, which was found in Sect. 7.6. Then $\langle\dot\varphi(t)\rangle$ is equal to

$$\langle\dot\varphi(t)\rangle = \int_0^{2\pi} (\Delta - \Delta_s \cos\varphi)p(\varphi)\,d\varphi.$$

7.7 Mean Frequency of Forced Oscillations with Noise 175

From (7.71) it follows that

$$\langle \dot{\varphi}(t) \rangle = \int_0^{2\pi} \left(J(\varphi) + \frac{D^2}{2} \frac{dp(\varphi)}{d\varphi} \right) d\varphi = \int_0^{2\pi} J(\varphi)\, d\varphi + \frac{D^2}{2} \int_0^{2\pi} dp$$

$$= \int_0^{2\pi} J(\varphi)\, d\varphi + \frac{D^2}{2} \underbrace{(p|_{\varphi=0} - p|_{\varphi=2\pi})}_{=0} = \int_0^{2\pi} J(\varphi)\, d\varphi.$$

From (7.74) it follows that since we are considering a stationary probability density distribution $p(\varphi)$ for which $dp(\varphi)/dt = 0$, then $dJ(\varphi)/d\varphi = 0$, i.e., $J = \text{const}$ and

$$\langle \dot{\varphi}(t) \rangle = 2\pi J, \tag{7.108}$$

where J is defined by

$$J(\varphi) = \frac{D^2}{2} \left[(Q - Q_s \cos \varphi)p - \frac{dp}{d\varphi} \right]. \tag{7.109}$$

The last expression is the same as (7.73), but with straight derivative of p. $p(\varphi)$ is known, and we only need to find $dp/d\varphi$ by differentiating (7.88), bearing in mind that N is a constant. In (7.88) denote

$$\widehat{I} = \int_\varphi^{\varphi+2\pi} \underbrace{e^{(-Q\psi + Q_s \sin \psi)}}_{=A}\, d\psi. \tag{7.110}$$

To take a derivative of $\widehat{I}$ whose limits of integration depend on φ, we use the fundamental theorem of calculus

$$\frac{d}{dx} \left(\int_0^x f(t)\, dt \right) = f(x). \tag{7.111}$$

We can represent an integral in (7.110) as

$$\widehat{I} = \int_0^{\varphi+2\pi} A\, d\psi - \int_0^\varphi A\, d\psi \tag{7.112}$$

and use (7.111) to obtain

$$\frac{d}{d\varphi} \widehat{I} = \frac{d}{d\varphi} \left(\int_0^{\varphi+2\pi} A\, d\psi \right) - \frac{d}{d\varphi} \left(\int_0^\varphi A\, d\psi \right)$$

$$= e^{(-Q(\varphi+2\pi)+Q_s \sin(\varphi+2\pi))} - e^{(-Q\varphi+Q_s \sin \varphi)}$$

$$= e^{(-Q\varphi+Q_s \sin \varphi)} \left(e^{-2\pi Q} - 1 \right).$$

Use the last result when differentiating (7.88)

$$\frac{dp}{d\varphi} = \frac{1}{N} e^{(Q\varphi - Q_s \sin \varphi)} (Q - Q_s \cos \varphi) \int_\varphi^{\varphi+2\pi} A\, d\psi$$

$$+ \frac{1}{N} e^{(Q\varphi - Q_s \sin \varphi)} e^{(-Q\varphi + Q_s \sin \varphi)} \left(e^{-2\pi Q} - 1 \right). \tag{7.113}$$

176   7 1:1 Forced Synchronization of Periodic Oscillations in the Presence of Noise

Substitute (7.113) into (7.109)

$$
\begin{aligned}
J &= \frac{D^2}{2} \left[(Q - Q_s \cos \varphi) \frac{1}{N} e^{(Q\varphi - Q_s \sin \varphi)} \int_\varphi^{\varphi + 2\pi} A \, d\psi \right. \\
&\qquad \left. - \frac{1}{N} e^{(Q\varphi - Q_s \sin \varphi)} (Q - Q_s \cos \varphi) \int_\varphi^{\varphi + 2\pi} A \, d\psi - \frac{1}{N} \left(e^{-2\pi Q} - 1 \right) \right] \\
&= \frac{D^2}{2} \frac{1 - e^{-2\pi Q}}{N}.
\end{aligned}
$$

Hence, from (7.108) $\langle \dot{\varphi}(t) \rangle$ is

$$
\begin{aligned}
\langle \dot{\varphi}(t) \rangle &= \frac{\pi D^2}{N} \left(1 - e^{-2\pi Q} \right) = \frac{\pi D^2}{N} \left(1 - e^{-2\pi Q} \right) \times \frac{2 e^{-\pi Q}}{2 e^{-\pi Q}} \\
&= \frac{2\pi D^2}{N} e^{-\pi Q} \frac{(e^{\pi Q} - e^{-\pi Q})}{2}.
\end{aligned}
$$

Finally,

$$
\langle \dot{\varphi}(t) \rangle = \frac{2\pi D^2}{N} e^{-\pi Q} \sinh \pi Q \tag{7.114}
$$

with N defined by (7.103) or (7.105), and Q by (7.72), in which Δ is defined by (3.20).

By analogy with forced oscillations without noise considered in Sect. 3.11, we can call $|\langle \dot{\varphi} \rangle|$ mean beat frequency. Consider forced van der Pol oscillator with noise (7.1) with $\lambda = 0.1$, $\omega_0 = 1$, $B = 0.01$ and the values of forcing frequency Ω close to 1, i.e., around the 1:1 locking region outlined in Fig. 3.5. In Fig. 7.5, $|\langle \dot{\varphi} \rangle|$ versus Ω estimated from (7.114) is shown by black lines for four different non-zero noise intensities D, from the smallest $D = 0.02$ given by the lowest curve, to the largest $D = 0.5$ (strong noise) given by the upper curve. Grey line shows analytical estimate of $|\langle \dot{\varphi} \rangle|$ without noise by formula (3.106). One can see that if noise is weak, the beat frequency demonstrates a plateau inside the locking region, just like in the case of noiseless oscillations (also compare with Fig. 3.16), although the slope of it is slightly non-zero. However, the stronger the noise, the larger the slope of the plateau becomes, and in the limit of very strong noise $(D \to \infty)^2$ the plateau vanishes completely and the dependence takes the form

$$
|\langle \dot{\varphi} \rangle| = |\omega_0 - \Omega|.
$$

Quantitatively the case of $D \to \infty$ looks almost indistinguishable from the case of $D = 0.5$ illustrated in Fig. 7.5.

Note that in experiments where only the forcing frequency Ω is being changed, we would normally detect synchronization by the presence of the plateau in the graph of beat frequency. If there is no plateau, we would say there is no synchronization. The strong noise destroys synchronization, in some sense overriding the effect of periodic forcing on the system, hence there will be no plateau.

---

[2] The case of an infinitely large noise is a mathematical abstraction here, since in reality when noise becomes too large, the whole dynamics is very smeared and it becomes impossible to introduce the phase of oscillations correctly.

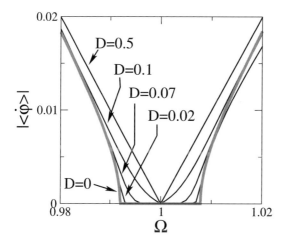

Fig. 7.5. Mean beat frequency $|\langle\dot\varphi\rangle|$ of forced oscillations with noise versus the forcing frequency Ω around 1 : 1 phase locking region for several different noise intensities D. In (7.1) the parameters are set as: $\lambda = 0.1$, $\omega_0 = 1$, $B = 0.01$. *Grey line* shows $|\langle\dot\varphi\rangle|$ without noise, i.e., at $D = 0$ (compare with Fig. 3.16). *Black lines* show $|\langle\dot\varphi\rangle|$ at non-zero noise intensities D, starting from $D = 0.02$ (*lowest curve*) and ending with $D = 0.5$ (*upper curves*)

7.8 Interpretation of Phase Dynamics

Consider (7.64) describing the dynamics of phase difference φ between the response and the forcing. One can interpret the behavior of φ as the behavior of a particle in a potential V that has the shape described by minus integral over φ of the deterministic part of the right-hand side $G(\varphi)$ in (7.64), (7.65) i.e.,

$$V(\varphi) = -\Delta\varphi + \Delta_s \sin\varphi. \tag{7.115}$$

This is illustrated in Fig. 7.6. In (a) the parameters of the forcing are chosen to be such that in the absence of noise the system is inside the synchronization (locking) region, i.e., $\Delta_s > \Delta$, meaning that the amplitude B of forcing is big enough to induce synchronization with the given detuning Δ. Three values of Δ are illustrated: the left part ($\Delta > 0$), the middle ($\Delta = 0$) and the right part ($\Delta < 0$) of the locking region. When $\Delta > 0$ and $\Delta_s > \Delta$ (upper panel of Fig. 7.6(a)), the particle under the influence of noise oscillates around one of the local minima of the potential well. If the applied noise has Gaussian distribution, i.e., is able to take any value at least occasionally, then whatever the barrier height, sooner or later the perturbation will achieve a value that would be enough to kick the particle over the barrier. The particle does jump to the neighboring well from time to time, and these jumps, or phase slips, represent a random process. With each jump to the right, φ increases by 2π, and with each jump to the left it decreases by the same value. It is much easier for the particle to overcome the lower barriers to the left than to the right, so the preferred direction of particle drift is to the right, although the jumps in the opposite direction are not impossible. The phase difference φ will in average decrease with

178   7 1:1 Forced Synchronization of Periodic Oscillations in the Presence of Noise

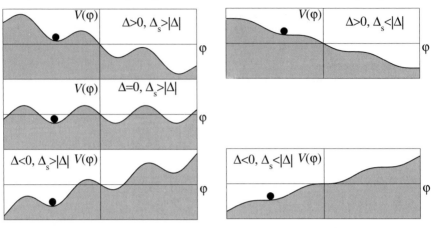

Fig. 7.6. Schematic illustration of the behavior of phase difference φ as a particle in an inclined potential $V(\varphi)$, see (7.64) and (7.115). **a** Inside the noise-free synchronization region, $\Delta_s > |\Delta|$. **b** Outside the noise-free synchronization region, $\Delta_s < |\Delta|$

time unboundedly. In Fig. 7.6(a), the second row illustrates the behavior of φ in the middle of synchronization region ($\Delta = 0$). Whatever the amplitude B of forcing is, the particle with equal probability jumps to the left or to the right, and in average φ does not change. Third row of Fig. 7.6(a) illustrates the case in the right-hand part of synchronization region, $\Omega > \omega_0$. In this case φ drifts preferably to the left. Generally, inside synchronization region phase difference displays plateaus of certain duration that correspond to the state of phase locking, interrupted by jumps by 2π. The average duration of staying within each potential well is proportional to the strength of synchronization.

In Fig. 7.6(b) the situation is illustrated schematically for the case when the forcing parameters are such that in the noise-free case the system is outside the locking region. In that case there are no potential wells for φ, and it slides down the surface in this or that direction, depending on the sign of detuning Δ.

Now let us follow the evolution of the phase difference in time in a full system describing forced periodic oscillations with noise. Consider van der Pol equations with external forcing in a slightly different form than (3.3), namely,

$$\begin{aligned}\dot{x} &= y, \\ \dot{y} &= \lambda(1-x^2)y - \omega_0^2 x + B\cos(\Omega t) + \tilde{D}\xi(t).\end{aligned} \quad (7.116)$$

Here as before, $\lambda = 0.2$ is non-linearity parameter, ω_0 is the frequency of self-oscillations just at birth (i.e., at $\lambda = 0$), B is the strength of external forcing and Ω is the forcing frequency. $\xi(t)$ is Gaussian white noise with zero mean and unity variance, and $\tilde{D}$ is the strength of applied noise. In the absence of noise, bifurcation diagram in the vicinity of 1:1 synchronization region looks as shown in Fig. 7.7.

7.8 Interpretation of Phase Dynamics

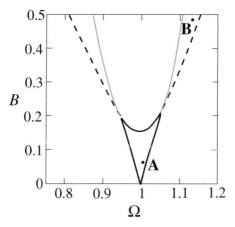

Fig. 7.7. (Color online) The vicinity of 1:1 synchronization region of a forced van der Pol system as described by (7.116) with $\lambda = 0.2$. *Solid lines*: saddle-node bifurcations; *dashed*: torus birth

We start from considering small amplitudes B of forcing in order to match the conditions of Stratonovich theory described in Sects. 7.4–7.7. In particular, consider point **A** inside locking region ($\Omega = 1.0118$ and $C = 0.06$).

When numerically calculating the phase difference between the forcing and the response, it is convenient to rewrite the original equations (7.116) in terms of amplitude A and phase ψ by introducing $x(t) = A(t)\cos\psi(t)$ and $y(t) = \dot{x}(t) = A(t)\sin\psi(t)$ and to arrive at the following equations:

$$\dot{A} = \lambda(1 - A^2\cos^2\psi)A\sin^2\psi + A\sin\psi\cos\psi - \omega_0^2 A\cos\psi\sin\psi$$
$$+ B\cos\Omega t\sin\psi + \tilde{D}\xi\sin\psi,$$
$$\dot{\psi} = \lambda(1 - A^2\cos^2\psi)\sin\psi\cos\psi - \sin^2\psi - \omega_0^2\cos^2\psi \qquad (7.117)$$
$$+ \frac{B}{A}\cos\Omega t\cos\psi + \frac{\tilde{D}}{A}\xi\cos\psi.$$

Without noise ($\tilde{D} = 0$), the difference φ between the phase ψ of the forced oscillations and that of the forcing Ωt calculated from the full system (7.117) oscillates with a small amplitude around a horizontal line, as compared to the straight line that results from the approximate truncated equations (7.64) for φ.

Note, that the noise strength D in (7.64) is not equal to $\tilde{D}$ in (7.117). When noise is included, the phase difference φ starts to jump both in the original equations (7.116)–(7.117) and in the truncated equations (7.64).

In Fig. 7.8, upper panel, we illustrate what happens when the phase slip occurs in (7.116): the phase difference gradually slips by 2π. It is interesting to observe what happens to the realizations of the response and of the forcing during the phase slip. In Fig. 7.8, lower panel, the realization $x(t)$ of (7.116) is given, on which its values are superimposed that are taken when the forcing is at the same (arbitrary) phase, i.e., in fact the values of the stroboscopic section of x. One can see that

180   7 1:1 Forced Synchronization of Periodic Oscillations in the Presence of Noise

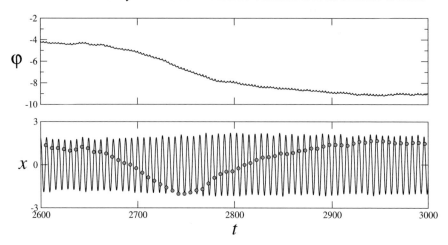

Fig. 7.8. Illustration of a phase slip in a forced van der Pol system with noise (7.116), (7.117) with $\lambda = 0.2$, $\Omega = 1.0118$, $B = 0.06$ and $\tilde{D} = 0.05$. **Upper panel**: phase difference φ gradually slipping by 2π. **Lower panel**: Realization x (*solid line*) and the values of x at the same phase of forcing, i.e., the values of its stroboscopic section (*circles*)

while phase difference is oscillating around the constant level around -4.2, each time the forcing makes one full oscillation, x is at approximately the same stage, in our case near its maximum. However during the phase slip the stroboscopic values of x run through all possible values, and after the phase slip ends, return to the original value around the maximum. Thus, phase slip can be noticed either in the plot of phase difference, or in the realization of the stroboscopic section. Another illustration of what happens when noise affects the phase-locked system is given in Fig. 7.9 (first column), where the phase differences are shown for three different values $\tilde{D}$ of noise intensity. It is clearly seen that as noise becomes stronger, the number of phase slips per time unit grows (first column). In the second column, the respective stroboscopic sections are given by black points, together with the manifold of the resonant torus of the noise-free system and the two cycles: stable (black circle) and saddle (white circle). Obviously, noise smears the stable cycle and the phase trajectories visit larger vicinities of it. But in addition to that, in terms of the phase space phase slip corresponds to the event when noise throws the phase point outside the delimiting stable manifold of the saddle cycle (not shown), so that the phase point makes full rotation along the surface of the torus before it comes back to the vicinity of the stable cycle.

7.9 Phase Diffusion

Strictly speaking, in the presence of unbounded noise there is no synchronization in its classical sense, because the phase difference φ does not oscillate around some fixed value, but rather occasionally jumps, or slips. However, the phase slips can

7.9 Phase Diffusion

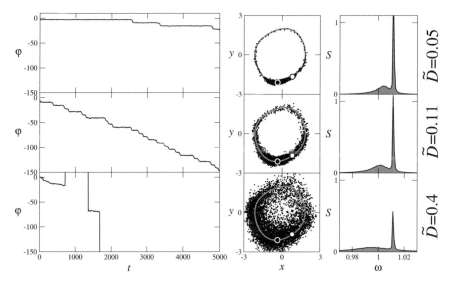

Fig. 7.9. Noise destroying a phase-locked periodic motion in a forced van der Pol oscillator (7.116). Different values of noise $\tilde{D}$ are given to the right of each row. **First column**: Phase difference φ between the system and the forcing. **Second column**: Stroboscopic section. **Third column**: Spectra of x_1

either occur quite often, or be rare events, and it is clear that in the former case the system is further away from its synchronized state than in the latter case. In [174] Malakhov has proposed the concept of effective synchronization. Following his terminology, we can say that the effectiveness of synchronization is associated with the frequency of phase slips: the more often phase slips occur, the further away the system is from the synchronized state. A general method to assess the effectiveness was described, e.g., in [153] which is based on the observation of the phase difference. Each panel of Fig. 7.10 shows 15 different realizations of the phase difference $\varphi/2\pi$ between the response and the forcing for a forced van der Pol system with noise (7.117) at $\Omega = 1.0018$ and $B = 0.06$, corresponding to 15 different realizations of noise $\xi(t)$ with the same intensity $\tilde{D}$ indicated to the right of each panel. For the convenience of comparison between different noise intensities, the ordinates of the two panels have the same scale. φ in these figures in normalized by 2π, so that one can clearly see that the size of each step is 1. The mean slope of φ is equal to the mean beat frequency $\langle\dot\varphi\rangle$, and it is evident that at smaller noise $\tilde{D} = 0.05$ it is smaller than at larger noise $\tilde{D} = 0.1$, as predicted by theory (compare with Fig. 7.5).

Using the terminology introduced in Sect. 7.1, in Fig. 7.10 there are 15 realizations of the same random experiment that was launched from the same initial conditions. If we look at these realizations carefully, we will notice that at some time $t_1 > 0$ there is a certain range of values that φ can take. At some larger time moment $t_2 > t_1$, the range of values taken by φ is larger than at t_1. At each time

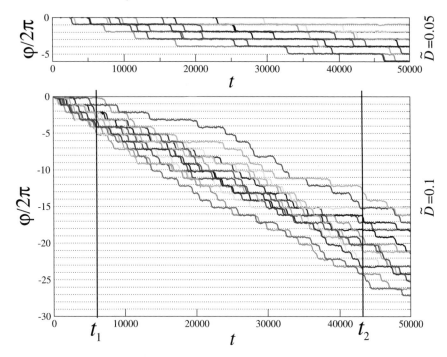

Fig. 7.10. 15 realizations of phase difference φ between the response and the forcing corresponding to 15 different realizations of noise $\xi(t)$ with the same intensity $\tilde{D}$ shown to the right of each panel. Data are obtained by numerical simulation of (7.117) at $\lambda = 0.2$, $\Omega = 1.0018$ and $B = 0.06$. Each time the system was launched from the same initial conditions

moment t, the random process governing the evolution of φ can be characterized by a probability density distribution $p_1(\varphi, t)$ and its moments: mean value $\langle\varphi(t)\rangle$ and variance $\sigma_\varphi^2(t)$. At the initial time moment $t = 0$ the probability density distribution (PDD) $p_1(\varphi, 0)$ was Dirac delta-function $\delta(\varphi - \varphi_0)$, where φ_0 is some initial phase difference, and the process had zero variance $\sigma_\varphi^2(0) = 0$. With the increase of time t, the mean value becomes negative in our case and decreases, while $p_1(\varphi, t)$ is smeared. This means that the variance $\sigma_\varphi^2(t)$ grows in time. Therefore, the PDD behaves like in Fig. 7.1(c), only it starts with a delta-function at $t = 0$.

Even from observing these realizations one can notice that at different $\tilde{D}$ the ranges of possible values of φ grow in time with different velocities: the larger the noise, the faster. The average velocity of growth of the variance $\sigma_\varphi^2(t)$ can serve a measure of how quickly the phase *diffuses*. A diffusion coefficient D_{eff} can be introduced as

$$D_{\text{eff}} = \lim_{t \to \infty} \frac{\sigma_\varphi^2(t)}{2t} = \lim_{t \to \infty} \frac{\langle\varphi^2(t)\rangle - \langle\varphi(t)\rangle^2}{2t}. \quad (7.118)$$

The larger the value of D_{eff}, the further the system is from the synchronized state.

7.10 Full-Scale Biological Experiment

We would like to illustrate the phenomenon of 1 : 1 phase synchronization in the presence of noise by a full-scale biological experiment. Consider a human cardiovascular system. It is well-known that our hearts are self-sustained systems that demonstrate non-damped oscillations. Moreover, it is equally well established that although the heart does have some well-defined time scale of its beatings, the time intervals between the successive beats are not the same and are distributed within a certain range of values. The interbeat intervals are often referred to as RR-intervals. Of course, there is a lot of random influence that is applied to the heart either from inside the rest of the human body, or from the outside.

In [22, 23] it was studied how weak non-invasive forcing in the form of a sequence of light and sound pulses could influence the heart rate of healthy volunteers. The main attention was paid to 1 : 1 synchronization by periodic forcing, and it has been demonstrated that most subjects were able to adjust their heart rhythm to a rhythmic signal whose frequency was close to that of the subjects. An easy and popular way to detect the time scale associated with the heart beats is to register an electrocardiogram (ECG)—a signal that reflects the electrical activity of the human heart. A typical ECG of a healthy human is shown in Fig. 7.11(a). One can notice that it has a very characteristic shape, and reproduces the same pattern again and again with a certain accuracy. The sharpest peaks in the ECG are the so-called R-peaks, and usually one introduces interbeat intervals as the time intervals T_i between the successive R-peaks. In Fig. 7.11(b) a forcing signal is represented schematically: at a certain time moment a red square appeared on the screen of the computer and simultaneously a "beep" signal was generated. The duration of this sound-and-light pulse was fixed at 0.1 sec. A volunteer was sitting comfortably in an arm-chair in front of the computer responsible for the generation of the forcing signal. The frequency of forcing was changed in the range ±25% of the average heart rate of the subject. The difference between the forcing frequency and the average heart rate at rest, relative to the average heart rate at rest, was called detuning Δ. The response to the forcing was recorded and processed for each value of Δ.

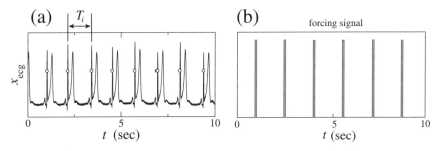

Fig. 7.11. (Color online) **a** A typical electrocardiogram (ECG) of a healthy human. T_i: interbeat interval. **b** A schematic representation of the forcing in the form of a sequence of light and sound pulses

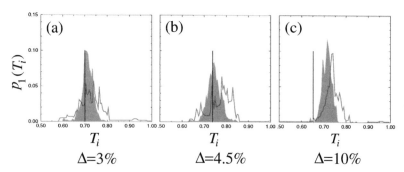

Fig. 7.12. (Color online) Probability density distributions $p_1(T_i)$ of interbeat intervals T_i: *solid line*—rest state immediately before the application of the forcing, *shaded*—in response to the forcing signal, *vertical lines* show the positions of forcing period. The respective values of frequency detuning Δ are given under the graphs

In Fig. 7.12 the probability density distributions $p_1(T_i)$ of interbeat intervals T_i are shown: solid line—rest state without forcing immediately before the application of the forcing, shaded—in response to the forcing signal, vertical lines show the positions of forcing period. First of all, one can notice that forcing makes the distribution sharper, more concentrated around some central value, which means regularization of heart beats. Now let us recall the meaning of synchronization by periodic forcing: the *basic* frequency of oscillations should coincide with the frequency of forcing. The basic frequency of heart beats can be associated with the most probable period of oscillations, which is the value of T_i at which the highest maximum of $p_1(T_i)$ occurs. From Fig. 7.12 one can see that before the forcing was applied, the most probable period of heart beats was different from the respective value of forcing period (compare the highest maximum of the solid line with the position of the vertical line). However, application of forcing can change that. Namely, at the values of the detuning Δ equal to 3% and 4.5% the most probable period of heart beats coincides with the period of forcing marked by the vertical line, and this implies that 1:1 synchronization takes place. However, when Δ is equal to 10%, i.e., the detuning between the forcing and the heart is too large, the forcing period is quite far away from the most probable period of heartbeats, and this means that there is no 1:1 synchronization.

In Fig. 7.13 two characteristics are given for the same healthy volunteer at different values of frequency detuning Δ between the forcing and the average heart rate at rest. (a) shows the ratio of the frequency of forcing f_f to the frequency of response f_r, the latter being the inverse of the average RR-interval T_i in a subject who is being subjected to the forcing. Compare this with Fig. 6.3(b) where the same dependence is given for van der Pol oscillator under periodic forcing to ensure that around $\Omega/\omega_0 = 1$ the similarity between the experimental function and the numerical one is quite remarkable. Namely, within the region of synchronization there are distinct plateaus of these graphs. However, Fig. 6.3(b) illustrates a noise-free case, and that is why all the plateaus there are strictly horizontal. In the full-scale experi-

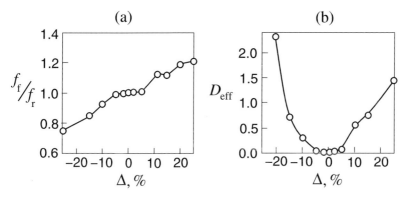

Fig. 7.13. Two quantities characterizing the response to a weak periodic forcing of the pattern of heart beats of a healthy volunteer. **a** The ratio of the forcing frequency f_f to the response frequency f_r, and **b** phase diffusion D_{eff} against the detuning Δ

ment there is plenty of noisy influence of various sorts, and in full agreement with Sect. 7.7 (see Fig. 7.5), the plateau in the experimental plot is slightly inclined.

In Fig. 7.13(a) forced frequency synchronization is illustrated, but what about the phase one? For both ECG and the forcing one can introduce phases using (8.10). Then one can calculate the phase difference and from that the effective phase diffusion D_{eff} that was introduced in Sect. 7.9. In Fig. 7.13(b) the phase diffusion D_{eff} is given for the same values of detuning Δ as in (a). Again, in full agreement with the theory, D_{eff} demonstrates a pronounced minimum inside synchronization region and almost reaches the value of zero. Outside synchronization region, phase diffusion is positive, and the further away from synchronization region, the larger D_{eff} is.

We believe that the experiments with the weak noninvasive forcing applied to healthy volunteers serve quite a good example of the Stratonovich's theory in action.

7.11 Effects of Noise on the Spectrum of a Synchronized System

In Sect. 7.7 we considered the effects of noise on the *mean* frequency of forced oscillations $\langle \dot{\varphi}(t) \rangle$. Following the analysis by Stratonovich, we have established that the beat frequency $\langle \dot{\psi}(t) \rangle$, by which the frequency of the synchronized system is shifted away from the forcing frequency Ω under the influence of noise, is defined by (7.114). Note, that the mean frequency $\langle \dot{\psi}(t) \rangle$ of forced oscillations can be interpreted in terms of their power spectral density $S_X(\omega)$, which has the meaning of the probability density distribution of the power of the process over all frequencies (see Sect. 7.1). Mean frequency $\langle \dot{\psi}(t) \rangle$ is the mean value of this distribution,

$$\langle \dot{\psi}(t) \rangle = \int_0^\infty \omega S_X(\omega)\, d\omega \times \left(\int_0^\infty S_X(\omega)\, d\omega \right)^{-1}. \tag{7.119}$$

However, (7.114) does not give us any clue as to how the change in $\langle \dot{\psi}(t) \rangle$ is linked to the change in the spectrum of the process.

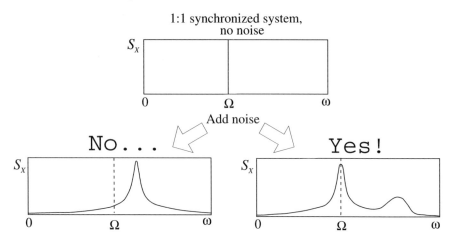

Fig. 7.14. Upper panel: spectrum of a 1 : 1 synchronized system without noise. **Lower panels**: the hypothesized changes of the spectrum due to noise, that could lead to the shift in the mean frequency $\langle\dot\psi(t)\rangle$ of forced oscillations. **Left**: The original spectrum peak is smeared and shifted (wrong). **Right**: The original spectrum peak is smeared but stays at the same position Ω. Another peak appears (correct)

When a distribution of some process changes its shape, its mean as well as other moments change, too. The power spectral density of the synchronized weakly non-linear system without noise in the first approximation has one delta-peak at the frequency Ω of forcing (upper panel of Fig. 7.14). Addition of noise smears the peak anyway, but what about its location? How should the spectrum change in order to cause the shift of its mean frequency? One possibility is that the only spectral peak is smeared and shifted in one direction (as shown in Fig. 7.14, lower left panel), and with that the mean value would shift in the same direction, too. Another possibility is that another peak appears to one side of the peak at Ω (as shown in Fig. 7.14, lower right panel). Then the mean frequency would correspond to a value somewhere in between the two peaks. What possibility is realized when noise is applied to a synchronized system?

Consider a very weak forcing. According to the linear response theory, the system perturbed periodically and weakly must contain a spectrum peak at the frequency of applied perturbation. Hence, in the limit of weak forcing the spectrum of the forced system must contain a peak at Ω. If the system is synchronized by forcing, there are no other peaks[3] besides the one at Ω, and it must exist even in the presence of noise. So the first hypothesis that the peak at Ω gives place to another peak does not seem plausible. Therefore, we need to abandon it, and to consider another option (Fig. 7.14, lower right panel).

---

[3] At least in the close vicinity of Ω. If either the forced oscillator, or forcing are not weakly non-linear, the spectrum peaks at the multiples of Ω might exist as well, but they are not taken into account when we discuss the immediate vicinity of the main peak at Ω.

7.11 Effects of Noise on the Spectrum of a Synchronized System 187

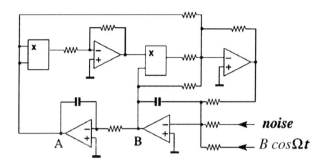

Fig. 7.15. Scheme of electronic circuit modelled by (7.116)

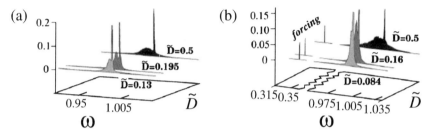

Fig. 7.16. Spectra measured in a full-scale experiment with van der Pol circuit (Fig. 7.15). **a** 1 : 1 phase locking; **b** 1 : 3 phase locking

Stratonovich [277] has shown by theoretical analysis that the picture in Fig. 7.14, lower right panel, is correct, but we do not repeat his calculations here because they are quite lengthy. Instead, we will illustrate the evolution of the spectra numerically. In Fig. 7.9 (third column) the spectra of the variable x of (7.116) are shown at three different values of noise intensity $\tilde{D}$ [42]. Note, that the forcing frequency $\Omega = 1.0118$ here is larger than the frequency of unforced oscillations which is approximately equal to 1. As predicted in [277], a new spectrum peak appears to the left-hand side of the peak at the forcing frequency, whose position, height and width depend on noise intensity. The mean frequency of oscillations is not equal to Ω, but is shifted towards the unperturbed frequency 1 due to the appearance of the additional peak induced by noise. In order to check the validity of theoretical and numerical predictions, in [42] an experiment with the electric circuit described by the van der Pol oscillator (3.3) was done. The scheme of the circuit is given in Fig. 7.15, and the resulting spectra for 1 : 1 phase locking in Fig. 7.16(a). The same parameters of the circuit and of the forcing were chosen as for illustrations in Fig. 7.9, only slightly different noise intensities are illustrated. One can make sure that in full agreement with the theory and with the numerical simulations, noise induces a new spectral peak to the left of the peak at Ω.

In addition to 1 : 1 synchronization, experiments were done for 1 : 3 phase locking as well, where the forcing frequency was set to $\Omega = 0.33216$ and forcing strength to $B = 0.3$. The spectra with three different noise intensities D are given in

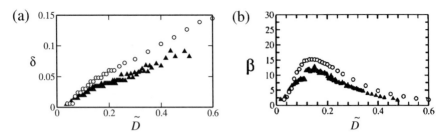

Fig. 7.17. Noise effect on the van der Pol oscillator synchronized by locking, parameters correspond to point **A** in Fig. 7.7. **a** Distance δ of the noise-induced peak from the main peak at the forcing frequency Ω. **b** Regularity β of the noise-induced peak. *Empty circles*: numerical simulation of (3.3); *black triangles*: experiment with the scheme in Fig. 7.15

Fig. 7.16(b) and testify to that the effect of the appearance of the noise-induced peak seems to be a universal one whenever phase locking is being destroyed by noise.

Let us discuss the physical meaning of the noise-induced spectral peak. Importantly, noise induces in the system the motion that was absent without external random fluctuations. This new motion is represented as excursions in the phase space which occur randomly, but with a certain mean frequency. Obviously, the frequency of the new spectrum peak is related to the frequency of these excursions. Also, the respective phase slips can occur more or less regularly, which is associated with the phenomenon of coherence resonance[4] [87, 191, 210, 241, 267]. The latter phenomenon means that (i) noise applied to a dynamical system can induce a new kind of motion that was not present without noise; (ii) there is a moderate value of noise intensity at which the regularity or coherence, of this motion takes its maximal value. So, counterintuitively, noise plays a constructive role by creating an additional relatively ordered motion in the system. A more detailed description of this phenomenon can be found in Sect. 9.1. It is interesting to characterize the parameters of the motion induced merely by external noise as it affects a synchronized system. The following characteristics were introduced: the distance δ of the noise-induced peak from the main one at Ω, and the regularity of the new motion. The first parameter is straightforward, and δ as a function of $\tilde{D}$ is shown in Fig. 7.17(a) for both numerical simulation and electronic experiment. At noise intensities close to zero, the position of the noise-induced peak almost coincides with that of the main one. As the noise intensity increases, the new peak moves away.

The parameter quantifying the regularity of the noise-induced motion needs to be defined carefully, since noise changes the properties of both kinds of motion in the system: oscillations in the phase-locked state and oscillations during the phase slip. In order to assess the changes in the noise-induced motion only, the latter has to be separated from the phase-locked one. This was done by artificially removing the main spectrum peak by means of a band-stop filter and, to avoid the resultant discontinuity in the spectrum, connecting the edges of the removed frequency range

---

[4] Noise-induced phenomena are discussed in Chap. 9.

7.11 Effects of Noise on the Spectrum of a Synchronized System   189

with a straight line. Then the coherence β of the noise-induced motion was estimated as the signal-to-noise ratio [87, 241] of the noise-induced peak using the method described in Sect. 9.3 with (9.3). Figure 7.17(b) shows the coherence β of the noise-induced peak as a function of noise intensity $\tilde{D}$ from numerical simulation (empty circles) and from the experiment (black triangles). Both functions have a resonant character characteristic of the phenomenon of coherence resonance, with β taking its maximal value at an optimal noise intensity $\tilde{D} \approx 0.2$. The latter could be treated as an evidence of noise-induced ordering.

7.11.1 Effect of Noise on the Spectrum of Oscillations Synchronized by Suppression

One might wonder if noise can induce a new ordered motion only in the system that is in the state of being phase-locked by an external forcing. What about the systems whose dynamics is synchronized by another mechanism, e.g., suppression? We remind the reader, that in the state of suppression the phase space no longer contains a resonant torus that was present in the lower part of synchronization tongue and illustrated in Fig. 7.9, second column. Inside the upper part of the tongue in Fig. 7.7 the phase space contains only a stable cycle. However, inside the tongue the properties of this cycle are not always the same: grey line signifies the transition between the two types of the cycle stability, i.e., from a node in the central part of the tongue to the focus in its peripheral part that contains point **B** corresponding to $\Omega = 1.129$ and $B = 0.48$. It can be mentioned in advance that when the vicinity of the cycle does not have any potential to oscillate, i.e., when the cycle is simply a node with all real Floquet multipliers, noise is not able to induce an ordered motion. However, the cycle stability can be such that its Floquet multipliers are complex-conjugate, and hence in the noise-free system the phase trajectories tend to the cycle while winding around it. In that case noise is capable of inducing these rotations on a regular basis. We illustrate the evolution of stroboscopic sections and spectra of oscillations at point **B** in Fig. 7.7 with different values of noise intensity. In Fig. 7.18 the first column shows the stroboscopic sections, while the second column the corresponding spectra of oscillations influenced by external noise in (3.3). By analogy with locking, noise appears to create a new peak to the right of the main one at Ω. However, unlike with locking, the stroboscopic section does not demonstrate large excursions in the phase space, since there are no guiding manifolds of the resonant torus here. The noise only initiates rotations around the stable cycle with a certain average frequency.

The distance δ of the new peak from the main one at Ω, and its regularity β, are given in Fig. 7.19 both for numerical simulation of (3.3) and for the experiment. As in case of locking, the system displays coherence resonance with the increase of noise intensity $\tilde{D}$.

To summarize, on the way of destroying synchronization of periodic oscillations by a periodic forcing, noise creates a new kind of motion whose regularity resonantly depends on the noise intensity. This phenomenon appears rather uni-

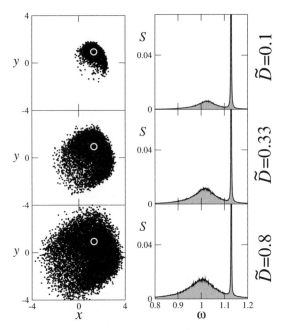

Fig. 7.18. Noise destroying a periodic motion in a forced van der Pol oscillator (7.116) that is synchronized by suppression. Different values of noise $\tilde{D}$ are given to the right of each row. **First column**: Stroboscopic section. **Second column**: Spectra of x_1

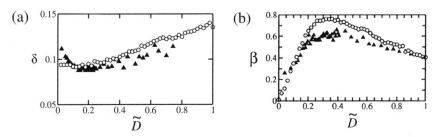

Fig. 7.19. Noise effect on the van der Pol oscillator synchronized by suppression, parameters correspond to point **B** in Fig. 7.7. **a** Distance δ of the noise-induced peak from the main peak at the forcing frequency Ω. **b** Regularity β of the noise-induced peak. *Empty circles*: numerical simulation of (3.3); *black triangles*: experiment with the scheme in Fig. 7.15

versal, since it occurs either in phase-locked states, or in suppressed ones near the boundary of synchronization region.

8 Chaos Synchronization

In the previous chapters we considered the synchronization of the simplest kinds of oscillations, namely, of periodic ones. In this chapter we examine synchronization of irregular oscillations which are associated with deterministic chaos. The concept of deterministic chaos is an important paradigm for the understanding of quite a broad range of phenomena, in which complex irregular oscillations play a crucial role. Examples of these phenomena are turbulence [121, 164, 254], rhythmical activity of living cells [92], population dynamics [182], charge transport in semiconductor devices [261], etc.

One might find it quite natural to extend the idea of synchronization from periodic to chaotic oscillations, but what would this imply? On the one hand, non-damped chaotic oscillations in dissipative systems are the rightful members of a family of self-sustained oscillations,[1] and as such are entitled to participate in synchronization phenomena in principle. On the other hand, the properties of chaos are so markedly different from the properties of regular oscillations, that were considered in this book so far, that the direct parallels are hardly possible. So what could be expected from the interaction of two or more oscillators, at least one of which being chaotic? In order to answer this question, in the first instance we need to reveal what it is that makes chaos so distinguishable from periodic oscillations. A reader familiar with the features of chaotic attractors can skip Sect. 8.1.

---

[1] See the definition of self-sustained oscillations in Sect. 2.3.

192   8 Chaos Synchronization

8.1 What Is Chaos?

Nowadays, theory of deterministic chaos is quite well developed, and there are a variety of textbooks devoted to this and related topics, e.g., [15, 101, 180, 199, 258, 262, 279]. This section explains the distinctive features of chaos that will be essential as one considers interaction between different oscillators with at least one chaotic among them. As before, we will describe the same phenomenon both in terms of the phase space and of power spectral density.

8.1.1 Exponential Divergence of Phase Trajectories

The complexity of oscillations can be of a deterministic origin and be caused by the sensitivity of the system to initial conditions. The latter means that if in the same dynamical system with the same set of control parameters one launches two trajectories from almost indistinguishably close initial conditions, with the course of time they would diverge exponentially fast.[2] Figure 8.1(a),(b) illustrates the sensitivity of the phase trajectories to initial conditions on a chaotic attractor, using as an example a famous paradigmatic system that can exhibit deterministic chaos, namely, a Rössler oscillator whose equations read

$$\dot{x} = -\omega y - z,$$
$$\dot{y} = \omega x + \alpha y, \qquad (8.1)$$
$$\dot{z} = \beta + z(x - \mu).$$

This is arguably the simplest chaotic system with continuous time,[3] because it has only one non-linear term zx in its equations. Moreover, its dimension is only three, which is the smallest dimension of the phase space of a system with continuous time in which a chaotic attractor can live. At $\alpha = \beta = 0.2$ and $\mu = 6.5$, the phase trajectory lies on a chaotic attractor which is born as a result of a sequence of period-doubling bifurcations of a stable limit cycle that existed in the system, e.g., at $\mu = 3$. In Fig. 8.1(a) two phase trajectories of Rössler system are shown that were launched from close initial conditions (I.C.) in the vicinity of the point marked by a filled circle. In Fig. 8.1(b) the corresponding realizations are given. The distance between the two sets of initial conditions was of the order of 10^{-2}. One can see that in the beginning the trajectories are very close to each other, but the discrepancy between them grows with the course of time.

---

[2] To be precise, the distance between the two nearby trajectories grows exponentially in time only when the trajectories remain close to each other. When the trajectories diverge too much, the distance between them cannot grow according to an exponent of time, if only because a chaotic attractor is of finite size and there is simply no room for separation that would be larger than the largest diameter of the attractor.

[3] There are simpler chaotic systems in the form of discrete maps, but they are beyond the scope of this book.

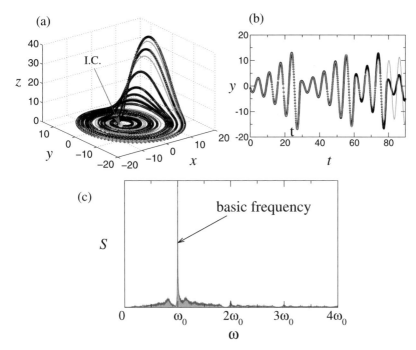

Fig. 8.1. (Color online) Illustration of deterministic chaos in Rössler system (8.1). **a** Two phase trajectories (denoted by *different colors*) that start from very close initial conditions (I.C. in the field of the figure) diverge with time. **b** Oscillations corresponding to the trajectories in **a**. **c** Spectrum of chaotic oscillations. In spite of complexity of oscillations, deterministic chaos can often be treated as a narrow-band process having a pronounced peak in the spectrum

8.1.2 Chaos Properties in Terms of Phase Space

In terms of the phase space, periodic oscillations are represented by stable (attracting) periodic orbits. Unlike them, chaotic oscillations are generally represented by a set with a fractal structure (chaotic attractor) whose dimension is non-integer. A typical example of a chaotic attractor is given in Fig. 8.1(a). There are several typical bifurcation scenarios that lead to the formation of chaos. This chapter will be devoted to chaos that is born as a result of an infinite cascade of period-doubling bifurcations of a limit cycle. This type of chaos is often called Feigenbaum chaos after a mathematician who was the first to reveal the universality of this scenario and has created its theory [77, 78]. For this type of chaos synchronization theory is currently developed to the largest extent, and the typical synchronization mechanisms are easier to understand using Feigenbaum chaos as an example.

Birth of Feigenbaum Chaos

In order to understand how chaos of Feigenbaum type can be born, let us perform an experiment in our mind. Out of all control parameters of the dynamical sys-

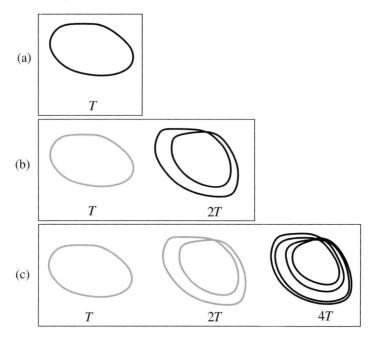

Fig. 8.2. (Color online) A schematic illustration of a cascade of period-doubling bifurcations leading to the birth of Feigenbaum chaos. *Black lines*: stable cycles; *grey lines*: saddle cycles; letters T, $2T$ and $4T$ indicate the approximate values of the periods of the cycles. **a** Before the first bifurcation at $\mu = \mu_0$; **b** after the first period-doubling bifurcation at $\mu = \mu_1$; **c** after the second period-doubling bifurcation at $\mu = \mu_2$, see text for reference

tem choose one and call it μ. Suppose at a certain $\mu = \mu_0$ the system demonstrates periodic oscillations with period T and therefore has a stable limit cycle in the phase space (Fig. 8.2(a)). Allow this parameter to increase gradually, and follow the evolution of regimes in the system. Assume that at $\mu = \mu_1$ the first period-doubling bifurcation takes place. As a result, at μ slightly exceeding μ_1 there are two periodic orbits (cycles) in the phase space of the system: one stable with period close to $2T$, and one saddle with period close to T (Fig. 8.2(b)). Assume that at $\mu = \mu_2$ the second period-doubling bifurcation occurs. As a result, at μ slightly larger than μ_2 there are three cycles in the phase space: one stable with period close to $4T$, and two saddle with periods close to $2T$ and T (Fig. 8.2(c)). At the end of an infinite sequence of period-doubling bifurcations, which by the way occurs within a finite range of the values of μ, there is an *infinite* number of saddle cycles in the phase space with periods close to, but not exactly equal to, nT, where n is integer number. And there are no stable limit cycles any longer. What remains in the phase space forms a chaotic attractor of Feigenbaum type.

Skeleton of Feigenbaum Chaos

Imagine an *infinite* (countable) number of saddle cycles that are packed in a *finite* volume of the phase space without intersecting each other. Bear in mind that this is possible only in a phase space whose dimension is three or larger, since there can be no saddle cycles in a two-dimensional space (plane). A collection of these cycles form a *skeleton of a chaotic attractor*.

Consider a single saddle periodic orbit (a saddle cycle), which is an intersection of two manifolds: a stable and an unstable one.[4] A sketch of a saddle periodic orbit in a three-dimensional phase space is given in Fig. 8.3. In the latter figure, a stable manifold is shown as a surface going sideways, and an unstable manifold as a cylindrical surface. However, the stability properties of these manifolds can be swapped, while their intersection will still be a saddle cycle.

For simplicity consider chaotic attractor in a three-dimensional phase space. One might argue that there is nothing really surprising about all these cycles being jammed into a finite volume, because each saddle cycle is a one-dimensional curve, and there is more than enough room for them to coexist comfortably in a three-dimensional space (3D), which is true. However, remember that the saddle cycles do not come on their own: they always carry a pair of their manifolds with them. Note, that in 3D these manifolds will be two-dimensional. Now imagine an infinite number of two-dimensional surfaces (manifolds) that are ought to fit within a finite volume of a 3D. Again, the dimension of each of these surfaces is less than the dimension of the phase space, so in principle this should not be a problem. However, these surfaces are not closed, i.e., for a single isolated saddle cycle they would either go to infinity, or to some attractor, or to another saddle object. In case of a chaotic

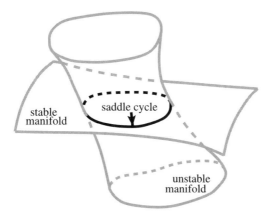

Fig. 8.3. (Color online) A schematic representation of a saddle cycle in a three-dimensional phase space. *Curvy cylinder*: unstable manifold; *surface going sideways*: stable manifold. Two manifolds intersect at a closed curve (*black line*) which is a saddle periodic orbit (*saddle cycle*)

---

[4] For properties of manifolds see Sect. 5.1.

196   8 Chaos Synchronization

attractor, there is no "other attractor" in the volume being discussed. The manifolds are still allowed to end at the infinity or at another saddle cycle, but imagine millions of them that start within a small volume of the phase space and try to find their ways. Whereas in fact there is much more of them that a million, or indeed than any other finite number whatever large! At least some of them will be bound to touch each other or to intersect, resulting in a hugely intricate homo(hetero)clinic structure.

It would be too difficult to illustrate even one intersection of two manifolds in the full 3D, so consider the Poincaré section instead. In Fig. 8.4 an intersection of two pairs of manifolds of two different saddle cycles is illustrated. We haste to note here that this illustration is *very* schematic and should not be regarded as mathematically accurate. The reason is that an accurate picture is immensely more complicated and is virtually impossible to sketch. Nevertheless, Fig. 8.4 does emphasize the main feature of manifolds' intersection, namely, two manifolds cannot intersect just once. If they intersect, they do so infinitely many times and make infinitely many horseshoes in the phase space.

This means that within a finite volume, intersection of only two manifolds will lead to a very intricate structure of the phase space. Now imagine that there are infinitely many manifolds intersecting with each other. Moreover, snake-like structures shown in Fig. 8.4 also intersect. In addition to that, it has been proved that in the vicinity of such a structure there exists an infinite number of periodic orbits among which there can be stable ones [15], and this structure is often referred to as non-hyperbolic attractor. It suffices to say that the structure of the phase space around the skeleton of the chaotic attractor is extremely tangled, apparently beyond imagination. The phase trajectories are winding around this skeleton, while obeying the ground rules: it never crosses manifolds, and if approaches one of them,

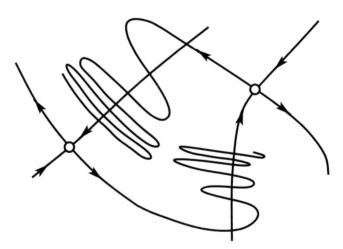

Fig. 8.4. A *very* schematic illustration of an intersection between the manifolds of two different saddle cycles. The picture is not accurate, since the real picture would be much more complicated

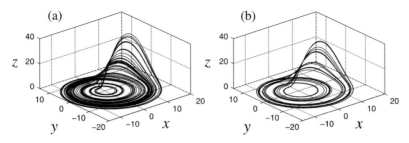

Fig. 8.5. **a** A fragment of a chaotic trajectory of Rössler system (8.1) and **b** seven unstable periodic orbits embedded into the attractor shown in **a** at $\alpha = 0.165$, $\beta = 0.2$, $\mu = 10$

follows it for a while. Consequently, the pathway of the phase point becomes very involved. What a drastic difference from a stable limit cycle! Note, that the skeleton defines the topological features of the chaotic attractor [89] and in many respects its dynamical properties as well [37, 99] (Fig. 8.5).

8.1.3 Chaos Properties in Terms of Spectra

What about the spectral properties of chaos? Feigenbaum chaos contains a countable set of saddle periodic orbits whose periods are approximately equal to the multiples of T. Therefore, the spectrum of chaotic oscillations of this type contains a pronounced peak close to $f = 1/T$, and possibly peaks at harmonics and subharmonics of f. However, oscillations are irregular, and their spectrum cannot be discrete. In fact, it is continuous and contains components at all frequencies. A spectrum of chaotic oscillations is illustrated in Fig. 8.1(c).

8.2 What Does Synchronization of Chaos Encompass?

Today synchronization of chaos is one of the central topics of contemporary nonlinear dynamics, which is confirmed by a large amount of publications devoted to this problem (see [24, 52, 185, 214] for reviews, and references within for more detail). In contrast to regular oscillations, the manifestations of chaos synchronization are not so pronounced and obvious. To a large extent this is because the chaotic oscillations have a continuous power spectral density, and therefore do *not* have a determined period which would unambiguously define the time scale of oscillations. Apparently, these factors would require different approaches to the understanding of chaos synchronization. Each of these approaches is based on its own phenomenological definition of chaos synchronization that reflects different aspects of ordering which can occur in cooperative dynamics of interacting chaotic systems.

8.2.1 Chaos Synchronization: Different Manifestations

For the first time the problems of interaction of the systems with chaotic dynamics were considered in [8, 84, 154, 155, 207] that examined the effects of coupling on

198   8 Chaos Synchronization

the dynamics of interacting *identical* systems, each demonstrating chaotic behavior. It was shown that for some sufficiently strong coupling the oscillations in interacting chaotic systems become completely identical, even if the subsystems start from different initial conditions. This phenomenon was called *complete synchronization* of chaos [206], and in fact is the strongest manifestation of chaos synchronization. However, complete synchronization is only possible in identical systems, while even the slightest difference between the oscillators leads to the disappearance of the exact coincidence of their phase trajectories. In order to extend this definition for non-identical systems, the concept of almost complete synchronization was introduced where synchronization was deemed to have been achieved if the distance between the phase trajectories in interacting systems did not exceed some small value [8]. Later it was found that at a certain sufficiently small mismatch, the interacting systems can demonstrate almost identical oscillations, but shifted in time with respect to each other by some time constant. This effect has acquired a name of *lag synchronization* [248].

Summarizing, we note that in different works synchronization of chaos has been associated with a series of effects, namely, the appearance of similarity in the dynamics of interacting oscillators [8], the decrease of the dimension of the attractor in the joint dynamical systems [51, 163, 223], the occurrence of functional dependence between the corresponding variables of coupled dynamical systems (*generalized synchronization*) [1, 255].

However, it was shown that synchronization of chaos can also be described in its classical sense, i.e., in terms of frequencies and phases of interacting oscillations [16–18, 208, 247].

8.2.2 Chaos Synchronization in a Classical Sense

Recall that with periodic forcing applied to a periodic oscillator, a synchronous regime is a periodic one with frequency equal to the frequency of external forcing, which is fairly simple (see Chap. 3). When noise comes into play and the spectrum of oscillations becomes continuous (see Chap. 7), synchronization needs a broader definition. Namely, the oscillations are regarded as $1:1$ synchronous with forcing if all three conditions below are satisfied simultaneously:

- The frequency of the highest spectral peak of forced oscillations coincides with the frequency of forcing.
- The graph of phase difference $\varphi(t)$ between the forcing and the response versus time demonstrates plateaus.
- These plateaus are sufficiently long.

A special term was introduced to characterize this phenomenon: *effective synchronization* [174].

Feigenbaum chaos which will be considered in this chapter is to some extent similar to a periodic regime smeared by noise, namely, (i) its spectrum has a pronounced peak, i.e., it is easy to single out a basic frequency of oscillations, and (ii) at the first glance its phase portrait looks like a limit cycle smeared by noise, although we know that its structure is much more complicated than that. With this in mind, it would be natural to define a synchronous chaos with the help of the same criteria. One can also suggest that for chaos the same mechanisms of synchronization might be valid, namely, phase (frequency) locking, suppression of natural dynamics, and possibly via crisis.[5] This approach was used in [16–18, 215] for the analysis of synchronization of chaos in a full-scale experiment and in numerical simulation.

8.3 Phase and Basic Frequency of Chaotic Oscillations

The concept of phase of oscillations is easy to understand if one considers harmonic oscillations $s(t) = A\cos(\Omega t + \varphi_0)$, also see Sect. 3.1. These oscillations are characterized by the amplitude A and have a period $T = 2\pi/\Omega$. The argument of the cosine $\Phi(t) = \Omega t + \varphi_0$ is called "phase." On the plane $(s, \Omega^{-1}\dot{s})$, where $\dot{s}$ is a derivative of s with respect to time t, these oscillations are represented by the motion of the state point along the circle with radius A and the current angle of rotation $\Phi(t)$. Thus one can single out the following main properties of the phase Φ:

1. Φ grows monotonously with time.
2. The increment of the phase by 2π corresponds to one full rotation of the state $(s, \dot{s})$.
3. The slope $\dot{\Phi}$ of the dependence Φ on time t is equal to the angular frequency Ω of oscillations.

In contrast to periodic oscillations, chaos does not have a unique period, which makes introduction of a phase for chaotic oscillations quite a non-trivial problem. At the moment there is no unique way to introduce a phase for deterministic chaos. However, quite often chaotic oscillations $x(t)$ can be regarded as a narrow-band (quasi)random process, for which it is known that it might be approximated by a signal with modulated phase and amplitude [153]

$$x(t) = A(t)\cos\Phi(t) = A(t)\cos(\omega^0 t + \varphi(t)), \qquad (8.2)$$

where $A(t) \geq 0$ is a random amplitude and $\varphi(t)$ is a random component of phase. In spite of its simplicity, such approximation has been shown to be quite accurate to describe the statistical properties of a wide class of chaotic oscillations [25, 26], e.g., of those born as a result of a cascade of period-doubling bifurcations. The spectra of chaotic oscillations quite often demonstrate pronounced peaks (Fig. 8.1(c)). Then a criterion for the definition of a narrow-band chaos could be, for example, the inequality $\Delta\omega \ll \omega^0$, where ω^0 is the central frequency of the main peak and $\Delta\omega$ is the width of the peak at the half of its height. The frequency ω^0 is sometimes called *basic frequency* of chaotic oscillations. Obviously, the approximation (8.2) is ambiguous, since for the given $x(t)$ there are arbitrarily many ways to define $A(t)$ and $\varphi(t)$, and, strictly speaking, even ω^0. For example, if for the fixed ω^0,

---

[5] This has been considered in Chap. 5.

$x(t)$ changes during some time interval, it is impossible to state unambiguously if this change was due to $A(t)$ or to $\varphi(t)$. Moreover, (8.2) does not change if $\varphi(t)$ is substituted by $\varphi(t) \pm 2\pi n$, $n = 0, 1, 2 \ldots$. In order to exclude the latter ambiguity, we consider φ wrapped into the interval $[-\pi, \pi)$.

One of the most popular ways to define the amplitude $A(t)$ and the phase $\Phi(t)$ in (8.2) involves Hilbert transform [85, 153]. Two signals $x(t)$ and $y(t)$ are connected via Hilbert transform, if

$$x(t) = -\frac{1}{\pi} \int_{-\infty}^{\infty} \frac{y(t')}{t - t'} dt', \quad y(t) = \frac{1}{\pi} \int_{-\infty}^{\infty} \frac{x(t')}{t - t'} dt', \quad (8.3)$$

where the integrals are taken in the sense of Cauchy principal values. Then we can formally introduce a complex signal $\eta(t)$ often called *analytic signal* as follows:

$$\eta(t) = A(t) \exp(i\Phi(t)) = A(t) \exp[i(\omega^0 t + \varphi(t))] = x(t) + iy(t). \quad (8.4)$$

We require that $y(t)$ be expressed as

$$y(t) = A(t) \sin \Phi(t) = A(t) \sin(\omega^0 t + \varphi(t)), \quad |A(t)| \geq 0, \; |\varphi| \leq \pi, \quad (8.5)$$

and is a Hilbert transform of $x(t)$, which is the real part of analytic signal $\eta(t)$. Using the functions $x(t)$ and $y(t)$, the instantaneous amplitude $A(t)$ and the instantaneous phase $\Phi(t)$ can be defined unambiguously as

$$A(t) = |\eta(t)| = \sqrt{x^2(t) + y^2(t)}, \quad (8.6)$$

$$\Phi(t) = \arg(\eta(t)) = \tan^{-1}(y(t)/x(t)). \quad (8.7)$$

In this case the phase Φ can be geometrically understood as an angle of rotation of a vector with the amplitude $A(t)$ in some projection plane (x, y) as shown in Fig. 8.6(a).

If the use of the narrow-band approximation is approved, one can consider synchronization in its classical sense similarly to how it was done in Chaps. 3, 4, 5 and 6

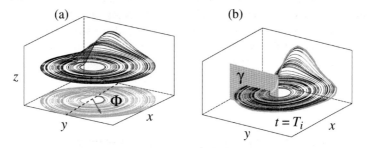

Fig. 8.6. (Color online) Two ways to introduce phase for chaotic oscillations: **a** as an angle of rotation of the phase trajectory in some projection; **b** via times T_i of trajectory return to a certain Poincaré secant surface γ

for periodic oscillations. That is, if there are two interacting systems that are characterized by triplets of variables $\{A_{1,2}(t), \omega_{1,2}^0, \Phi_{1,2}(t)\}$, where subscripts mean the numbers of the respective systems, then a criterion of synchronization can be the following restriction imposed onto the phase difference:

$$\Delta\Phi_{nm}(t) = \left| \Phi_1(t) - \frac{n}{m}\Phi_2(t) \right| < 2\pi, \tag{8.8}$$

or a rational ratio of the basic frequencies that can be formally expressed as

$$n\omega_1^0 - m\omega_2^0 = 0. \tag{8.9}$$

Remarkably, due to the ambiguity in the definition of the basic frequency ω^0 for chaotic oscillations, the criteria (8.8) and (8.9) do not always coincide. For example, often as a characteristic frequency the mean frequency $\omega^0 = \overline{\dot{\Phi}(t)}$ of oscillations is taken, where the overline denotes averaging over time and which is calculated as in (3.89). Obviously, the frequency ω^0 defined this way might not be equal to the position of the most pronounced peak in the spectrum of chaotic oscillations. Hence, the phase and frequency synchronization are sometimes regarded as different phenomena in literature [131, 283, 284].

If one presumes that the 2π growth of phase corresponds to one full rotation of the phase trajectory around some center, Poincaré section technique can sometimes be used for the introduction of the phase. As illustrated in Fig. 8.6(b), one can define a Poincaré secant surface γ and collect time moments T_i at which the phase trajectory crosses the surface γ in the chosen direction, e.g., from left to right. Between two successive intersections, i.e., during the time interval $(T_{i+1} - T_i)$ the phase grows linearly by 2π. Then we can express phase Φ at any time moment t as

$$\Phi(t) = \frac{t - T_i}{T_{i+1} - T_i} + 2\pi i. \tag{8.10}$$

Note that in order to be able to introduce phase like this, it is important to define the section γ to be transversal to *all* phase trajectories on the attractor and in its vicinity. Otherwise some rotations of the phase trajectory might not be taken into account, and the quantity Φ might have essentially non-monotonous character, which would contradict property 1 of the phase mentioned in the beginning of this section.

In cases when chaotic oscillations cannot be treated as narrow-band signals, the correct introduction of the phase becomes quite a complicated problem. Although the general concept of phase synchronization for such oscillations is still under development, there are several approaches which might be useful in particular situations [123, 198, 245, 250].

8.4 Forcing Chaos Periodically: What to Expect?

Perhaps the most obvious question that comes into one's mind when thinking about synchronization in connection with chaos would be: "If periodic oscillations can be

202   8 Chaos Synchronization

synchronized by periodic forcing, what about chaotic ones? And generally, what can we expect from forcing chaos periodically?" In this section we will again use the power of our mathematical imagination in order to predict the possible outcomes.

Consider a Feigenbaum chaotic attractor described in Sect. 8.1.2. We remind the reader that the finite volume of the phase space that contains this attractor, also contains a countable number of saddle cycles that form the attractor skeleton. Also, it is good to bear in mind that if we are discussing only systems with continuous time, the smallest dimension of the phase space where chaos can fit in, is three. Consider the simplest case: Feigenbaum chaos in a three-dimensional phase space, to which a periodic forcing of the simplest shape is applied. Note that periodic excitation increases the dimension of the phase space of the system being forced by, at least, one. Therefore, the smallest dimension of the phase space of a chaotic periodically forced system is four, which is really difficult to visualize. By analogy with the case of two reactively coupled periodic oscillators considered in Sect. 4.4, we will visualize the whole picture in the Poincaré section, thus reducing the dimension of the space to three.

Let us try to understand what effect can the periodic perturbation cause. Single out just one saddle cycle out of the attractor skeleton. Its Poincaré section is shown schematically in Fig. 8.7 (upper right panel), together with a projection (not a section!) of a stable cycle which represents periodic forcing (upper left panel). A *weak* periodic perturbation should make the saddle cycle undergo Neimark–Sacker bifurcation, which would have two implications. First, the cycle will acquire two more unstable directions and become a thrice saddle cycle (empty circle in lower panel of Fig. 8.7)—which in 4D means that it will become a repeller, since one direction of any periodic orbit is always neutral. Secondly, a saddle torus might be born out of the saddle cycle. We have already discussed a resonant saddle torus in Sect. 4.4 (Fig. 4.8(b)). A sketch of a Poincaré section of an ergodic saddle torus together with its manifolds is shown in the lower panel of Fig. 8.7.

Note that in the unforced Feigenbaum chaos there is not just one, but infinitely many saddle cycles. Therefore, in the phase space of a chaotic system that is forced by a weak periodic signal there are

- infinitely many thrice saddle cycles (repellers in 4D)
- possibly infinitely many saddle tori packed within the finite volume[6]

For comparison and to remind you, in a similar situation the phase space of a periodically forced periodic system would contain only one twice saddle cycle (a repeller in 3D) and one stable torus.

What possibilities would this structure of the phase space provide with regard to synchronization? We can try to exploit our background in synchronization of periodic oscillators and maybe draw some analogies. It would be natural to ask ourselves: "If synchronization of chaos can take place by analogy with synchronization

---

[6] Interestingly, the qualitative picture of the phase space structure in the Poincaré section of the forced chaotic system would then be the same as the one in the *full* phase space of the unforced system, if we ignore the repellers. Therefore, this is a considerably more complicated chaos.

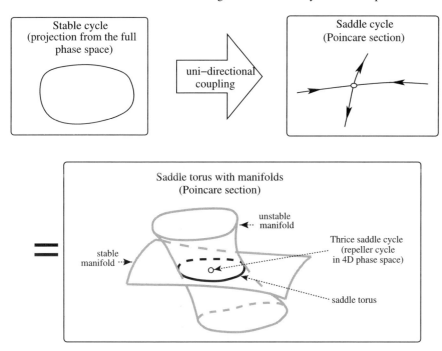

Fig. 8.7. (Color online) A schematic illustration of a periodic forcing applied to a system with a saddle cycle. The objects involved are indicated in the field of the figure

of periodic oscillations, can we expect the same mechanisms of it?" Consider the feasibility of phase (frequency) locking and of suppression of chaotic dynamics in turn.

8.4.1 Phase Locking of Chaos

In periodic oscillations, phase locking is associated with the birth of a pair of cycles on the surface of the stable torus. For chaos we can suggest that a similar bifurcation can take place on the surfaces of the saddle tori, as a result of which a pair of cycles would be born: saddle and twice saddle. Then these saddle tori will become resonant with the structure shown in Fig. 4.8(b). If each of the infinite number of the saddle tori becomes resonant, then the phase space acquires an infinite number of saddle cycles packed within a finite volume. The latter could form a skeleton of a chaotic attractor of Feigenbaum type, i.e., of the same type as the unforced chaos. All the other objects in the phase space might continue to exist, and these would be: twice saddle cycles, thrice saddle cycles (repellers in 4D), and resonant saddle tori with their manifolds, with an infinite number of each. The latter objects would of course complicate the general picture, but they would be less stable than saddle cycles with a single unstable direction and therefore less visible in an experiment. If we ignore them, we can say that the (almost) original chaos is reincarnated in the system as a result of phase locking on the saddle tori!

What should we expect for the bifurcation diagram on the plane of the forcing parameters "frequency detuning"–"strength of forcing"? In a periodically forced periodic oscillator there is just one line of a saddle-node bifurcation on the torus, which forms the lower boundary of the synchronization tongue (see Fig. 4.14, first row and first column). In a periodically forced chaotic system there are infinitely many tori, and to expect them to become resonant at the same values of the control parameters would really be to expect too much. Therefore, we could predict an infinite number of lines of saddle-node bifurcations on the plane of control parameters. If we are lucky, they are going to lie not too far away from each other. And if we are extremely lucky, all of them would fit within a finite area of the parameter plane, like the lines of period-doubling bifurcations do on the route to chaos. In that case, by crossing the full bunch of these lines we would fall into the region of a phase-locked chaos! This is where our imagination has taken us so far.

8.4.2 Suppression of Chaos

Recall that in periodically forced periodic oscillations, suppression of natural dynamics is associated with a torus death[7] bifurcation of a stable torus, as a result of which the torus shrinks to become a stable cycle. For a periodically forced chaotic system one can envisage the same bifurcation occurring to the saddle torus: it can shrink and finally merge with the repeller, i.e., black line in the lower panel of Fig. 8.7 can shrink and merge with the empty circle. This bifurcation would take away the two unstable directions of the cycle and make the cycle once saddle, just like before the forcing was applied. However, the shape of this cycle might not be exactly the same as without forcing. If all saddle tori undergo the torus death bifurcation, then there again will be infinitely many saddle cycles in a finite volume of the phase space, which could make the skeleton of a chaotic attractor of the same type that existed in the unforced system. This would imply that chaos is reborn via suppression of natural dynamics! As to the bifurcation lines in the plane of parameters "frequency detuning"–"strength of forcing," we can again expect one for each saddle torus, i.e., an infinite number of them. If they all fit within a finite area of the parameter plane, crossing of this area would lead us to chaos synchronized via suppression.

8.4.3 Any Other Options?

Yes, of course. We cannot help mentioning here that in addition to bifurcations of regular solutions like cycles and tori, there are more bifurcations that can occur in a chaotic system. Afraimovich and Shilnikov have proved [7], and later Anishchenko et al. have confirmed in an experiment [19], that a torus can undergo a transformation called "torus breakdown" by losing its smoothness. As a result of that, torus will

---

[7] Inverse torus birth, or Neimark–Sacker, bifurcation.

be destroyed and instead a chaotic attractor will appear. Obviously, the same transition can occur in the opposite direction: chaos can regain smoothness (or rather loose its non-smooth fractal structure) and turn into a torus. This has to be considered as a possibility.

What would it have to do with synchronization? If under certain parameters of forcing chaos would turn into a stable torus, then this torus might obey the same laws than the one in a periodically forced periodic system. Namely, it might undergo torus death and turn into a stable cycle which would constitute a synchronized, but non-chaotic regime.

8.4.4 Interacting Chaotic Systems

Here, it is pertinent to mention what one can expect in a more involved case when one simplest chaotic system is forced by another simplest chaotic system. In this case an infinite number of saddle periodic orbits, that represent the system being forced, will be under the influence of another infinite set of saddle periodic orbits representing the forcing system. By analogy with the above, we can assume that this would mean the birth of an infinite number of non-stable tori, which will now be twice saddle rather than simply saddle. These tori can be either ergodic, or resonant.

What if two chaotic systems are coupled mutually? We can also expect the existence of an infinite number of saddle tori in their joint phase space. However, from Chap. 4 we already know that mutual coupling would make the phase space structure more complicated, as compared to the case of unidirectional coupling. To summarize, interaction of chaotic systems should allow for the phenomena approximately similar to the one occurring at periodic forcing of chaos, however, even more options can be anticipated. Running a few steps ahead, note that two interacting chaotic systems are analyzed in detail in Sect. 8.7, while a detailed experimental and numerical study of unidirectionally coupled chaotic systems is made in [18, 165, 222, 247, 270].

In this section we have allowed ourselves to indulge in speculations about the possible outcomes that might emerge out of periodically perturbing a chaotic attractor. It is time to stop and to check if anything of the above is true.

8.5 Synchronization of Chaos by Periodic Forcing

8.5.1 Experiment

In [16–18, 215] several cases of chaos synchronization were studied by means of both full-scale experiment and numerical simulation. A model system used as a chaos generator was the Anishchenko–Astakhov oscillator [15]. Periodically forced chaos was modelled by coupling two of these generators unidirectionally, so that the forcing unit was in a periodic regime, and the unit being forced in a chaotic one.

8 Chaos Synchronization

The equations read

$$\dot{x}_1 = (m_1 - z_1)x_1 + y_1,$$
$$\dot{y}_1 = -x_1, \quad (8.11)$$
$$\dot{z}_1 = g(f(x_1) - z_1),$$

$$\dot{x}_2/p = (m_2 - z_2)x_2 + y_2 - B(x_2 - 3x_1) + B(y_2 - 3py_1),$$
$$\dot{y}_2/p = -x_2, \quad (8.12)$$
$$\dot{z}_2/p = g(f(x_2) - z_2),$$

where

$$f(x) = \begin{cases} 0, & x < 0, \\ x^2, & x \geq 0. \end{cases}$$

Equations (8.11) represent the oscillator that generates the forcing signal, and (8.12) represent the oscillator being forced. A scheme of an experimental setup for these oscillators is shown in Fig. 8.8.

At $m_1 = 0.7$ and $g = 0.3$ the first system (8.11) is in a periodic regime whose spectrum on the screen of an oscilloscope is given in Fig. 8.9, upper panel. At $m_2 = 1.19$ and $g = 0.3$ the second oscillator demonstrates chaotic oscillations that correspond to a well-developed Feigenbaum chaos formed as a result of period-doubling bifurcations, and its spectrum is shown in Fig. 8.9, lower panel. Parameter p in (8.12) defines the frequency detuning between the subsystems as follows: all the time derivatives of the system variables x_2, y_2 and z_2 are divided by p, and thus the time in this subsystem is slowed down if $p > 1$, and is sped up if $p < 1$. p is equivalent to the frequency detuning denoted by the same symbol in Sect. 4.8. B is the strength of forcing.

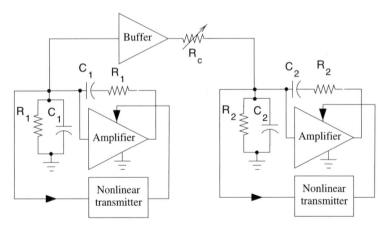

Fig. 8.8. Scheme of an experimental setup used to illustrate the phenomena of chaos synchronization. Two Anishchenko–Astakhov oscillators are coupled unidirectionally via a buffer

8.5 Synchronization of Chaos by Periodic Forcing   207

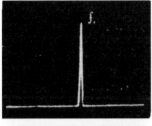

Spectrum of the forcing signal

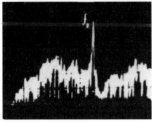

Spectrum of oscillations in the system to be forced

Fig. 8.9. Upper panel: spectrum of periodic oscillations in (8.11) (forcing system) at $m = 0.7$ and $g = 0.3$. **Lower panel**: spectrum of chaotic oscillations in (8.12) (the system to be forced) at $m = 1.19$ and $g = 0.3$ without forcing ($B = 0$).

Note that, unlike in Chaps. 3, 7, 6 and Sect. 5.2, forcing in (8.11)–(8.12) is included not by simply adding the signal from the forcing system to one of the system's equations. Instead, the forcing term has the form of a difference between the variable of the forced system and the scaled variable of the forcing one, which is rather by analogy with mutually coupled oscillators considered in Chap. 4. Also, the coupling term actually consists of two components in the first of (8.12): both the x_1 and y_1 variables from the forcing system are included. This particular form of unidirectional coupling arises in the experiment as one tries to implement the simplest and most natural way of connecting one electronic system to another.

In Fig. 8.10 the vicinity of 1 : 1 synchronization tongue is shown that was obtained in a full-scale experiment. The lines in this figure were obtained as follows: the physical features of the oscillatory regimes occurring in the experiment were observed, such as phase portraits, realizations and spectra. When these features changed drastically, on a qualitative level, it was believed that a qualitative transition similar to a bifurcation has occurred, and the respective set of forcing parameters (p, B) was attributed to a certain line. The designations are indicated in figure caption.

The lower boundary of the tongue is the line at which a non-synchronous chaos becomes synchronous via frequency locking. This transition is illustrated with the evolution of spectra in the first row of Fig. 8.12 as one follows route **A** in Fig. 8.11. Without forcing, the system demonstrates Feigenbaum chaos with a single well-defined peak in the spectrum (first column). When forcing is applied (second column), one can see that another peak appears very close to the main one: but because this is chaos whose spectrum is continuous, and also because the detuning p is very

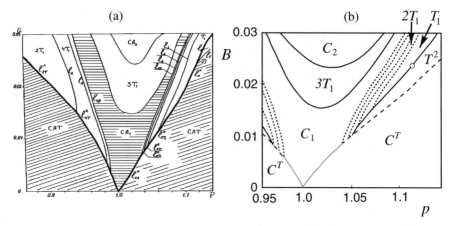

Fig. 8.10. (Color online) The vicinity of 1 : 1 synchronization region for a chaotic oscillator forced by a periodic signal, see (8.11)–(8.12) with $m_1 = 0.7, m_2 = 1.19$ and $g = 0.3$. **a** Original drawing from [215], **b** a revised version with only most essential lines. Designations in **b**: C_1 and C_2 are synchronous chaotic regimes, C^T is a non-synchronous chaos, T, $2T$ and $3T$ are stable cycles with periods approximately equal to T_1, $2T_1$ and $3T_1$, respectively. *Grey line*: borderline between C_1 and C^T; *dashed line*: inverse torus breakdown transition; *solid black*: torus death (inverse Neimark–Sacker) bifurcation; *dotted*: period-doubling bifurcation; *empty circle*: Takens–Bogdanov point (see Chap. 5)

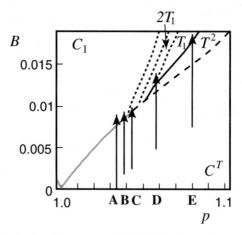

Fig. 8.11. (Color online) The enlargement of Fig. 8.10(b). Several routes are marked by letters, namely, **A**: $p = 1.034$, **B**: $p = 1.039$, **C**: $p = 1.043$, **D**: $p = 1.058$, **E**: $p = 1.08$. Illustrations of these transitions are given in Fig. 8.12

close to 1, we can notice this mainly from the thickening of the main peak. Inside synchronization tongue, there is again only one spectral peak which is at the position of the forcing frequency. This is how frequency locking has taken place.

A few more transitions that can occur on the way inside synchronization region via different routes are illustrated in Fig. 8.12: different rows correspond to routes

8.5 Synchronization of Chaos by Periodic Forcing

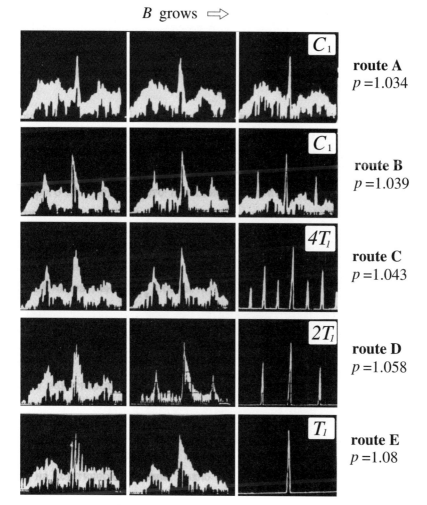

Fig. 8.12. Illustration of several synchronization transitions in unidirectionally coupled Anishchenko–Astakhov oscillators (8.11)–(8.12) with $m_1 = 0.7$, $m_2 = 1.19$ and $g = 0.3$, i.e., when a chaotic oscillator is forced by a periodic signal. Spectra of forced oscillations are shown, all pictures being the snapshots of the screens of oscilloscopes in a full-scale experiment. Each row corresponds to a route marked from **A** to **E** in Fig. 8.11 at different values of detuning p provided to the right of the row

from **A** to **E** indicated in Fig. 8.11. In all these routes, the detuning p is fixed at a certain value and the forcing strength is increased. One can note that the end points of these routes are different: routes **A** and **B** end at synchronized chaos C_1, route **C** ends at the period-four limit cycle $4T_1$, route **D** ends at the period-two limit cycle $2T_1$, and route **E** ends at the period-one limit cycle T_1. These transitions are associated with locking in [16–18, 215].

210   8 Chaos Synchronization

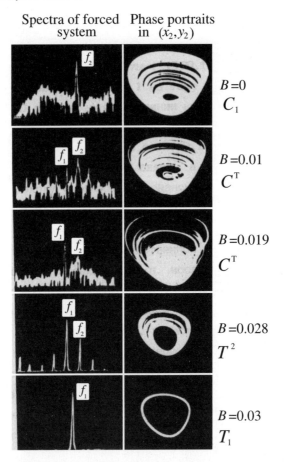

Fig. 8.13. Transition to chaos synchronization by suppression in (8.11)–(8.12) with $m_1 = 0.7$, $m_2 = 1.19$ and $g = 0.3$, i.e., when a chaotic oscillator is forced by a periodic signal. Detuning is set to $p = 1.13$. Spectra and phase portraits of forced oscillations are shown, all pictures being the snapshots of the screens of oscilloscopes in a full-scale experiment. Each row corresponds to a different value of forcing strength B provided to the right of the row. For reference, see Fig. 8.10

Consider a larger value of the detuning $p = 1.13$ and enter the synchronization region in Fig. 8.10(b) by increasing B from zero to 0.03. The evolution of spectra and of the phase portraits is illustrated in Fig. 8.13. To the right of each row the respective values of B are given together with the designations for the regime observed. This transition is associated with the suppression of natural dynamics. Namely, without forcing $B = 0$ the regime was chaotic C_1, with the basic frequency f_2. As forcing grows from zero to $B = 0.01$, its frequency f_1 appears in the spectrum at a certain distance from f_2, while the observed regime is a non-synchronous chaos C^T. At $B = 0.019$ we still observe the non-synchronous

8.5 Synchronization of Chaos by Periodic Forcing 211

chaos C^T, but now the peak at f_1 corresponding to the forcing becomes higher than the peak at f_2 that is associated with the natural dynamics. At $B = 0.028$ the chaos has turned into a torus T^2 via the inverse torus breakdown transformation, its spectrum is discrete and contains peaks at f_1, f_2, and at their combinations. Note, that the peak at the forcing frequency f_1 is higher than the one at the natural frequency f_2. At $B = 0.03$ the torus has died and the observed regime is a stable cycle with the frequency f_1 of external periodic forcing. Thus, suppression of natural dynamics has taken place.

Is anything of the observed in the course of these experiments similar to the picture we have created in our minds in Sect. 8.4? First of all, at stronger forcing an inverse torus breakdown bifurcation has turned chaos into a stable torus. And then this torus has shrunk to become a stable cycle inside the tongue. So, suppression here is in fact the elimination of chaos. Although, remarkably, deeper inside the tongue chaos is reestablished again.

As to the weaker forcing, frequency locking has indeed taken place, as it follows from the spectra. But did it take place as a result of saddle-node bifurcations as we have imagined in Sect. 8.4? This is impossible to tell yet, because the saddle-node bifurcations are expected from saddle cycles that are invisible in an experiment. In order to check this, a numerical analysis is required.

8.5.2 Numerical Analysis

In [16–18, 215] it was first suggested that the transition to chaos synchronization via phase locking is associated with the saddle-node bifurcations of the saddle cycles. There, the vicinity of the 1 : 1 forced synchronization of chaos in (8.11)–(8.12) was obtained numerically with all parameters as in Sect. 8.5.1, only m_1 was set to 0.6 instead of 0.7—but the forcing system still demonstrated a periodic regime. The bifurcation diagram on the plane of forcing parameters "detuning p"–"forcing strength B" is given in Fig. 8.14: the original drawings from [215] are shown together with the clearer modern version. Solid grey lines in (b) denote saddle-node bifurcations for a saddle and a twice saddle cycles, and one can see three such lines that were revealed numerically. Solid black lines in (b) show saddle-node bifurcations for a stable and a saddle cycle, and dotted lines show period-doubling bifurcations for stable cycles.

It is of course impossible to plot an infinite number of bifurcation lines by means of a numerical simulation. Moreover, in the late 1980s when this result was being obtained, powerful computers were not easily available and calculation of even those saddle-node lines shown in Fig. 8.14 presented a significant challenge for a researcher. And on top of that, unfortunately, even with the modern numerical techniques available it is still impossible to prove or to refute the existence of the infinite number of saddle tori in the phase space, and to check whether these bifurcations do take place on their surfaces. But the numerical evidence at least does not contradict the hypothesis of Sect. 8.4. For a more accurate numerical evidence we refer the reader to Sect. 8.6.

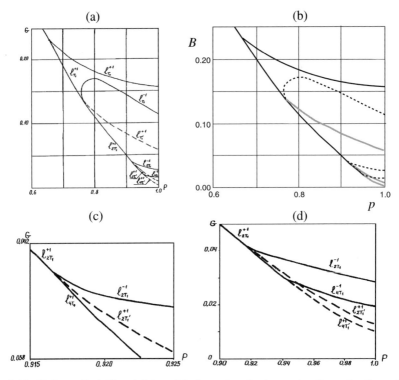

Fig. 8.14. (Color online) The vicinity of 1 : 1 synchronization region for a chaotic oscillator forced by a periodic signal (8.11)–(8.12)—numerical simulation. Parameter values are: $m_1 = 0.6$, $m_2 = 1.19$ and $g = 0.3$. **a** The original drawing from [215] and **b** revised clearer version. **c** and **d** are enlargements of two parts of the diagram in **a**, **b**. Lines in **b** are denoted as: *solid grey*—saddle-node bifurcations involving saddle cycles, *solid black*—saddle-node bifurcations of stable cycles, *dotted*—period-doubling bifurcations of stable cycles

8.6 Synchronization of Periodic Oscillations by Chaos

In this section we consider a different situation: assume that the system being forced is periodic, but the forcing is chaotic. As an example, let the Rössler system [252] drive the paradigmatic van der Pol oscillator. This way, the forced van der Pol system would always have a chaotic attractor with one positive Lyapunov exponent, and any bifurcations evoked by coupling would not affect the existence of chaos. Let us write down the equations of our systems in the following form:

$$\begin{aligned}\dot{x}_1 &= -\omega_1 y_1 - z_1, \\ \dot{y}_1 &= \omega_1 x_1 + \alpha y_1, \\ \dot{z}_1 &= \beta + z_1(x_1 - \mu),\end{aligned} \quad (8.13)$$

$$\begin{aligned}\dot{x}_2 &= y_2 + C(x_1 - x_2), \\ \dot{y}_2 &= \varepsilon(1 - x_2^2)y_2 - \omega_2^2 x_2,\end{aligned} \quad (8.14)$$

8.6 Synchronization of Periodic Oscillations by Chaos

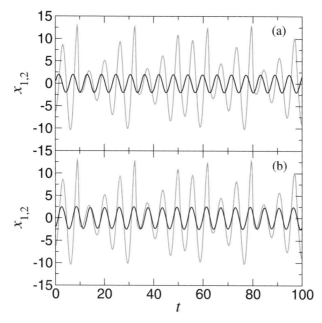

Fig. 8.15. (Color online) Oscillations in the periodic oscillator forced chaotically (8.13)–(8.14) at $\omega = 1.08$ and two different values of forcing strength C: **a** $C = 0.002$, **b** $C = 0.02$. *Grey line* denotes x_1 of driving Rössler system (8.13), *black line* denotes x_2 of the driven van der Pol oscillations (8.14)

where (8.13) represent the forcing Rössler system and (8.14) the van der Pol oscillator being forced. Parameters α, β and μ determine the dynamics of Rössler system, ε is the non-linearity parameter of the van der Pol oscillator, ω_1 and ω_2 define the time scales of two oscillators, respectively, and the parameter C is the forcing strength. We fix $\alpha = \beta = 0.2$, $\mu = 6.5$, and $\omega_1 = 1$, at which Rössler oscillator demonstrates a well-developed one-band chaos. ε was set to 0.2.

In Fig. 8.15 oscillations of driving (grey) and of driven (black) systems are presented for $\omega_2 = 1.08$ and two different values of C. Already from this figure one can conclude that at larger C the systems oscillate more synchronously. For example, the maxima of oscillations in the partial systems occur almost simultaneously at larger C.

8.6.1 Spectra

In order to better understand the effect of coupling on cooperative dynamics in interacting systems, let us examine how the spectra of oscillations evolve with variation of forcing strength C. This evolution for $\omega_2 = 1.08$ is illustrated in Fig. 8.16. The spectra of the forcing chaotic Rössler system are denoted by grey shaded areas, while the spectra of the driven van der Pol system are represented by black shaded

214   8 Chaos Synchronization

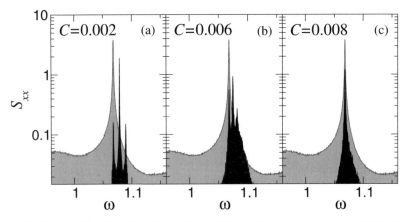

Fig. 8.16. (Color online) Locking of periodic oscillations by chaotic forcing in (8.13)–(8.14) at $\omega_2 = 1.08$. Spectra of forcing (*grey*) and of response (*black*) for different values of forcing strength C: **a** $C = 0.002$, **b** $C = 0.006$, **c** $C = 0.008$

areas. The driving system has a continuous spectrum with a distinct peak at the frequency $\omega \approx 1.07$. At small C (Fig. 8.16(a)) the oscillations of the driven van der Pol system have a spectrum with two characteristic peaks. One of them is at $\omega \approx 1.07$ and is the result of driving from the Rössler system, whereas the second one with frequency $\omega \approx \omega_2 = 1.08$ corresponds to the own dynamics of the van der Pol oscillator. Note that the third peak present in the spectrum of the van der Pol system appears at the combination frequency. It reflects the fact of non-linear interaction between the two dynamical systems, and its position is completely defined by a linear combination of two basic frequencies. As C grows, the spectral peak corresponding to van der Pol dynamics moves towards the characteristic frequency of the Rössler system (Fig. 8.16(b)), and at $C \approx 0.007$ coincides with it. Further increase of C does not change the characteristic frequencies of oscillators (Fig. 8.16(c)). If we forget that the spectra of coupled oscillators are continuous, the scenario considered here reminds of the forced synchronization of periodic oscillations via frequency *locking* described in Sects. 3.9–3.10.

The system (8.13)–(8.14) demonstrates a very different behavior at $\omega_2 = 1.2$ (see Fig. 8.17). As before, the spectrum of the chaotically driven van der Pol oscillator has two characteristic peaks. However, in contrast to the previous case, the increase of the forcing strength C practically does not change the position of the peak related to the own dynamics of van der Pol oscillator. Instead, we can see that the power of oscillations around ω_2 drops with the increase of C, whereas the peak at the forcing frequency grows. This can be explained as *suppression* of natural dynamics of the van der Pol oscillator by chaotic force. The reported behavior of spectra is very similar to that typical of synchronization via suppression of natural dynamics occurring in a forced periodic oscillator (Sects. 3.9–3.10). Hence, the transformations of spectra that accompany the onset of synchronization in chaotic

8.6 Synchronization of Periodic Oscillations by Chaos

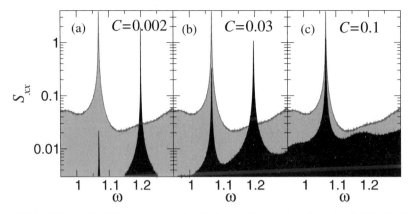

Fig. 8.17. (Color online) Suppression of periodic oscillations by chaos in (8.13)–(8.14) at $\omega = 1.2$. Spectra of oscillations at different values of forcing strength C: **a** $C = 0.002$, **b** $C = 0.03$, **c** $C = 0.1$

systems, are quite similar to what is observed for forced synchronization of periodic oscillations.

In spite of the continuity of the spectrum of chaotic oscillations, one can still single out two basic scenarios of synchronization. Namely, in the first case the main peak of the forced system moves closer to the frequency of forcing with the increase of coupling. In the second case the forcing suppresses the oscillations at frequencies that are close to the natural frequency of the oscillator being forced.

8.6.2 Poincaré Sections

The analogy with the classical locking and suppression may become clearer, if we consider the evolution of the attractors of (8.13)–(8.14). Figure 8.18 illustrates topological changes of the attractors corresponding to the onset of synchronization when frequency locking is realized at $\omega_2 = 1.08$. The projections (x_2, y_2) of the Poincaré section defined by $x_1 = 0$ are shown for different values of C. At small C, the section looks like a smeared closed curve with homogeneously distributed points (Fig. 8.18(a)). With the increase of C, this distribution becomes more and more inhomogeneous (Fig. 8.18(b)). Starting from a certain value of C, all section points are grouped in a segment of the what was a smeared closed curve (Fig. 8.18(c)). This looks similarly to what happens at frequency locking of periodic oscillations (see Fig. 3.11): the onset of synchronization corresponds to a saddle-node bifurcation on the torus.

The evolution of Poincaré sections at $\omega = 1.2$ with the change of forcing strength C is shown in Fig. 8.19. Here, the increase of forcing strength makes the initial smeared closed curve shrink along x_2 directions until it collapses into a homogeneous cloud of points. Such a behavior is to some extent similar to the torus death bifurcation that is responsible for synchronization of periodic oscillations via suppression of natural dynamics (see Fig. 3.12). Thus, we can conclude that the

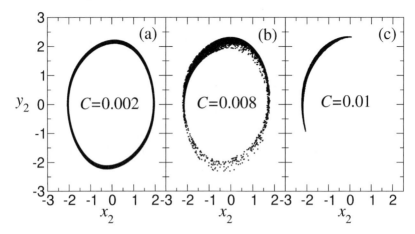

Fig. 8.18. Locking of periodic oscillations by chaos in (8.13)–(8.14) at $\omega_2 = 1.08$. Evolution of Poincaré section defined by $x_1 = 0$. **a** $C = 0.002$, **b** $C = 0.008$, **c** $C = 0.01$

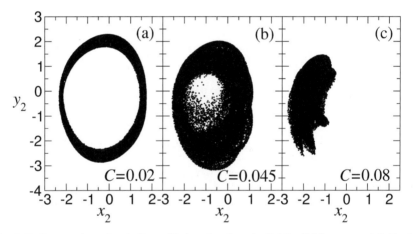

Fig. 8.19. Suppression of periodic oscillations by chaos in (8.13)–(8.14) at $\omega_2 = 1.2$. Evolution of Poincaré section defined by $x_1 = 0$: **a** $C = 0.02$, **b** $C = 0.045$, **c** $C = 0.08$

mechanisms of frequency locking and of suppression of natural dynamics are general and common both for periodic and chaotic oscillations. Moreover, the shape of the synchronization region on the parameter plane "coupling strength"–"frequency detuning" has a similar tongue-like structure as shown in Fig. 8.22 below.

8.6.3 Phase Difference

Once we have revealed that synchronization of chaos could be understood in terms of basic frequencies corresponding to the main (independent) peaks in the spectra, it would be interesting to find out if this phenomenon can be interpreted in terms of the phase of oscillations. For this purpose, we make use of the fact that as it is seen

8.6 Synchronization of Periodic Oscillations by Chaos   217

from the spectra, the oscillations of both coupled systems can be treated as narrow-band signals. Then we can introduce the instantaneous amplitudes $A_{1,2}$, and phases $\Phi_{1,2}$ as follows:

$$x_1(t) = A_1(t)\cos\Phi_1(t), \qquad y_1(t) = A_1(t)\sin\Phi_1(t),$$
$$x_2(t) = A_2(t)\cos\Phi_2(t), \qquad y_2(t) = -A_2(t)\sin\Phi_2(t).$$

Signs before sines and cosines above where chosen in order to ensure that phases in both systems are growing, rather than decreasing, with time. Then we can rewrite (8.13) and (8.14) in terms of amplitudes and phases

$$\dot{A}_1 = \alpha A_1 \sin^2\Phi_1 - z_1\cos\Phi_1,$$
$$\dot{\Phi}_1 = \omega_1 + \frac{z_1}{A_1}\sin\Phi_1 + \alpha\sin\Phi_1\cos\Phi_1,$$
$$\dot{z}_1 = \beta + z_1(A_1\cos\Phi_1 - \mu),$$

$$\dot{A}_2 = \left(\omega_2^2 - 1\right)A_2\sin\Phi_2\cos\Phi_2 + C(A_1\cos\Phi_1 - A_2\cos\Phi_2)\cos\Phi_2$$
$$+ \varepsilon\left(1 - A_2^2\cos^2\Phi_2\right)A_2\sin^2\Phi_2,$$

$$\dot{\Phi}_2 = \sin^2\Phi_2 + \omega_2^2\cos^2\Phi_2 - C\left(\frac{A_1}{A_2}\cos\Phi_1 - \cos\Phi_2\right)\sin\Phi_2$$
$$+ \varepsilon\left(1 - A_2^2\cos^2\Phi_2\right)\sin\Phi_2\cos\Phi_2,$$

and analyze the phase difference $\Delta\Phi = \Phi_1 - \Phi_2$.

The dynamics of $\Delta\Phi$ for different values of C is presented in Fig. 8.20(a) to illustrate frequency locking. At $C = 0.001$ the phase difference changes almost linearly in time. However, as C increases, the realization of $\Delta\Phi(t)$ starts to demonstrate plateau-like segments, on which the phases of oscillations in the partial systems grow with the same velocity. With the further increase of the forcing strength C, the plateaus become longer, and starting with some value of C, the phase difference simply oscillates around some mean value, demonstrating no linear trend. The latter implies the onset of synchronization. This behavior of the phases can also be illustrated by the distribution p of the phase difference $\Delta\Phi$ "wrapped" inside the interval $[-\pi, \pi)$, see Figs. 8.20(b)–(d). At small C the values of $\Delta\Phi$ are distributed almost homogeneously (Fig. 8.20(b)). A larger C evokes heterogeneity in the distribution; namely, some values of the phase difference become more probable than the others (Fig. 8.20(c)). Finally, at a sufficiently large forcing C the distribution of the wrapped $\Delta\Phi$ becomes narrow, implying that $\Delta\Phi(t)$ never achieves some values in $[-\pi, \pi)$ (Fig. 8.20(d)). The latter corresponds to phase synchronization.

As seen from Fig. 8.21, qualitatively the same phase behavior is observed when synchronization is realized via suppression of natural dynamics. Hence, strictly speaking, by looking at the evolution of phase difference alone one cannot determine exactly what mechanism of synchronization is being realized in the given case.

It is important to note that if we define the synchronization condition using, for example, (8.8) or (8.9), the region of synchronization on the parameter plane (ω_2, C)

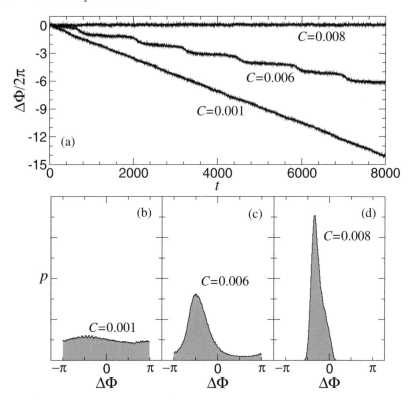

Fig. 8.20. (Color online) Locking of periodic oscillations by chaos in (8.13)–(8.14) at $\omega_2 = 1.08$. **a** Phase difference $\Delta\Phi$; distribution of wrapped $\Delta\Phi$: **b** $C = 0.001$, **c** $C = 0.006$, **d** $C = 0.008$

has the classical tongue-like shape, as shown in Fig. 8.22(a), where the condition of synchronization was used in the form of the rational connection between the basic frequencies in coupled oscillators (8.9).[8]

Now, let us find out what bifurcations are associated with the onset of synchronization of chaotic oscillations. In Fig. 8.22(a) the vicinities of three different synchronization tongues are illustrated on the plane of parameters "frequency ω_2 of forced system"–"strength C of forcing," and in Fig. 8.22(b) the enlargement of 1 : 1 region is given. Black solid and dashed lines are the lines of bifurcations of some unstable periodic orbits of a chaotic attractor. Namely, solid lines represent saddle-node bifurcations, and dashed lines denote Neimark–Sacker (torus birth/death) bifurcations. These lines form synchronization regions for individual unstable limit cycles. As seen from the figure, accumulation of bifurcation curves for different synchronization regions for different saddle cycles leads to the occurrence of chaos synchronization. With this, locking of chaos results from the accumulation of saddle-

---

[8] For the case considered, synchronization criteria for the phase and for the basic frequencies are satisfied almost at the same values of C and ω_2.

8.6 Synchronization of Periodic Oscillations by Chaos 219

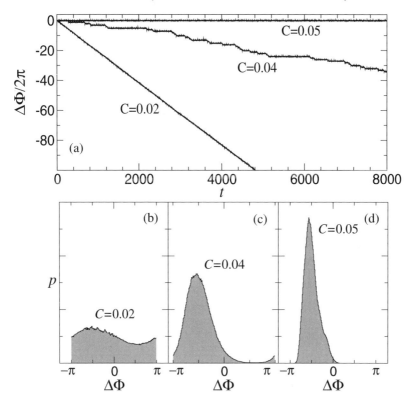

Fig. 8.21. (Color online) Suppression of periodic oscillations by chaos in (8.13)–(8.14) at $\omega_2 = 1.2$. **a** Phase difference $\Delta\Phi$; and distribution of wrapped $\Delta\Phi$: **b** $C = 0.02$, **c** $C = 0.04$, **d** $C = 0.05$

node bifurcations, whereas suppression of chaos is associated with accumulation of Neimark–Sacker bifurcations[9] [17, 18]. Thus, with variation of the parameters, the transition to chaos synchronization is realized gradually, unlike in periodic oscillations. Moreover, from the analysis of Fig. 8.22(a) it also follows that both mechanisms of synchronization seem to be general not only for synchronization 1 : 1 (for which the conditions (8.9) and (8.8) are valid with $n = m = 1$), but also for other ratios $n : m$. In particular, in the figure we can see accumulation of bifurcation lines for unstable limit cycles that form the regions of synchronization of the order 3 : 4 and 5 : 4. The latter regions have qualitatively the same structure as that for 1 : 1 synchronization.

Note that the tips of synchronization tongues for periodic orbits embedded into a chaotic attractor do not necessarily coincide (Fig. 8.22(b)) as discussed in Sect. 8.4, since the periods of unstable cycles in unperturbed chaos are usually not rationally related, but are characterized by some distribution [303]. Synchronization of chaos occurs in the parameter region where synchronization tongues of individual unsta-

---

[9] Compare with the discussion in Sect. 8.4.

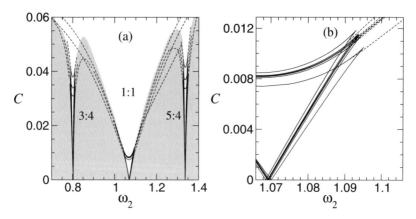

Fig. 8.22. a Region of synchronization of periodic oscillations by chaotic forcing (*white area*) in (8.13)–(8.14), which is defined from (8.9). **b** Zoomed part of **a**. Lines are bifurcations of unstable limit cycles embedded in the chaotic attractor: *solid lines*—saddle-node bifurcations, *dashed*—Neimark–Sacker (torus birth/death) bifurcations

ble cycles overlap. Therefore, strictly speaking, in contrast to synchronization of periodic oscillations, synchronization of chaotic oscillations can be achieved only at some non-zero forcing.

For chaos that is not a narrow-band process, the distribution of periods of unstable cycles can be quite broad, and the complete overlapping of synchronization regions of unstable periodic orbits might not occur. In this case, if the phase of chaotic oscillations is formally introduced using the approximation (8.2), the synchronization condition (8.8) for the fixed n and m is valid only for finite time intervals. This effect is sometimes called *imperfect phase synchronization of chaos* [201, 303].

Synchronization of chaos via phase/frequency locking, which is associated with accumulation of saddle-node bifurcations of unstable periodic orbits, can be regarded as a crisis of chaotic sets. Namely, while moving towards synchronization region in the parameter plane, each saddle-node bifurcation implies the birth of a pair of unstable cycles. In the system being considered, these are periodic orbits with one unstable direction (saddle orbit) and with two unstable directions (twice saddle orbits). Accumulation of these bifurcations creates a pair of synchronous chaotic sets—a stable and an unstable ones, whose skeletons are formed by the saddle and the twice saddle limit cycles, respectively [211, 246].

8.6.4 Lyapunov Exponents

It is clear that bifurcations of unstable limit cycles should somehow influence the stability properties of chaotic attractors. In order to illustrate this, consider evolution of Lyapunov exponents of a chaotic attractor for two mechanisms of synchronization: for phase/frequency locking and for suppression of natural dynamics. Three largest Lyapunov exponents are given in Fig. 8.23 as functions of the forcing strength C. Both for locking (a) and for suppression (b), chaos outside the synchro-

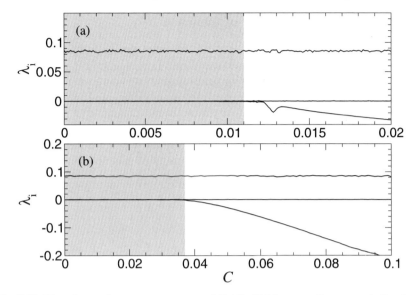

Fig. 8.23. Three largest Lyapunov exponents of (8.13)–(8.14) vs coupling strength C: **a** locking with $\omega_2 = 1.08$; **b** suppression with $\omega_2 = 1.2$. *White areas* outline synchronization region

nization region is characterized by one positive and two zero Lyapunov exponents. With the increase of C one of zero exponents becomes negative, which reflects accumulation of bifurcations of unstable orbits, i.e., occurrence of synchronization. Such behavior of Lyapunov exponents is quite typical of synchronization of a narrow-band chaos [225, 247]. However, for more complex chaotic oscillations evolution of Lyapunov spectrum on the way to synchronization can be different [198, 289].

So far we have considered synchronization of chaotic oscillations by periodic forcing, and of periodic oscillations by chaotic forcing. A natural question to ask would be whether chaotic oscillations can be synchronized by chaotic forcing. Forced synchronization of chaos by chaos was studied both in an experiment and in numerical simulations in [17, 18, 215].

Summarizing the main results of this section, we can conclude that the mechanisms of synchronization of chaos are similar to the ones of synchronization of periodic oscillations. Actually, there are two basic scenarios for the onset of synchronization—frequency locking and suppression of natural dynamics, which are well distinguished through the evolution of spectra. However, the analogy between synchronization of chaotic and of periodic oscillations is even deeper. Namely, synchronization is realized as a result of bifurcations which are the same for periodic and chaotic systems. The frequency locking is associated with saddle-node bifurcations of periodic orbits, whereas the suppression is due to Neimark–Sacker bifurcations. However, if for periodic oscillations synchronization is achieved as a result of only one bifurcation, the transition to chaos synchronization is accompanied by an infinite cascade of bifurcations of unstable orbits embedded in the chaotic attractor.

222   8 Chaos Synchronization

8.7 Mutual Synchronization of Chaos

In Sect. 8.6 we considered the simplest case of chaos synchronization, when the coupling between the systems was unidirectional. However, in realistic situations oscillators would often be coupled reciprocally. In this section we are going to investigate a more complicated case, when the chaotic oscillators are coupled mutually, i.e., when the dynamics of *each* system effects the oscillations in the other system through coupling. The cooperative dynamics in such systems was previously considered, e.g., in [18, 165, 222, 247, 270]. For our study we use the following model equations that describe the dynamics of two mutually coupled Rössler systems:

$$
\begin{aligned}
\dot{x}_1 &= -\omega_1 y_1 - z_1 + C(x_2 - x_1), \\
\dot{y}_1 &= \omega_1 x_1 + \alpha y_1, \\
\dot{z}_1 &= \beta + z_1(x_1 - \mu), \\
\dot{x}_2 &= -\omega_2 y_2 - z_2 + C(x_1 - x_2), \\
\dot{y}_2 &= \omega_2 x_2 + \alpha y_2, \\
\dot{z}_2 &= \beta + z_2(x_2 - \mu).
\end{aligned}
\tag{8.15}
$$

Here $x_{1,2}$, $y_{1,2}$, $z_{1,2}$ are dynamical variables of the first and of the second oscillators; α, β, and μ are the parameters governing the individual dynamics of the systems; C defines the strength of coupling between the oscillators, and $\omega_{1,2}$ determine the main frequencies of oscillations in the respective subsystems. We choose the parameter values of α, β, and μ to be such that at $C = 0$ both systems demonstrate chaotic oscillations, namely: $\alpha = 0.165$, $\beta = 0.2$, $\mu = 10$. For convenience, we represent $\omega_{1,2} = \omega_0 \pm \delta$, where δ determines frequency detuning between the two systems.

8.7.1 Phase/Frequency Locking

Figure 8.24 illustrates how spectra of oscillations behave with variation of coupling strength C at $\delta = 0.02$. Their evolution looks like a typical evidence of synchronization via phase/frequency locking. Actually, at $C = 0$ (Fig. 8.24(a)) spectra of oscillations in each of the two subsystems have quite sharp peaks at the frequencies defined by, but not equal to, ω_1 and ω_2. However, at $C \neq 0$ the oscillators start to interact. This interaction manifests itself via the appearance of the second characteristic peak in the spectra of each oscillator (Fig. 8.24(b)). With the increase of interaction strength C, two characteristic peaks in the spectra approach each other (Fig. 8.24(b)), and starting with some C coincide (Fig. 8.24(c)). Such evolution of spectra is very similar to what we have observed in Sect. 8.6 for forced synchronization of chaos when the frequency locking was being realized with the only difference that here *both* spectral peaks move towards each other. This conclusion is also confirmed by the behavior of Poincaré sections shown in Fig. 8.25. The growth of C makes the motion of phase points in the Poincaré section more inhomogeneous (Fig. 8.25(a), (b)), and once synchronization is achieved, the phase points concentrate only in a small fragment of the initial (for $C = 0$) Poincaré section.

8.7 Mutual Synchronization of Chaos   223

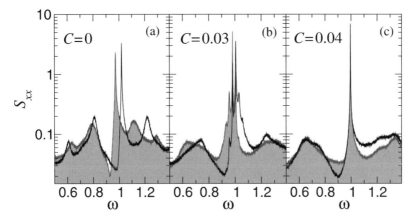

Fig. 8.24. (Color online) Locking in mutually coupled chaotic systems (8.15) at $\delta = 0.02$. Spectra at different values of the coupling strength C: **a** $C = 0$, **b** $C = 0.03$, **c** $C = 0.04$. *Shaded areas*: first subsystem; *black lines*: second subsystem

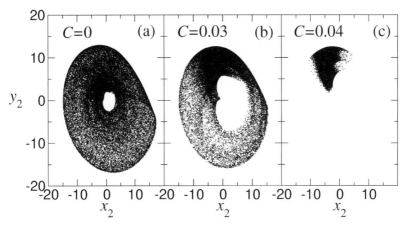

Fig. 8.25. Locking in mutually coupled chaotic systems (8.15) at $\delta = 0.02$. Poincaré section defined by $y_1 = 0$ at different values of C: **a** $C = 0$, **b** $C = 0.03$, **c** $C = 0.04$

8.7.2 Suppression

Now consider how synchronization is developed for relatively large detuning $\delta = 0.07$. Evolution of spectra is illustrated in Fig. 8.26. At small coupling C (a) spectra of both subsystems demonstrate two pronounced peaks that are associated with the natural time scales of the interacting oscillations. With the increase of C, low-frequency peak decreases, while the high-frequency peak grows (b), and at C large enough, only one characteristic peak is left in the spectrum (c). This transformation of the spectra is characteristic of synchronization via suppression of natural dynamics. It is also in line with suppression of mutually coupled periodic oscillators considered in Chap. 4. However, in contrast to what we have seen in the case of forced

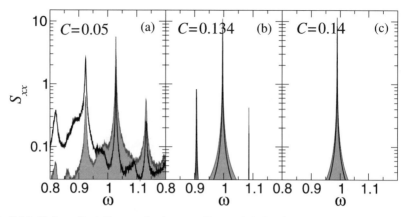

Fig. 8.26. (Color online) Suppression in mutually coupled chaotic systems (8.15) at $\delta = 0.07$. Spectra at different values of the coupling strength C: **a** $C = 0.05$, **b** $C = 0.134$, **c** $C = 0.14$

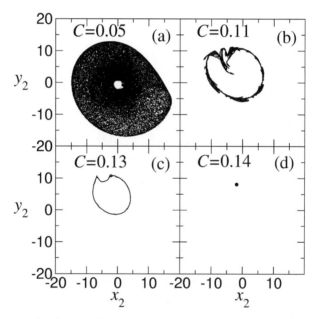

Fig. 8.27. Suppression in mutually coupled chaotic systems (8.15) at $\delta = 0.07$. Poincaré section defined by $y_1 = 0$ at different values of C: **a** $C = 0.05$, **b** $C = 0.11$, **c** $C = 0.13$, and **d** $C = 0.14$

synchronization in Sect. 8.6, for mutually coupled systems the increase of C leads to the disappearance of chaos in both subsystems. As demonstrated in Fig. 8.27, while the systems are approaching the state of synchronization, the chaotic attractor (a), (b) is first transformed into an (non-smooth) ergodic torus (c), which then collapses into a limit cycle (d) as a result of an inverse Neimark–Sacker bifurcation. The latter corresponds to the onset of synchronization.

8.7.3 Phase Behavior

Synchronization in (8.15) can also be described in terms of phases. Let us introduce substitutions $x(t) = A(t) \cos \Phi(t)$, $y(t) = A(t) \sin \Phi(t)$. Then model equations read

$$\dot{A}_1 = a A_1 \sin^2 \Phi_1 + C(A_2 \cos \Phi_2 - A_1 \cos \Phi_1) \cos \Phi_1 - z_1 \cos \Phi_1,$$

$$\dot{\Phi}_1 = \omega_1 - \frac{C}{A_1}(A_2 \cos \Phi_2 - A_1 \cos \Phi_1) \sin \Phi_1 + a \sin \Phi_1 \cos \Phi_1$$
$$+ \frac{z_1}{A_1} \sin \Phi_1,$$

$$\dot{z}_1 = b + z_1(A_1 \cos \Phi_1 - m), \tag{8.16}$$

$$\dot{A}_2 = a A_2 \sin^2 \Phi_2 + C(A_1 \cos \Phi_1 - A_2 \cos \Phi_2) \cos \Phi_2 - z_2 \cos \Phi_2,$$

$$\dot{\Phi}_2 = \omega_2 - \frac{C}{A_2}(A_1 \cos \Phi_1 - A_2 \cos \Phi_2) \sin \Phi_2 + a \sin \Phi_2 \cos \Phi_2 + \frac{z_2}{A_2} \sin \Phi_2,$$

$$\dot{z}_2 = b + z_2(A_2 \cos \Phi_2 - m).$$

The dynamics of phase difference $\Delta \Phi = \Phi_1 - \Phi_2$ and the evolution of the distribution of wrapped phase difference are presented in Figs. 8.28 and 8.29. For both frequency locking and suppression, synchronization manifests itself as localization of $\Delta \Phi$ that occurs starting with $C = 0.05$.

It is important to note here, that when suppression is being realized, an interesting effect occurs near the boundary of synchronization. Namely, there is a range of the parameter values at which the phase difference $\Delta \Phi$ is localized, although the basic frequencies of oscillations are still incommensurate. This situation is illustrated in Fig. 8.30 where a Poincaré section and a distribution of wrapped phase difference are presented for $\delta = 0.07$ and $C = 0.13$. In spite of the fact that (8.15) demonstrate quasiperiodic oscillations, which are represented by ergodic torus (a), the corresponding (non-wrapped!) phase difference is limited and well localized within an interval $[-\pi, \pi)$ (b). This effect was also mentioned in Sect. 3.8 where forced synchronization of periodic oscillations was considered. Thus, the use of the inequality (8.8) alone as a synchronization criterion does not always lead to the appropriate results.

The relationship between the bifurcations of unstable periodic orbits embedded into the chaotic attractor and the mutual synchronization of chaos is elucidated in Fig. 8.31. The region of synchronization for which both criteria (8.8) and (8.9) hold true, is indicated as a white area. Because of symmetry in (8.15), the figure is symmetric with respect to $\delta = 0$, and therefore for simplicity all bifurcation lines are given only for positive δ. Similarly to the case of forced synchronization, here the region of synchronization has a tongue-like structure, and locking is realized via accumulations of saddle-node bifurcations of unstable limit cycles (black dashed lines). Due to the symmetry of (8.15), the tips of the tongues for unstable periodic orbits meet at one point $\delta = 0$, thus forming a nested structure. With this, suppression occurs as a result of a single inverse Neimark–Sacker bifurcation (black

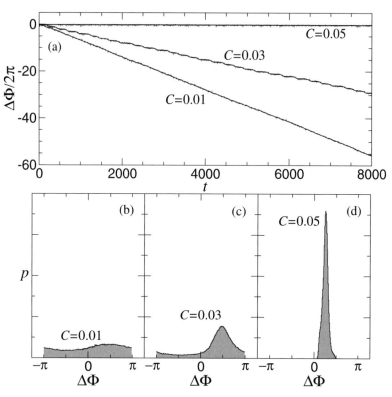

Fig. 8.28. (Color online) Phase locking in (8.16) at $\delta = 0.02$. **a** Phase difference $\Delta\Phi$ and distribution of wrapped $\Delta\Phi$ for: **b** $C = 0.01$, **c** $C = 0.03$, **d** $C = 0.05$

dot-dashed line), which transforms an ergodic torus into a stable limit cycle. The inside of synchronization region has a very complicated bifurcation structure that reflects a variety of transitions between regular and chaotic attractors. In particular, period-doubling bifurcations (grey lines) play an important role in the variety of the phenomena that take place inside the synchronization region. For example, these bifurcations are crucial for complete and lag synchronizations. They are also responsible for the development of phase multistability, which is considered in Chap. 12.

Reorganization of different attractors that is related to the onset and the development of chaos synchronization is illustrated in Fig. 8.32. In this figure four largest Lyapunov exponents are given as functions of coupling strength C for two characteristic values of detuning δ corresponding to locking and to suppression, respectively. Note that in both cases the transition to synchronization is associated with the transformation of an attractor with two zero Lyapunov exponents into an attractor with one zero Lyapunov exponent. However, these transformations are different for two mechanisms. Consider first how the stability of the chaotic attractor changes when locking mechanism is realized (Fig. 8.32(a)). At small value of C the chaotic attractor has two positive Lyapunov exponents, implying that the system demonstrates

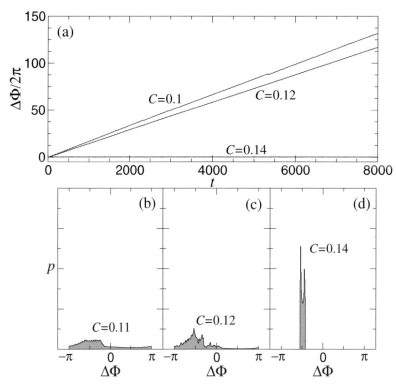

Fig. 8.29. (Color online) Suppression in (8.16) at $\delta = 0.07$. **a** Phase difference $\Delta\Phi$ and distribution of wrapped $\Delta\Phi$ for **b** $C = 0.11$, **c** $C = 0.12$, **d** $C = 0.14$

hyperchaos [253]. The onset of synchronization does not change the number of positive Lyapunov exponents. However as C increases further, the stability of the synchronous attractor changes, and one of the positive Lyapunov exponents becomes negative. This transition leads to the occurrence of a specific correlation between the amplitudes $A_{1,2}$ of oscillations that are associated with a phenomenon known as "lag synchronization" [248, 270].

For suppression, the increase of C first of all changes the number of positive Lyapunov exponents. Namely, first hyperchaos becomes just chaos with one positive Lyapunov exponent, and then the chaotic attractor is transformed into a torus with two zero and four negative Lyapunov exponents. The transition to synchronization is associated with bifurcation of ergodic torus, which collapses into a stable limit cycle (see Fig. 8.32(b)).

8.8 Homoclinic Synchronization of Chaos

In Sects. 8.6 and 8.7 we have established that two kinds of local bifurcations of the saddle cycles embedded into a chaotic attractor are responsible for the realization of

228   8 Chaos Synchronization

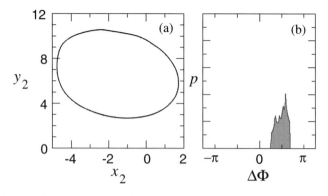

Fig. 8.30. (Color online) Phase versus frequency synchronization near suppression boundary in (8.15) at $\delta = 0.07$ and $C = 0.13$. **a** Projection (x_2, y_2) of Poincaré section defined by $y_1 = 0$; and **b** the corresponding distribution $p(\Delta\Phi)$ of wrapped phase difference

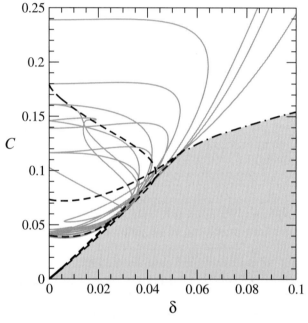

Fig. 8.31. (Color online) The vicinity of synchronization region for two coupled Rössler oscillators (8.15). *White area* outlines synchronization region. Bifurcations of unstable limit cycles are indicated by lines: *black dashed lines*—saddle-node bifurcations; *black dot-dashed line*—Neimark–Sacker bifurcation; *grey solid lines*—period-doubling bifurcations

two classical mechanisms of synchronization of chaos, namely, frequency (phase) locking and suppression of natural dynamics. However, *non-local* (global) bifurcations can also lead to the onset of synchronization [218, 220]. Synchronization of periodic oscillations via homoclinic bifurcation is considered in Chap. 5. In or-

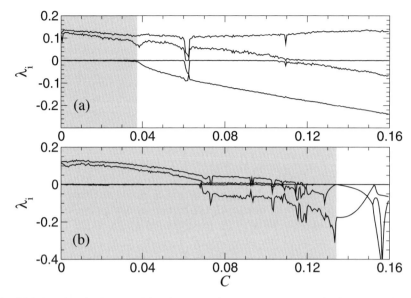

Fig. 8.32. Synchronization transitions in terms of Lyapunov exponents in (8.15). Four largest Lyapunov exponents are given as functions of C at **a** $\delta = 0.02$ (locking), **b** $\delta = 0.07$ (suppression). *White areas* correspond to synchronization regions

der to study the mechanism of chaos synchronization involving global bifurcations, consider the following model equations:

$$\begin{aligned}
\frac{dB_i}{dt} &= \nu B_i S_i \frac{1}{S_i + K} - B_i(\rho - \alpha \omega P_i), \\
\frac{dI_i}{dt} &= \alpha \omega B_i P_i - \rho I_i - \frac{I_i}{\tau}, \\
\frac{dP_i}{dt} &= -P_i\big(\rho + \alpha(B_i + I_i)\big) + \frac{\beta I_i}{\tau}, \\
\frac{dS_i}{dt} &= \rho\big(F_i(t) - S_i\big) - \gamma \nu B_i S_i \frac{1}{S_i + K}, \quad i = 1, 2, 3.
\end{aligned} \quad (8.17)$$

These equations describe the dynamics of populations of viruses and bacteria in three pools coupled via nutrition flow. This system can be obtained by the appropriate modification of (5.2). Here, as before B_i, I_i and P_i are the concentrations of non-infected bacteria, infected bacteria and viruses in ith pool, respectively, S_i represents the local concentration of nutrients inside the ith pool. All these concentrations are given in 10^6 ml^{-1}. The parameter values were chosen as follows: $\nu = 0.024$ min^{-1}, $K = 10\,\mu\text{g ml}^{-1}$, $\tau = 30$ min, $\omega = 0.8$, $\gamma = 0.01$ ng, $\beta = 100$ [41, 183] and are in general agreement with experimentally estimated values [167]. We define the inlet concentrations $F_{1,2,3}(t)$ as

$$F_1(t) = \sigma_1, \quad F_2(t) = S_1 + \sigma_2, \quad F_3(t) = S_2 + \sigma_3, \quad (8.18)$$

where σ_i, $i = 1, 2, 3$, are the values of the nutrient inlet concentrations given in mg/ml. We fix $\sigma_2 = 0$, and then setting the value of σ_1 we can vary the type of forcing signal applied to the third population from regular to chaotic oscillations. Then using an appropriate value of σ_3 we can observe either a synchronous or a non-synchronous response of the third system to the forcing signal.

The main bifurcations and regimes that are induced by variation of σ_1 and σ_3 are indicated in Fig. 8.33(a). The area of existence of synchronous attractors is shown by white, non-synchronous regimes occur inside the grey area. At small σ_1, a synchronous regime is a stable limit cycle which undergoes a period-doubling bifurcation as σ_1 grows and crosses the line *PD*. As a result of this bifurcation, the initial limit cycle loses its stability, but another stable limit cycle of doubled period is born in its vicinity. Further increase of σ_1 induces a cascade of period-doubling bifurcations, as a result of which the synchronous cycles of higher periods are consecutively born and then lose their stability. Finally, we cross the line l_{cr} on which a chaotic synchronous attractor appears in the phase space.

Variation of σ_3 changes timescale of third unit, and can thus lead to desynchronization. Boundaries of the region inside which periodic synchronous oscillations exist are torus birth bifurcation lines T_i. However, no stable tori appears on these lines, only the periodic solutions near the saddle cycle loses its stability. As it was shown for the two-dimensional system [31, 132], this kind of transition is accompanied by a global bifurcation involving the homoclinic orbit of the

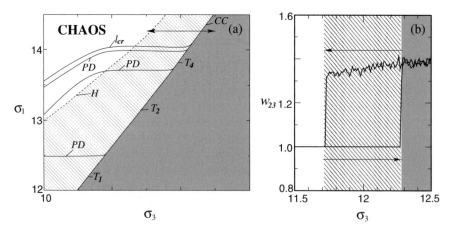

Fig. 8.33. (Color online) **a** Bifurcation diagram of coupled bacteria–viruses population (8.17) in the parameter plane (σ_1, σ_3). Inside *white area* only a synchronous attractor exists. Inside *grey area* only a non-synchronous attractor exists. In the *hatched area* both synchronous and non-synchronous attractors coexist. *PD*: lines of period-doubling bifurcations; T_i, $i = 1, 2, 4$: lines of Neimark–Sacker bifurcations; H: line of homoclinic bifurcation on which an non-synchronous attractor collides with a saddle limit cycle. l_{cr}: transition to chaos inside synchronization region; *CC*: crisis of the synchronous chaos. **b** Hysteresis in the dependence of the winding number w_{23} on σ_3 at $\sigma_1 = 14.25$. *Arrows* indicate the directions of change of the parameter σ_3

8.8 Homoclinic Synchronization of Chaos

saddle. Above $\sigma_1 \approx 14.07$ the variation of σ_3 produces a transition between synchronous and non-synchronous *chaotic* oscillations. Consider this transition in more detail. The difference between two types of chaos involved is illustrated in Fig. 8.34, where the Poincaré sections and the respective distributions of phase difference $\Delta\Phi = \Phi_3 - \Phi_2$ are shown. Here $\Phi_{2,3}$ are phases of oscillations in the second and the third pool, respectively, which are introduced by means of (8.10). In order to classify different chaotic regimes, we can use the mean return time[10] to a Poincaré secant surface. For this purpose, we define the Poincaré secant surface for each subsystem as $B_{2,3} = P_{2,3}/5$. The ratio of the mean return times provides the so-called winding number[11] w_{23}, which is a rational number in the case of a synchronous chaos, and an irrational one when the chaos is non-synchronous. w_{23} as a function of σ_3 is given in Fig. 8.33(b). Remarkably, this dependence has two overlapping branches. One of them lies on the line $w_{23} = 1$, while the other slightly changes with σ_3, assuming a range of values between 1.3 and 1.4 down to $\sigma_3 \approx 11.73$, where w_{23} drops abruptly

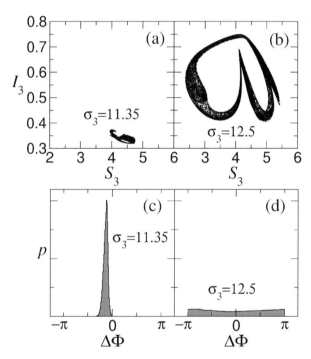

Fig. 8.34. Two different attractors in the phase space of the dynamical system (8.17) at $\sigma_1 = 14.25$, $\sigma_2 = 0$. **a, b** Poincaré sections defined by $B_3 = 0.11$: **a** of a synchronous chaotic attractor at $\sigma_3 = 11.35$; **b** of a non-synchronous chaotic attractor at $\sigma_3 = 12.5$. **c, d** Distributions of phase difference corresponding to: **c** synchronous chaos in **a**, **d** non-synchronous chaos in **b**

---
[10] Poincaré return times are defined in Sect. 6.1.
[11] For the definition of winding (rotation) number see Sect. 6.1.

232   8 Chaos Synchronization

down to unity. Thus, the value σ_3 at which the transition between synchronous and non-synchronous chaos occurs depends on the direction of variation of σ_3; i.e., there exists a range of parameter values at which a synchronous and a non-synchronous attractors coexist in the phase space. In Fig. 8.33(a), (b) this range is denoted as a hatched area.

This observation is consistent with the homoclinic synchronization mechanism that we have already illustrated for the regular oscillations in Chap. 5. Hence, we can assume that a similar mechanism is also realized in case of chaos. In order to verify this assumption, we fix $\sigma_1 = 14.25$ and calculate the Poincaré section at three different values of σ_3: on the homoclinic bifurcation curve H, in the middle of the coexistence area, and on the line of chaos crisis CC. The result is shown in Fig. 8.35(a)–(c). In the middle of coexistence area $\sigma_3 = 12.14$ two different chaotic attractors are quite well separated (b). However, with the decrease of σ_3, the non-synchronous (large) chaotic attractor approaches a saddle cycle separating two chaotic attractors in the phase space,[12] and on line H $\sigma_3 \approx 11.73$ these two objects collide as it is shown in (a). After this crisis the non-synchronous chaos becomes unstable. Namely, any trajectory that finds itself in the vicinity of this collision escapes towards the attractor corresponding to the synchronous chaos. A similar picture can be observed with the increase of σ_3. However, in this case the collision happens between the synchronous (small) attractor and the saddle cycle, see Fig. 8.35(c).

In order to examine these bifurcations more rigorously, we calculated the distance between the specified objects in the phase space. In Fig. 8.36 the smallest distance between each of the two chaotic attractors and the saddle cycle is plotted as a function of σ_3. With this, Fig. 8.36(a) illustrates the collision between the saddle limit cycle and the non-synchronous chaos, whereas Fig. 8.36(b) reflects the crisis of the synchronous chaos. In the inset the distance profile along the saddle cycle in the vicinity of the collision is given. 5000 points were recorded along the saddle cycle, and the smallest distance is shown for each point. As one can see, in both cases the chaotic attractors approach the saddle cycle and touch it at the point of bifurcation.

8.9 Effects of Noise on a Synchronized Chaos

In Sect. 7.11 we described the effects produced by noise that is applied to a periodic oscillator synchronized by periodic forcing. It has been demonstrated that while destroying either of two main types of synchronization, locking or suppression,[13] noise produces a new ordered motion whose regularity resonantly depends on noise intensity. A natural question to ask in this respect is: "What about noise effects on a synchronized chaos?" In this section we will try to answer this question.

---

[12] The stable manifolds of this saddle limit cycle form a boundary separating the basins of attraction of the synchronous and of the non-synchronous chaos.

[13] Provided that we are sufficiently close to the boundary of suppression region.

8.9 Effects of Noise on a Synchronized Chaos   233

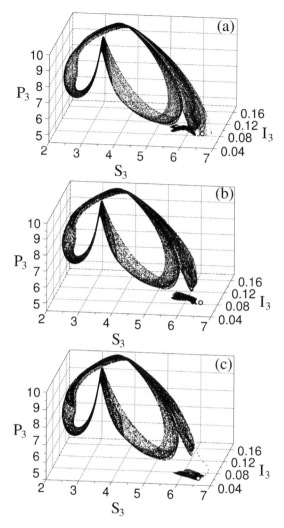

Fig. 8.35. Poincaré sections $B_3 = P_3/5$ of the two coexisting chaotic sets for **a** $\sigma_3 = 11.73$, **b** $\sigma_3 = 12.14$ and **c** $\sigma_3 = 12.25$. $\sigma_1 = 14.25$, $\sigma_2 = 0$. *Black dots* correspond to chaotic attractors, *grey empty circle* is a saddle periodic orbit involved in the bifurcation

8.9.1 Chaotic System Frequency-Locked by a Harmonic Signal

First, consider deterministically chaotic oscillations which are synchronized via frequency-locking by an external periodic forcing. As shown in [18] and in Sect. 8.6, this type of synchronization is associated with an accumulation of saddle-node bifurcations of unstable periodic orbits embedded in the chaotic attractor. As an example, we examine the Rössler oscillator under external harmonic forcing, described by the

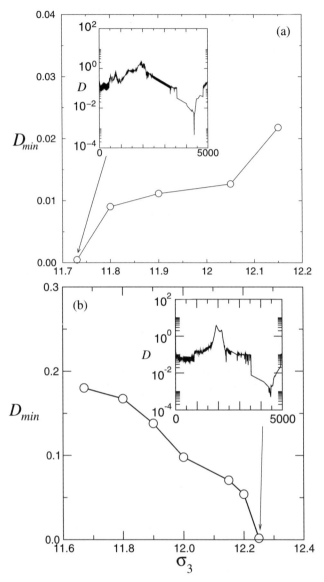

Fig. 8.36. a Smallest distance D_{min} between the saddle cycle and the **a** non-synchronous chaotic attractor and **b** synchronous chaotic attractor for $\sigma_1 = 14.25$ as a function of σ_3. The *inset* shows the distance profile along the saddle cycle in **a** for non-synchronous chaos at $\sigma_3 = 11.73$; in **b** for synchronous chaos at $\sigma_3 = 12.25$

equations

$$\dot{x} = -y - z + B\sin(\Omega t) + \tilde{D}\xi(t),$$
$$\dot{y} = x + \alpha y, \quad (8.19)$$
$$\dot{z} = \alpha + z(x - \mu).$$

8.9 Effects of Noise on a Synchronized Chaos   235

Here, α and μ are some parameters, $\xi(t)$ is Gaussian white noise of zero mean and unity variance, and $\tilde{D}$ is the noise intensity. Without noise ($\tilde{D} = 0$) and forcing ($B = 0$), at $\alpha = 0.2$ and $\mu = 6.5$ this system demonstrates chaotic oscillations with the attractor of Feigenbaum type shown in Fig. 8.1. When forcing with $B = 0.1$ and $\Omega = 1.061$ is applied, the system is $1:1$ frequency-locked. The Fourier spectrum of the oscillations is continuous, but contains one peak at the forcing frequency (Fig. 8.37(a)).

As noise intensity increases from zero, a new peak appears in the close vicinity of the first one. In Figs. 8.37(b), (c) and (d), spectra are shown for gradually increasing noise intensities $\tilde{D}$. It is evident that a peak appears on the right-hand side of the main one, grows (Fig. 8.37(b)), narrows (Fig. 8.37(c)), and then widens and decreases in amplitude again (Fig. 8.37(d)).

By analogy with periodic oscillations, the distance δ between the noise-induced peak and the one at forcing frequency Ω was estimated (Fig. 8.38(a)), showing a monotonic increase of the absolute value of δ with the increasing $\tilde{D}$. This means that the noise-induced peak gradually moves away from the main one as noise becomes stronger. The coherence β of the noise-induced motion was estimated using the approach explained in Sect. 7.11. Figure 8.38(b) shows β as a function of noise intensity $\tilde{D}$. It has a resonant character, with β taking its maximal value at an optimal noise intensity $\tilde{D} \approx 0.2$.

Based on these observations, we conclude that noise destroys the frequency-locked state of Feigenbaum chaos in a manner that is in many respects the same as in the case of periodic oscillations. This can be accounted for empirically in the following way. It is known that for Feigenbaum chaos, which is characterized by the presence of a well-resolved peak in the spectrum, the instantaneous amplitude $A(t)$ and phase $\Phi(t)$ of the oscillations can be introduced according to (8.2). Substituting the latter into (8.19) and making transformations similar to those given in [277, 289] and also considered in Chap. 7, one can rewrite (8.19) as follows:

$$\dot{A} = \alpha A + \frac{C}{2} \sin\phi + \Psi_a(\phi, z, A, t) + \xi_a, \tag{8.20}$$

$$\dot{\phi} = \Delta - \frac{C}{2A} \cos\phi + \Psi_\phi(\phi, z, A, t) + \xi_\phi, \tag{8.21}$$

$$\dot{z} = \alpha + z\big(A\cos(\phi + \Omega t) + \mu\big).$$

Here $\phi = \Phi - \Omega t$ is the phase difference between the forced oscillations and the forcing, Δ is some effective detuning, $\Psi_{a,\phi}$ are certain non-linear functions, and $\xi_{a,\phi}$ are two independent sources of Gaussian white noise whose intensities are completely defined by the intensity of the original noise ξ. Without the loss of generality, one can treat $\Psi_{a,\phi}$ as additional limited "random" forces that are defined by the chaotic dynamics of the system. Then the system of equations (8.20)–(8.21) could be considered as those describing the dynamics of a limit cycle oscillator under periodic external perturbation in the presence of *two* kinds of random forces: $\Psi_{a,\phi}$ which would be a *bounded* correlated random force, and the external noise $\xi_{a,\phi}$ which is *unbounded*. The term "bounded" means literally that the given force

236 8 Chaos Synchronization

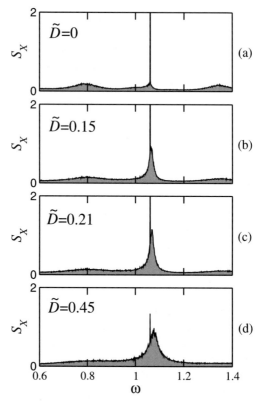

Fig. 8.37. Spectra of oscillations in Rössler system (8.19) in a chaotic regime, which is synchronized by external periodic forcing via phase (frequency) locking mechanism, at different noise intensities $\tilde{D}$. Note that the main peak is exactly at the frequency Ω of forcing

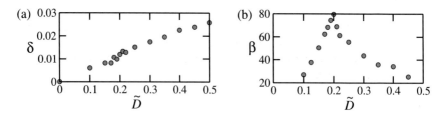

Fig. 8.38. Characteristics of a Rössler system (8.19) frequency-locked by harmonic forcing, plotted as functions of noise intensity $\tilde{D}$: **a** distance δ between the noise-induced and main peaks; **b** coherence β of noise-induced peak. For details of computation, see text

can take values only within a certain finite interval, while "unbounded" means that the force can take any values, whatever large, although perhaps with different probabilities.

Even in the absence of external noise $\xi_{a,\phi}$, the effective limit cycle in (8.20)–(8.21) will be smeared by the random forces $\Psi_{a,\phi}$. Synchronous chaos occurs when the largest possible values $\Psi_{a,\phi}$ are still not large enough to induce phase slips (or such that phase slips are rare). It is the addition of external noise ξ that induces phase slips. Hence, the effect of noise on a synchronized Feigenbaum chaos is at least qualitatively similar to its effect on periodic oscillations.

8.9.2 Periodic System Suppressed by Chaotic Forcing

In Sect. 8.6 it was described how chaotic forcing applied to a periodic oscillator can suppress natural oscillations in the latter leading to a synchronized state. Consider (8.14) that describe the van der Pol oscillator that are forced by Rössler system (8.13). The parameters of Rössler system are fixed as $\alpha = \beta = 0.2$, $\mu = 6.5$ and $\omega_1 = 1$, at which it demonstrates a well-developed one-band chaos. Non-linearity ε (equivalent to λ in (7.116)) in van der Pol oscillator was set to 0.1. The parameters of unidirectional coupling between the two systems were set as $\gamma = 0.04$ and $\omega_2 = 1.16$, at which oscillations in van der Pol system are synchronized by external chaotic forcing via suppression mechanism illustrated in Figs. 8.17 and 8.19. The spectrum of x_2 in the synchronized state is given in Fig. 8.39(a). Note that the frequency of the unforced oscillations in the van der Pol system was $\omega \approx 1.16$, which is larger than the frequency of chaotic forcing ≈ 1.079 at which the system has a peak in the synchronized state.

Let us add a random term $\tilde{D}\xi(t)$ into the first equation of (8.14) for $\dot{x}_2$ and follow the evolution of spectra as one increases the noise intensity $\tilde{D}$ from zero to some finite value (Fig. 8.39). Again, a new peak appears to the right of the main one, which initially grows with noise and becomes sharper, reaches the narrowest width at $\tilde{D} = 0.32$, and then widens and decreases in height. Interestingly, unlike in the case of locking illustrated in Fig. 8.38, the peak due to noise appears at a finite distance from the main one, and with noise growing stronger *approaches* the main peak instead of moving further away, see Fig. 8.40(a) where the distance δ between the peaks is shown. At the same time, the coherence β estimated from the noise-induced peak as described in Sect. 7.11 displays a resonant behavior taking its maximal value at $\tilde{D} = 0.4$.

8.10 Summary

It appears that in spite of the significant complexity of chaotic oscillations as compared to periodic ones, their synchronization obeys the same fundamental laws. Namely, the mechanisms of chaos synchronization are the same as those of synchronization of periodic oscillations: frequency (phase) locking, suppression and crisis (homoclinics). Moreover, the effect produced by the external noise applied to a chaotic system in a synchronized state is very much the same as the effect of

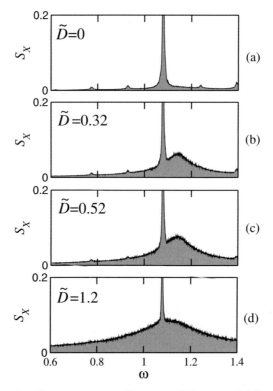

Fig. 8.39. Spectra of oscillations in a periodic van der Pol system (8.14), which is synchronized by external *chaotic* forcing coming from Rössler system (8.13), by suppression mechanism, at different noise intensities $\tilde{D}$. The frequency of unforced oscillations in van der Pol system (8.14) was $\omega \approx 1.16$

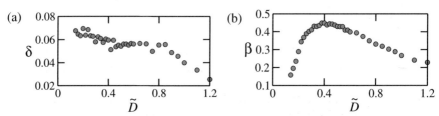

Fig. 8.40. Characteristics of a van der Pol oscillator (8.14) synchronized via the suppression mechanism, by chaotic forcing from a Rössler system, plotted as functions of noise intensity $\tilde{D}$: **a** distance δ between the main peak at the frequency of forcing and the noise-induced one; **b** coherence β of noise-induced peak

noise on a synchronized periodic system. In both cases, noise creates a new ordered motion that has the largest regularity at some moderate value of noise intensity.

9 Synchronization of Noise-Induced Oscillations

Today it is widely recognized that noise can induce oscillations. But what exactly does one mean by that?

Suppose we apply Gaussian white noise[1] to a *linear* oscillator with dissipation.[2] In terms of radioelectronics one would say that we allow a signal with continuous spectrum to pass through a band-pass filter. As a result, the spectrum of the signal at the output will have a peak, which will be the more pronounced, the less the power losses are in the oscillator. A realization of these oscillations will look similar to the ones in noisy self-oscillations. Can we say that these oscillations are *induced* by noise?

The answer is "no," because a linear filter, which in our example is represented by an oscillator with dissipation, is only able to weaken the components with different frequencies which are already present in the original signal. With this, the power of the signal at the output is always less than at the input.

In *non-linear* systems the situation is crucially different. There, noise is able to influence the way the power is taken from its source,[3] and also to play the role of the energy source itself, either completely, or partly. As a result, oscillations *induced* by noise acquire the properties that are similar to the ones of self-oscillations. Namely,

---
[1] See definition of Gaussian white noise in Sect. 7.1.
[2] See discussion on dissipation in Sect. 2.3.2.
[3] See Sect. 2.3.2 for the discussion on the power source in an oscillating system.

240   9 Synchronization of Noise-Induced Oscillations

their amplitude and frequency depend non-linearly on the noise intensity, and are also defined by the properties of this particular system.

With this, one can single out two special cases:

1. *Noise-activated oscillations.* Without noise the system is in principle capable of oscillating on its own, although perhaps its oscillations will decay in time. This is possible thanks to a special structure of trajectories in its phase space. Normally, when the system is in its relaxed state these regions of the phase space are not visited, and the oscillations are not observed. However, noise can kick the system towards the respective region of the phase space thus *activating* its oscillatory properties which were already there.

   An example is a system just below a saddle-node bifurcation of a stable and a saddle periodic orbits, when there is no pair of cycles in the phase space yet (or already), but there is a condensation of phase trajectories (a "ghost" of these cycles) on which the phase point can spend a significant amount of time after having been thrown there by noise.

   Importantly, the time scale of these oscillations depends weakly on noise intensity because it is defined mostly by the inherent properties of the system. At the same time, the regularity of these oscillations, i.e., the degree of their closeness to periodic oscillations is controlled directly by the noise intensity.

2. *Noise-induced oscillations.* Without noise no repetitive oscillations in the system are possible even in the form of a transient process. The trajectories in the phase space do not form loops or closed unstable trajectories which could be highlighted (activated) by noise. However, the structure of the phase space is such that relatively small fluctuations can push the phase trajectory on a pseudo-orbit, which would be an almost closed trajectory [127].

   In terms of the realization of the process, it looks like a single splash induced by noise. Upon the return to the original state the system can again be thrown onto the pseudo-orbit. Thus, a sequence of pulses arise which occur irregularly in time. It is important that the frequency of splashes is directly controlled by the intensity of noise, although it does depend on the time it takes the phase point to return to its initial state. Thus noise gives birth to a new time scale which was absent in the deterministic (noise-free) system. Such oscillations can be classified as *noise-induced*, and the systems where they arise are called excitable [170].

Like any classification, the one given above is an idealization. A particular system can behave in a complex manner combining the features of both mechanisms. However, the partition described above is useful for understanding of different manifestations of the dynamics that arises due to noise.

Non-linear effects caused by noise have a short, but impetuous history. The first kind of systems where the effects arising due to noise were studied were bistable systems, and the phenomenon of interest was stochastic resonance (SR) [21, 130]. The process of switchings between the two states of a bistable system does not have a pronounced maximum in the spectrum. In spite of that, the study of this process in

9 Synchronization of Noise-Induced Oscillations 241

terms of the phase of switchings has allowed one to discover both the mutual synchronization of switchings in coupled bistable systems [189], and the effect which looks like phase locking of switchings by an external forcing [266]. Further studies have confirmed that the SR effect is indeed accompanied by synchronization, and it can be characterized in terms of phase diffusion[4] [82, 83, 192, 193, 268].

Somewhat later another effect was identified, namely, that of coherence resonance (CR). CR consists in that the degree of regularity (closeness to a periodic process) of oscillations induced by noise has a maximum at some optimal noise intensity. It was first discovered in a situation when a pair of fixed points, a stable and a saddle, lie on the limit cycle [87, 241]. Later on CR was studied in the systems close to local bifurcations[5] [42, 166, 191]. In [210] a simple explanation of CR was proposed based on an excitable system. In [221] CR was studied in an electronic model of a monovibrator which has no oscillatory solutions in the absence of noise, whatever the values of its parameters are. CR has a special value for neurodynamics, since the excitable regime is one of the main regimes in which nervous cells operate.

CR is accompanied by a formation of a pronounced peak in the spectrum of oscillations[6] induced by noise. This allows one to consider the problem of synchronization of such oscillations in classical terms by analyzing phase (frequency) locking and suppression of oscillations. In this chapter we do this based on the concept of a stochastic limit cycle introduced below.

At present synchronization of stochastic oscillations is a hot topic in non-linear dynamics [170], and its applications embrace synchronization of applause [196] to synchronization of tunneling in quantum systems [96].

Stochastic Limit Cycle

Although no deterministic periodic orbits are involved in the formation of noise-induced trajectories in the phase space, the phase portrait itself may look like a smeared-out limit cycle. Moreover, the notion of a "stochastic limit cycle" was proposed in [287, 288]. A stochastic limit cycle can be formally introduced if one considers an appropriate projection of the phase portrait on some manifold (plane or surface), and calculates a two-dimensional probability distribution density on this manifold. If this distribution has a shape reminiscent of a crater, at least qualitatively, one can define a closed curve through its ridge (highest points), and call this a stochastic limit cycle.

One can also introduce an average period for such a limit cycle. Of course, both the shape and the period of a stochastic limit cycle will be defined only in a statistical, averaged, sense. In addition, the motion around the stochastic limit cycle can be smeared out to a smaller or larger extent, and also the instantaneous periods of oscillations can deviate from the average period more or less. This means that the

---

[4] See definition of phase diffusion in Sect. 7.9.
[5] An example is given in Sect. 7.11.
[6] For the definition of a spectrum see Sect. 7.1.

noise-induced motion can have different degrees of regularity. Hence, noise-induced motion does possess a characteristic shape and time scale of its oscillations.

9.1 Noise-Induced Oscillations

Non-linear systems perturbed by noise display a wide spectrum of complex phenomena, ranging from noise-induced chaos [76, 177] and noise-induced order [114, 238] to stochastic ratchets [106, 129].

From the point of view of non-linear dynamics, one of the interesting effects of noise is to wash out some of the detailed structures in the bifurcation diagrams [66, 286]. Application of noise to a period-doubling system will truncate the bifurcation sequence by opening a so-called bifurcation gap around the accumulation point for the period-doubling cascade. As soon as the noise amplitude becomes comparable with the trajectory splitting for a (high-periodic) orbit, the subsequent bifurcations can no longer be observed.

However, noise can also play a constructive role by activating dynamics that is not observed in a noise-free system. The simplest example of this phenomenon is a linear damped oscillator. While being forced by noise, it exhibits a sequence (superposition) of relaxation processes converging towards the equilibrium point. Schematically, this mechanism can be depicted as illustrated Fig. 9.1(a). The resulting behavior is characterized by a pronounced global maximum in the power spectrum. Hence, there is a frequency that can be assigned to noise-induced behavior. We emphasize that a damped linear oscillator only acts a filter for the noisy forcing signal. Thus, there is no self-sustained dynamics.

Quite a different situation can be observed for non-linear systems. Noise can have different effect when acting on oscillatory or on excitable non-linear systems. In the deterministic case the oscillatory system already possesses an eigenfrequency that can be modified by the random forcing. For example, the power spectrum displayed by the system after a bifurcation may be visible even before the bifurcation if noise is applied [166, 191, 298]. Thus, noisy precursor of the bifurcation, i.e., a noise-activated time scale is observed. The phase portrait for noise-activated dynamics near an Andronov–Hopf bifurcation is similar to the portrait in Fig. 9.1(a) but the power of the output signal can be larger than the input noise intensity.

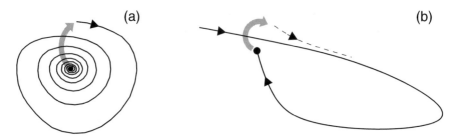

Fig. 9.1. (Color online) Two main mechanisms for noise-induced oscillations

The influence of noise on excitable systems is even more dramatic. According to the definition by Izhikevich [126]: A dynamical system having a stable equilibrium is excitable if there is a large amplitude periodic pseudo-orbit passing near the equilibrium as shown in Fig. 9.1(b). If system is excited beyond the pseudo-orbit (the excitation threshold) it will perform an excursion in the phase space of the system (a spike) and return into the vicinity of its stable equilibrium point. In this case, the pseudo-orbit plays the role of separatrix between the subthreshold behavior (inside the loop) and excited behavior (beyond the loop).

It was found that the rhythmicity of noise-induced events (spikes in neural systems, for example) depends significantly on the noise intensity. There is an optimal noisy level at which the regularity (closeness to periodic processes) is maximal. Such a non-linear effect is known as coherence resonance and will be considered in detail below. In many ways, systems with noise-induced oscillations behave like noisy limit cycle oscillators. In the following sections we shall study the different types of synchronization that can arise between such systems using the concept of coherence resonance oscillator [105, 224, 271].

9.2 Models

To examine the synchronization of noise-induced oscillations we shall consider two representative models, namely, the Morris–Lecar model that describes the spiking and refractory dynamics of a nerve cell, and the electronic circuit (monovibrator) that likewise belongs to the class of excitable systems.

9.2.1 Morris–Lecar Model

The Morris–Lecar (ML) model [181] is a simplification of the original Hodgkin–Huxley model [116] which describes the spiking and refractory properties of biological neurons. The Morris–Lecar model includes a calcium current generating fast action potentials and a delayed rectifier potassium current. To maintain a constant potential in the resting state, a leak current is also taken into account. The ML-model is two-dimensional and does not display bursting dynamics, period doublings or chaos. However, application of noise of a proper magnitude can bring the system across a separatrix in the phase space, upon which it spikes and returns to the stable equilibrium point.

To study the synchronization phenomena, we consider two diffusively coupled ML models. Equations may be written as

$$
\begin{aligned}
\frac{dv_{1,2}}{dt} &= I_{\text{ion}}(v_{1,2}, w_{1,2}) + I + D_{1,2}\xi_{1,2}(t) + g(v_{n,1} - v_{1,2}), \\
\frac{dw_{1,2}}{dt} &= \epsilon \frac{w_\infty(v_{1,2}) - w_{1,2}}{\tau_\infty(v_{1,2})},
\end{aligned}
\tag{9.1}
$$

244   9 Synchronization of Noise-Induced Oscillations

where

$$I_{\text{ion}}(v, w) = g_{\text{Ca}}m_\infty(v)(v_{\text{Ca}} - v) + g_{\text{K}}w(v_{\text{K}} - v) + g_{\text{L}}(v_{\text{L}} - v),$$
$$m_\infty(v) = \left[1 + \tanh\{(v - v_a)/v_b\}\right]/2,$$
$$w_\infty(v) = \left[1 + \tanh\{(v - v_c)/v_d\}\right]/2,$$
$$\tau_\infty(v) = 1/\cosh\{(v - v_c)/(2v_d)\}.$$

Here, v denotes the transmembrane voltage of the neuron and w represents the activation of the potassium current. I is the external stimulus current and $\xi_{1,2}$ denote uncorrelated sources of Gaussian noise with intensities $D_{1,2}$. The last term in the first line of (9.1) represents the diffusive interaction between the two cells with a coupling strength g. The parameter set used in our simulations is: $I = 0.23$, $v_a = -0.01$, $v_b = 0.15$, $v_c = 0.0$, $v_d = 0.3$, $g_{\text{Ca}} = 1.1$, $g_{\text{K}} = 2.0$, $g_{\text{L}} = 0.5$, $v_{\text{Ca}} = 1.0$, $v_{\text{K}} = -0.7$, $v_{\text{L}} = -0.5$, and the time separation parameter $\epsilon = 0.02$. v_{Ca}, v_{K}, and v_{L} represent the reversal potentials associated with the different currents, and g_{Ca}, g_{K}, and g_{L} are the corresponding conductances. For a detailed explanation of the remaining parameters we refer the reader to the original literature [181]. The subscript n in (9.1) determines the different types of interaction. If $n = 1$, a unidirectional interaction is realized; the first system being the "master" and the second the "slave." If $n = 2$, the systems are mutually coupled.

9.2.2 Monovibrator Circuit

Our experimental studies are performed with a monovibrator circuit that generates a single electric impulse whenever the external signal exceeds a threshold level [221]. The electric scheme of the two coupled monovibrator circuits is shown in Fig. 9.2. This system is described by the following dynamical equations:

$$\epsilon \frac{dx_{1,2}}{dt} = \chi\left(x_{1,2} - y_{1,2} - \left(D_{1,2}\xi_{1,2}(t) + \alpha x_{1,2} + \gamma v_b\right)\right) - y_{1,2},$$
$$\frac{dy_{1,2}}{dt} = x_{1,2} - y_{1,2} + g(x_{1,2} - y_{1,2} - x_{2,1} + y_{2,1}),$$

(9.2)

where $x_{1,2}$ are voltages at the output of the operational amplifier and $y_{1,2}$ are voltage drop across the capacitor C. The constants α and γ are positive and defined by the value of resistors R_1, R_2, R_3, and R_f. v_b represents the normalized threshold voltage. The function χ is a sign function which takes values of $+1$ and -1 for positive and negative arguments, respectively. The independent noise sources $\xi_{1,2}$ with noise intensities $D_{1,2}$ are introduced.

9.3 Coherence Resonance Oscillator

Being forced by white Gaussian noise, both of the above models manifest excitable dynamics. Namely, there is a continuous sequence of spikes for the ML model and a sequence of electric impulses for the monovibrator circuit (Fig. 9.3). Remarkably,

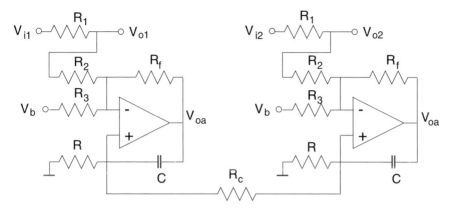

Fig. 9.2. The electrical scheme for the coupled monovibrator circuit. Both units are identical but the noise sources (V_{i1} and V_{i2}) are independent

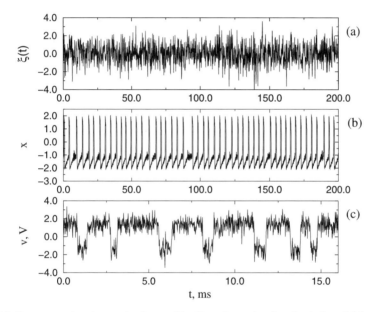

Fig. 9.3. Representative time series for **a** white Gaussian noise, **b** noise-induced firing in the Morris–Lecar model, and **c** noise-induced oscillations in an electronic circuit

the intervals between the noise-induced events seem to be quite regular rather than random.

Figure 9.4(a) displays the typical power spectra observed for the relaxation-type neuron model (9.1) with vanishing interaction $g = 0$ at different noise intensities D. Each spectrum possesses a well-defined global maximum which may be associated with the natural frequency of the noise-induced oscillations. The regularized behavior is observed within a finite range of noise intensities.

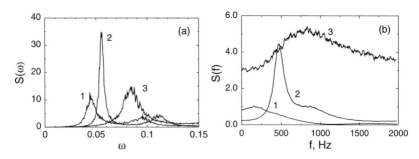

Fig. 9.4. The evolution of power spectra **a** for the noise-driven Morris–Lecar model (curves *1*, *2*, and *3* in **a** correspond to $D = 0.0001, 0.001$, and 0.01 V^2, respectively) and **b** for the noise-driven monovibrator circuit (curves *1*, *2*, and *3* in **b** correspond to $D = 0.015, 0.1$, and 0.6 V^2, respectively)

It is interesting to observe a similar influence of noise on the features of the power spectrum for the monovibrator system (9.2). For small noise intensity ($D \ll 0.1$ V^2), the monovibrator generates impulses of duration $\tau \approx \tau_0 = -RC \ln\{(V_b/E + 1)/2\}$. The time intervals between the impulses are much longer than τ. Thus, the respective power spectrum results from a superposition of randomly appearing impulses. The smooth and broad peak at low frequency can be observed in this case (Fig. 9.4(b), curve 1).

For an optimal noise strength $D \approx 0.1$ V^2 the pauses between impulses are approximately equal to their duration. The corresponding peak in the power spectrum is sharp and relatively high (Fig. 9.4(b), curve 2). Finally, for the strong noise, the pauses between impulse onsets tends to zero because the monovibrator is immediately pushed out from the equilibrium state. The peak in power spectrum is absorbed by the increasing level of noise background (Fig. 9.4(b), curve 3).

Thus, for both systems we observe a noise-induced time scale of the system (pronounced peak in power spectrum) which is not a noisy precursor of deterministic behavior. The described non-linear effect is known as coherence resonance [87, 241] and manifests itself in a rather regular oscillatory response of an excitable system to the application of noise of a proper magnitude. In contrast to stochastic resonance, there is no external forcing involved. However, the excitable system exhibits a characteristic time constant associated with the duration of a spike (or impulse) when the system is excited. Pikovsky and Kurths [210] used this observation to explain the coherence resonance in terms of a different noise dependence of the activation (or excitation) and excursion (or relaxation) times.

To characterize the coherence behavior (i.e., the degree of its regularity) one uses a quantity that can be interpreted as the signal-to-noise ratio [87, 241]:

$$\beta = h\omega_p / \Delta\omega_p. \quad (9.3)$$

Here, ω_p is the peak frequency in the power spectrum of the noise excited system, $\Delta\omega_p$ is the width of the peak, and $h = H_p/H_b$ is the peak height normalized with respect to the noise background (Fig. 9.5). Note that $\omega_p/\Delta\omega_p$ is the familiar inverse

9.3 Coherence Resonance Oscillator

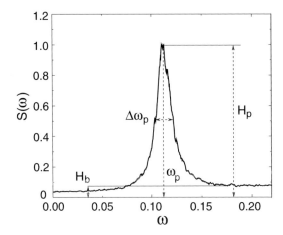

Fig. 9.5. How to measure regularity

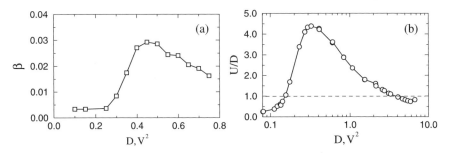

Fig. 9.6. Electronic experiment on monovibrator circuit. **a** Regularity β of noise-induced oscillations and **b** output power vs noise intensity D

quality factor Q of a signal [278]. In the following sections, β will be referred to as a measure of regularity.

As a function of noise intensity, the regularity β (Fig. 9.6(a)) clearly demonstrates a coherence resonance maximum at a finite noise intensity. As discussed above, this can be explained in terms of an optimal balance between the mean duration of a impulse generated by the monovibrator and the mean duration of a pause [210, 221]. For strong noise the pauses between impulse onsets tend to zero because the monovibrator is immediately pushed out from the equilibrium state. Strong noise can also disrupt the recharging process of the capacitor C. Thus, the impulse duration attains a random value. This leads to decreasing measure of regularity β when the noise intensity D increases beyond 0.1 V^2.

Being related to the dynamics of excitable system at an optimal noise intensity, coherence resonance can be regarded as activating non-linear properties of the system. Let us examine the corresponding aspects of noise-induced oscillations. Figure 9.6(b) shows the relation between the output signal power U and the input noise intensity D. The dashed line at 1.0 indicates the equal output and input power. It is

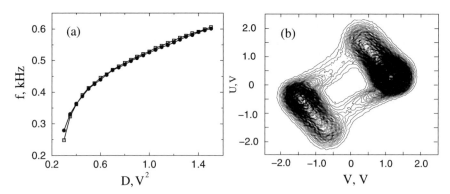

Fig. 9.7. a Peak and mean frequencies of noise-induced oscillations vs noise intensity D; **b** two-dimensional probability density distribution for noise-induced oscillations shows the ring structure similar to noisy limit cycle

clearly seen that there is an interval of noise intensity where the U/D ratio exceeds one. Hence, the non-linear system not only transforms the input noise signal into impulses (spikes) but also spends some internal energy (for the electronic circuit this is provided by the power supply). This is similar to self-sustained system with one important difference: the power release is controlled by the noise intensity.

Figure 9.7(a) illustrates that the peak and mean frequencies of the signal coincide and grow as the noise strength increases. Thus, we observe a noise-induced time scale of the system but not a noise activated deterministic behavior. Let us consider the geometrical image of such behavior. The two-dimensional probability density distribution has a clear ring-like structure (Fig. 9.7(b)). This is very similar to the case of noisy self-sustained relaxation oscillations with segments of fast and slow motion. Such a structure is particularly pronounced when the noise intensity is in the optimal range and disappears both for too weak and for too strong noise. Since the observed structure reveals the geometry of "stochastic limit cycle" [287, 288], one can introduce a phase of noise-induced oscillations via a position on the cycle.

We can conclude that a noise-driven excitable system in the regime of coherence resonance can be considered as a "coherence resonance oscillator" whose behavior is characterized (i) by a peak frequency governed by the noise intensity and (ii) by a phase defined as the position on a stochastic limit cycle. Hence, the question naturally arises [105]: To what extent interacting nonidentical coherence resonance oscillators can adjust their motion in accordance with one another so as to attain a form of synchronization?

9.4 Frequency and Phase Locking

In this section, we shall study the synchronization of coupled *non-identical excitable systems* each operating in a regime of coherence resonance. The noise intensity governs the frequency of the noise-induced oscillations and can, therefore, be considered as a frequency mismatch parameter. The transition from the non-synchronous to the synchronous state is signaled by the merging of the peak frequencies in the power spectra and also by an evolution of the distribution of instantaneous phase differences. With a small mismatch, the transition occurs via a frequency locking of noise-induced oscillations. For large mismatch, the transition is related to the suppression of the peak frequency.

9.4.1 Frequency Locking: Electronic Experiment

Let us now analyze synchronization of two diffusively coupled coherence resonance oscillators. To investigate the effect of frequency mismatch on the synchronization of CR oscillators, the noise intensity of the second oscillator is chosen to be different from that of the first system. We refer to D_2 as a mismatch parameter.

In Fig. 9.8 (left panels), the evolution of the power spectra is plotted as a function of the noise intensity D_2 for the coupled electronic monovibrators. For $D_2 = D_1 = 0.45$ V^2 both excitable units are identical and peaks in their power spectra coincide (top left panel). With increasing D_2 (middle and bottom left panels) peak in power

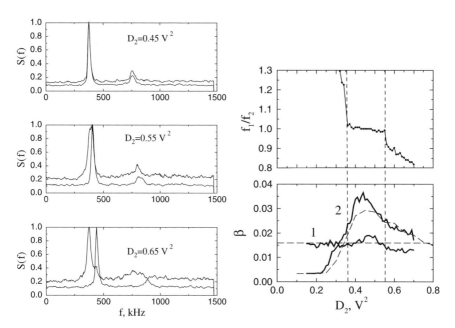

Fig. 9.8. Frequency locking observed in the electronic experiment. **Left panels**: Evolution of the normalized power spectra as the D_2 increases. **Right top panel**: The ratio of the peak frequencies f_1/f_2 (winding number) stabilizes near 1.0 within a range of D_2. **Right bottom panel**: The degree of regularity β for the second system (solid curve 2) displays the maximum in the frequency locking region of D_2. Note, the regularity for the first system (solid curve 1) also weakly increases even though D_1 is fixed. $D_1 = 0.45$ V^2 and $g = 0.0125$

spectra of the second monovibrator moves to the right from the initial position. Frequencies become unlocked.

In Fig. 9.8 (right top panel) a frequency-locked region is easily identified within a certain range of the noise intensity D_2, where the ratio of the peak frequencies f_1/f_2 (winding number) stabilizes near 1.0. In the right bottom panel the degree of regularity β is plotted versus the D_2 for both monovibrators. It is clearly seen, that for the second system (curve 2) β displays a maximum in the frequency locking region of D_2. The regularity for the first system (curve 1) also increases weakly although of D_1 was fixed. Note, the $D_2 \approx 0.45$ V^2 is the optimal noise intensity for the second subsystem. Thus, the observed gain of regularity when D_2 approaches that value is expected. However, Fig. 9.8 indicates an important effect: the maximal achieved degree of regularity in coupled subsystems is higher that the maximal individual value for each uncoupled monovibrator (as indicated by the dashed curves in right bottom panel). The latter means that the noise-induced oscillations are more regular in the regime of stochastic frequency locking [226]. This can be regarded as an example of array-enhanced coherence resonance.

The frequency-locked interval tends to become broader as the coupling strength g is increased. In Fig. 9.9 it is shown that there is the triangular-shaped zone in (D_2, g) parameter plane where the frequencies of noise-induced oscillations are locked. The latter is determined by the condition that winding number f_1/f_2 has to

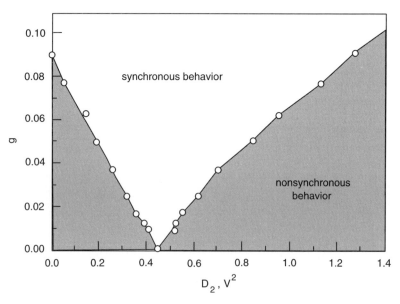

Fig. 9.9. Synchronization region for two coupled monovibrators. The noise intensity D_2 effectively plays the role of a frequency mismatch ($D_1 = 0.45$ V^2). *White triangular-shaped area* corresponds to the frequency locked regime determined by the condition $|f_1/f_2 - 1| < 0.002$

be sufficiently close to unit. More precisely, the expression $|f_1/f_2 - 1| < 0.002$ was used to diagnose whether the frequencies were locked.

The resulting area of stochastic synchronization is resembles the well-known Arnold tongue. This provides one more evidence that the behavior of the noisy excitable systems is in many ways similar to the self-sustained dynamics.

9.4.2 Phase Locking: Coupled Morris–Lecar Models

Similar results were observed in numerical simulations of the coupled ML system ((9.1) with $n = 2$). In Fig. 9.10, we have plotted the phase diagram in the two-dimensional parameter space spanned by the coupling strength g and the mismatch parameter D_2. The synchronization region which clearly resembles the Arnold tongue was obtained by the condition of closeness of the peak frequencies $|\omega_1 - \omega_2| < \text{const} = 0.0002$. We analyze the instantaneous phases of the two ML oscillators to provide an alternative diagnostics of synchronization.

Neiman et al. [192] and Rosenblum et al. [247] showed how the instantaneous phases of stochastic oscillations can be locked. Once instantaneous phases are defined for the CR oscillators, they can be used to detect synchronization of two coupled CR oscillators. According to Refs. [287, 288] the stochastic limit cycle was defined by connecting the most likely escape trajectory out of a stationary point with the most likely return trajectory back to that point. The system's state on this circular trajectory could be described in terms of phase-like variables. The instanta-

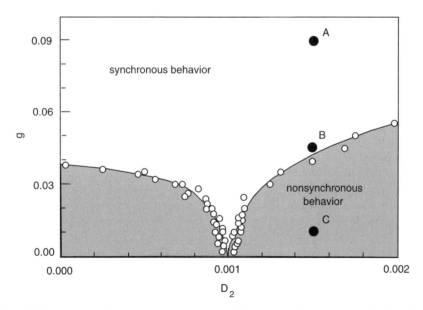

Fig. 9.10. Synchronization region for two coupled ML models. The noise intensity D_2 effectively plays the role of a frequency mismatch ($D_1 = 0.001$).

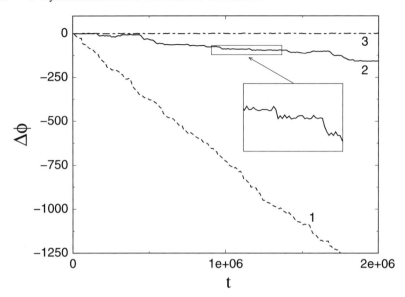

Fig. 9.11. Variation of the phase difference in two coupled Morris–Lecar models as a function of time for non-synchronous ($g = 0.02$), nearly synchronous ($g = 0.035$) and synchronous ($g = 0.08$) states. $D_2 = 0.00075$. The phase slips of 2π for the nearly synchronous regime are clearly seen in the enlarged *inset*

neous phase can be defined as [247]: $\phi(t) = 2\pi(t - \tau_k)/(\tau_{k+1} - \tau_k) + 2\pi k$, where τ_k is the time of the kth firing.

Based on the phase variable for each ML system, the instantaneous phase difference is specified as $\Delta\phi = \phi_1 - \phi_2$. As the coupling is increased, for a given frequency mismatch $\Delta\Omega$, we observe a transition from a regime where phase difference grows ($\Delta\phi \sim \Delta\Omega t$) to a synchronous state where the phase difference remains bounded, but oscillates around some mean value (Fig. 9.11). Hence, there is no average (or long term) phase drift. Phase locking for noisy systems can be observed during a long but *finite* time [192, 278]. Therefore, it has to be determined a priori how long the phases should be locked (in average) to assert that a noisy system is effectively synchronized. We assume that the stochastic oscillations are synchronous if no 2π phase slip occurs during 50 000 periods.

Figure 9.12 illustrates the distribution function of the phase differences (measured during 50 000 time periods) and the Poincaré section for three discernible regimes (corresponding to the points A, B and C in Fig. 9.10, respectively). Inside the synchronization region (point A), the Poincaré section is concentrated in a small area (Fig. 9.12(a)) and the distribution density of $\Delta\phi$ appears to be limited to a finite range near a vanishing phase difference. But outside the synchronization region (point C), the Poincaré section is completely different and takes the form of a ring in the phase space of the system (Fig. 9.12(c)). Moreover, the distribution of the phase differences is nearly homogeneous over 2π. At the boundary of synchroniza-

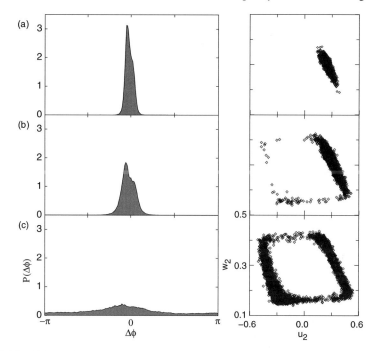

Fig. 9.12. (Color online) The distributions of the phase difference and Poincaré sections (*insets*) for two coupled ML oscillators: **a** inside the synchronization region ($g = 0.09$), **b** near the boundary ($g = 0.045$), and **c** outside this region ($g = 0.01$). $D_1 = 0.01$ and $D_2 = 0.0015$. The Poincaré section is specified by the condition $\omega_1 = 0.35$. From these plots, one can draw an analogy with the transition from a torus to a limit cycle in the deterministic case

tion (point *B*), the Poincaré section indicates a closed curve, but it is not equally dense everywhere (Fig. 9.12(b)). These results clearly allow us to draw an analogy between the transition from an ergodic torus to a limit cycle in the deterministic case and the evolution observed in the stochastic oscillations. In this way, we can complement the term "stochastic limit cycle" [287, 288] with the notion of a "stochastic torus."

9.4.3 Phase Dynamics Inside the Synchronization Region: Electronic Experiment

Further inspection of Fig. 9.12 shows that the peak of phase difference distribution is located close to zero but not precisely at zero. The value of mean phase difference characterizes the synchronization phenomena as well as the temporal behavior of the phase difference.

For the regular oscillations it is known that inside the Arnold tongue the phase difference increases with increasing frequency mismatch, taking the zero value at vanishing mismatch. Outside the Arnold tongue the mean phase difference decreases

9 Synchronization of Noise-Induced Oscillations

because the phase distribution tends to uniform distribution with zero mean value. A similar behavior is observed for chaotic oscillations in spite of the phase difference distribution having a finite width in the synchronized state. However, for stochastic synchronization of noise-induced jumps in bistable systems the phase difference behaves in a rather different way [263].

Does this reflect the essentially different nature of stochastic synchronization? We study this problem by means of an electronic experiment with unidirectionally coupled monovibrators. D_2 is fixed at 0.9 V^2 while D_1 changes within the interval [0.5; 1.3] V^2 providing a frequency mismatch of the noise-induced oscillations. Figure 9.13 shows the evolution of phase difference distribution (left panel) when D_1 increases. It is clearly seen that for minimal ($D_1 = 0.5$ V^2) and maximal ($D_1 = 1.3$ V^2) values of the noise intensity, the distribution covers all interval of $\Delta\phi$. This corresponds to the asynchronous regime. In the intermediate panels, the phase difference distribution seems to be limited and shifted to larger

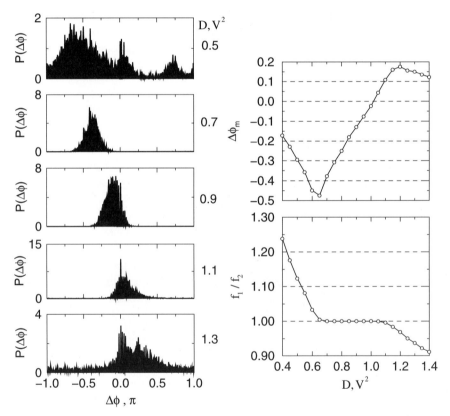

Fig. 9.13. Stochastic synchronization in unidirectionally coupled monovibrators (electronic experiment). **Left panels**: The phase difference distribution with increasing of $D = D_1$. **Right top panel**: Mean phase difference. **Right bottom panel**: The evolution of frequency ratio

$\Delta\phi$ with increasing D_1. The changes of mean phase difference is illustrated in top right panel while the bottom right panel represents the corresponding changes of the frequency ratio. The stochastic frequency locking is observed approximately from $D_1 = 0.6$ V² till $D_1 = 1.1$ V² where $f_1/f_2 \approx 1.0$. In the same range of D_1, the mean phase difference $\Delta\phi_m$ increases roughly linear. But outside the frequency locking region $\Delta\phi_m$, as before, approaches zero.

9.5 Synchronization via Suppression

Let us now focus on the synchronization phenomena observed in the unidirectionally coupled ML model ($n = 1$ in (9.1)). In this case, the first and the second subsystems play the role of "master" and "slave," respectively. Noise intensity of the master system is taken at the optimal value ($D_1 = 0.001$) while D_2 is varied as a mismatch parameter.

Unidirectionally interacting CR oscillators bear many common features to the behavior observed in forced self-sustained systems. When the coupling is introduced various patterns of response, depending on the time scales of the two subsystems are elicited. Figure 9.14 displays the 1 : 1 synchronization region. At the boundary of this region the mean frequencies of the noise-induced oscillations coincide ($|\omega_1 - \omega_2| < \mathrm{const} = 0.0002$), and remain constant within a range of mismatch parameter (i.e., inside the synchronization region).

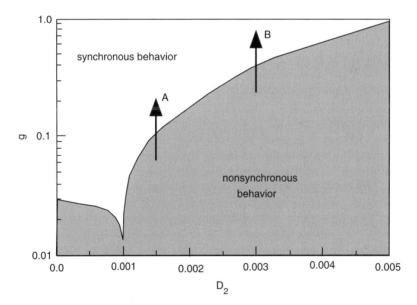

Fig. 9.14. The synchronization region for unidirectionally coupled ML models. Directions A and B indicate different transitions to synchronized behavior

256   9 Synchronization of Noise-Induced Oscillations

The transitions to synchronous regime are realized through locking of the peak frequencies of the interacting units (direction A in Fig. 9.14) as we discussed above for mutually coupled systems, or via suppression of noise-induced oscillations in one of the coupled subsystems by the signal from the other (direction B in Fig. 9.14).

Figure 9.15 shows how the transition to synchronous behavior for large mismatch parameters occurs. The peak frequency of the noise-induced oscillations in the driven system keeps its constant value while its height decreases and the width becomes broader until it becomes difficult to resolve the peak in the noise background. At this moment suppression of this frequency takes place. When the coupling is further increased, the height of the peak frequency which is equal to the frequency of the driving system can be distinguished in the power spectrum. It grows, and becomes narrower.

For coupled deterministic self-sustained systems such a spectral evolution reflects the transition from a two-dimensional torus (two distinct peaks at different frequencies) to a limit cycle (peaks in both systems are at the same position). At the same time, the Poincaré section changes from an invariant curve for the torus section to a single point representing the limit cycle. Let us check what can be observed in the Poincaré section of unidirectionally coupled noisy ML models. In Fig. 9.16 the phase projection of the slave subsystem is shown for the selected moments of time when v_1 increases through the value 0.35. Thus, such a phase projection is an analog of the Poincaré section (it is not a true Poincaré section because one can not be sure that all noisy trajectories are transversal to the secant). At weak coupling $g = 0.01$ one can observe that points form the ring like structure that actually shows the geometry of the stochastic limit cycle for the slave ML model. This means that the firing processes in the master system and in the slave system are uncorrelated, i.e., asynchronous. This corresponds to the regime illustrated in Fig. 9.12(c). For stronger coupling $g = 0.2$ the ring-like structure still exists, but in contrast to Fig. 9.12(b) its size is considerably reduced. Finally, at strong coupling $g = 0.6$ the discussed structure converges to a small spot of points. It is clear that the observed evolution is qualitatively equivalent to the transition from a closed curve in the torus section to a single point in the section of the limit cycle. Thus, we conclude, that suppression synchronization mechanism is observed for the unidirectionally coupled ML models in terms of the evolution both of the power spectra and of Poincaré sections. How does this relate to the regularity of noise-induced oscillations?

The β evolution for the "slave" system is shown in Fig. 9.17, where the horizontal solid and dashed lines indicate the regularity level for the "master" system and for the "slave" system without coupling, respectively. From figure it is seen, that for weak coupling the regularity in "slave" system remains almost the same up to $g \approx 0.01$. This means that the "master" system does not significantly influence its dynamics. When g values exceed 0.01, the measure of regularity in the driven system sharply falls and reaches a minimum value at $g \approx 0.08$. When g increases further, β rapidly rises up to the constant value that corresponds to β in the driving system with optimal noise intensity. All changes described are in a clear relation with the above discussed spectral evolution. It is clear that the drop of regularity

9.5 Synchronization via Suppression   257

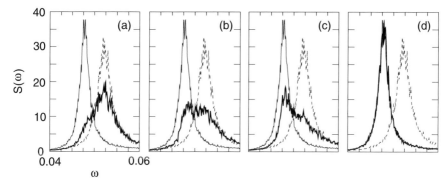

Fig. 9.15. Evolution of the power spectrum along the direction B (Fig. 9.14) at increasing coupling strengths **a** $g = 0.05$, **b** $g = 0.08$, **c** $g = 0.10$, and **d** $g = 0.40$. The transition via the suppression of the noise-induced frequency in the driven system takes place. *Dashed curve* corresponds to the driving system with vanishing coupling

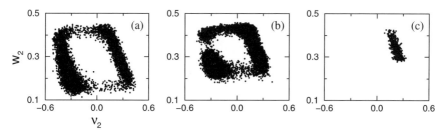

Fig. 9.16. Evolution of Poincaré section (with condition $v_1 = 0.35$) for the slave subsystem. Coupling strength increases from left to right: **a** $g = 0.01$, **b** $g = 0.20$, **c** $g = 0.60$

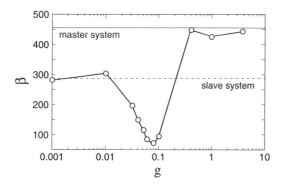

Fig. 9.17. Measure of regularity vs coupling strength for unidirectionally coupled Morris–Lecar models: $D_2 = 0.003$ (along the direction B). *Horizontal solid* and *dashed lines* indicate the maximum regularity level for "master" and "slave" systems respectively, with vanishing coupling

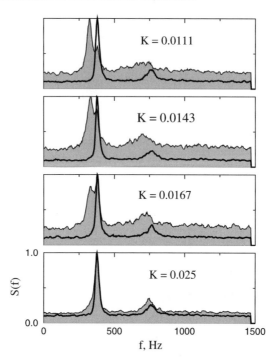

Fig. 9.18. (Color online) Evolution of the power spectrum obtained from electronic experiment. This figure illustrates a transition via the suppression of the noise-induced frequency in the driven system

is provided by the suppression of noise-induced oscillations in the "slave" system, while the subsequent sharp rise reflects the onset of firing regime that just repeats the firing in the "master" system.

To be sure that observed mechanism of stochastic synchronization is not specific for the model considered, we also made experiments on electronic monovibrators. The results are shown in Fig. 9.18. In spite of the different non-linear properties of the monovibrator (there are no bifurcations for deterministic monovibrator model, just hook-like trajectory that converges to the stable node), the observed spectral evolution is the same. It is clearly seen that the peak in the slave system (given in grey) decreases in height without changing its position, while there is the new peak growing precisely at the frequency of the master system.

To conclude, we have demonstrated, that both synchronization mechanisms known for periodic oscillations can be also detected for interacting noisy excitable systems. This supports the generality of the synchronization mechanism for any kind of oscillations that have a pronounced peak in the power spectrum, regular, chaotic or noise-induced.

10 Conclusions to Part I

The main message that we were keen to convey in this part was as follows. Whatever the nature of self-sustained oscillations we are faced with, be it perfectly periodic, deterministically chaotic, influenced by noise or even noise-induced, and no mat-

260   10 Conclusions to Part I

ter what the exact form of coupling between them is, they all obey the same laws dictated by the fundamental and universal phenomenon of synchronization. There are three mechanisms of synchronization: phase (frequency) locking, suppression of natural dynamics, and crisis, or homoclinic bifurcation.

We would also like to emphasize an important practical aspect of synchronization, namely, that it can be used as a tool for the control of virtually all kinds of oscillations. Indeed, from the beginning of the 20th century the problem of synchronization was studied with regard to the problem of control, when the need arose to stabilize the frequency of a powerful generator of electromagnetic waves by means of applying weak external perturbation. In addition to that, synchronization allows one to manipulate the frequencies of oscillations in the system, as well as their amplitudes, shapes and degree of regularity. Sometimes, in engineering applications one needs to stop oscillations altogether, and oscillation death is one of many phenomena that can occur in coupled oscillating systems and can be helpful in this respect.

Part II

Case Studies in Synchronization

*One day Alice came to a fork in the road and saw a Cheshire cat in a tree.
"Which road do I take?" she asked. "Where do you want to go?" was his response.
"I don't know," Alice answered. "Then," said the cat, "it doesn't matter."*
Lewis Carroll, "Alice in Wonderland"

Logic will get you from A to B. Imagination will take you everywhere.
Albert Einstein

This part offers an exciting excursion into the complex web of synchronization phenomena. There are systems of various origins, there are connections of various forms. Real living systems are still beyond our imagination. But there is universality in their behavior.

11 Synchronization of Anisochronous Oscillators

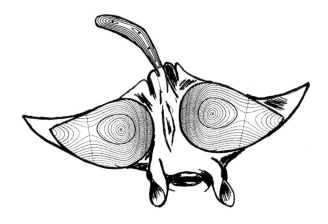

Chapter 3 of this book introduces some theoretical approaches that can be helpful as one studies the synchronization of periodic oscillations. It provides generic and useful tools that allow one to predict the behavior of coupled oscillators at different values of the control parameters, as long as the basic assumptions of the theory are valid.

One of the most important assumptions used there was that oscillations should be close to harmonic ones. Thus, the geometry of the cyclic motion in the phase space of the oscillator was predefined, requiring the shape of the limit cycle to be close to an ellipse. Note that for a weakly non-linear oscillator the phase velocity of a point on the limit cycle is approximately proportional to the distance between this point and unstable equilibrium point inside this limit cycle. As a result, if the limit cycle grows in size preserving its elliptic shape, the period remains almost constant. This approach allows one to separate phase and amplitude dynamics, and also to consider the amplitude of oscillations as a slow variable. With this, there are no physical reasons why a different geometry of a limit cycle should lead to considerably different synchronization features.

If a limit cycle has a shape different from an elliptic one, the spectrum of corresponding oscillations consists not only of the single frequency that is inversely proportional to the cycle period, but also of the multiples of this frequency that are

266   11 Synchronization of Anisochronous Oscillators

called "higher harmonics." We expect that *anharmonicity* (the presence of higher harmonics) of oscillations alone is not crucial from the view point of synchronization.[1] However, it becomes important when it is related to a special feature that is often found in periodic oscillations: *anisochronicity*.

Generally anisochronicity means that in different fragments of a limit cycle the phase point moves with substantially different velocities. In other words, a phase trajectory slows down in some parts of the limit cycle and accelerates in others.[2] It is clear that the response of such an oscillator to perturbations should depend on the current position of the phase point on the limit cycle.

If we consider two periodic oscillators, each being only weakly anisochronous, a small external perturbation will only slightly change the shape and amplitude of the existing limit cycle, and will also adjust its phase. The character of these changes does not depend on the exact form of nonlinearity in the interacting systems. The situation is completely different if partial oscillations in coupled units are anisochronous. In this case, the non-linearity of the systems plays an essential role.

In this chapter we discuss non-linear effects that appear due to anisochronicity of individual oscillatory units and introduce some useful methods for their analysis.

11.1 Phase Velocity Field and Coupling Vector

First of all, let us define the phenomenology and introduce some quantities that are helpful for the analysis of anisochronous oscillations. In this section we focus on the oscillators whose phase space is two-dimensional (2D), i.e., is a phase plane. However, the approaches being discussed here can be also applied to complex higher-dimensional systems.

A set of ordinary differential equations for a 2D dynamical system reads

$$\begin{aligned} \dot{x} &= f(x, y), \\ \dot{y} &= g(x, y), \end{aligned} \tag{11.1}$$

and the respective vector field of phase velocity $\vec{v}$ is given by

$$\vec{v} = \{\dot{x}, \dot{y}\}. \tag{11.2}$$

Vector $\vec{v}$ provides one with information about the direction of motion of the phase point. Moreover, if one introduces its absolute value $|v|$ as follows:

$$|v| = \sqrt{\dot{x}^2 + \dot{y}^2}, \tag{11.3}$$

---

[1] This statement was partly confirmed when we considered chaos in terms of periodic orbits (see Sects. 8.4, 8.5.2 and 8.6), which often have quite complex geometry, but nevertheless obey the phenomenology introduced for ellipse-like limit cycles.

[2] Similar behavior was also considered in Sect. 6.4.

the derivative of $|v|$ with respect to time reflects the acceleration or deceleration of the phase point along its trajectory, and thus defines the degree of anisochronicity of oscillations.

Equations (11.1) can be regarded as equations of motion for a phase point, with the right-hand parts playing the role of *internal* forces. Any *external* perturbation can be considered as an additional force that can change either the direction, or the velocity of the motion of the phase point.

Consider two identical oscillators and assume that they are coupled weakly. Then we can represent the behavior of coupled oscillators using a *superposition of their phase planes* (Fig. 11.1). This representation provides us with a convenient method to estimate the phase shift between the oscillators, and also to analyze the effect of coupling geometrically.

Let us consider a particular kind of coupling between the two subsystems that would lead to the following effect: each variable of the first subsystem tends to be equal to the respective variable of the second subsystem, and vice versa. This can be realized if the coupling is introduced into the system as follows:

$$\begin{aligned}
\dot{x}_1 &= f_1(x_1, y_1) + \gamma_x(x_2 - x_1), \\
\dot{y}_1 &= g_1(x_1, y_1) + \gamma_y(y_2 - y_1), \\
\dot{x}_2 &= f_2(x_2, y_2) + \gamma_x(x_1 - x_2), \\
\dot{y}_2 &= g_2(x_2, y_2) + \gamma_y(y_1 - y_2),
\end{aligned} \quad (11.4)$$

where subscripts denote the first and second coupled subsystems, and $\gamma_{x,y}$ are the coupling constants. This coupling will be referred to as *diffusive coupling*. The term "diffusive" naturally arises as one describes chemical or biological processes with $x_{1,2}$ and $y_{1,2}$ being the concentrations of some quantities in connected chambers. In these cases $\gamma_{x,y}$ are the diffusion coefficients.

On the superimposed phase plane such diffusive coupling is represented by the vectors $\vec{d}_1$ and $\vec{d}_2$ attached to the phase points of the first and second subsystems, respectively (Fig. 11.2). In the case of "true" diffusion, $\gamma_x = \gamma_y$ and coupling vectors

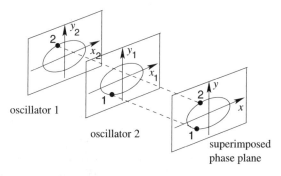

Fig. 11.1. In order to analyze the cooperative behavior of two identical oscillators, one can superimpose the phase trajectories of the two systems onto one phase plane, which we will hereafter refer to as "superimposed phase plane"

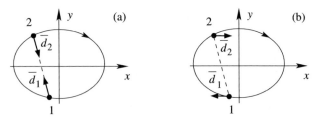

Fig. 11.2. Diffusive coupling on the superimposed phase planes. **a** All-variable coupling with $\gamma_x = \gamma_y$. **b** One-variable coupling with $\gamma_x > 0$, $\gamma_y = 0$

$\vec{d}_1$ and $\vec{d}_2$ are always directed towards each other (Fig. 11.2(a)). It looks like points 1 and 2 attract each other in order to minimize the distance between them.

Generally, coupling can be anisotropic, for instance a membrane that is semipermeable to chemicals, and $\gamma_x \neq \gamma_y$. In this case, $\vec{d}_1$ and $\vec{d}_2$ are not collinear any longer. In extreme cases, γ_x or γ_y can be equal to zero, and we arrive at the *one-variable diffusive coupling* that is illustrated in Fig. 11.2(b): geometrically, only horizontal component of the coupling force is present. This kind of coupling can arise for example in interacting electric circuits that are coupled via a resistor r. Then the coupling current i_c is equal to

$$i_c = (u_1 - u_2)/r, \qquad (11.5)$$

where u_1 and u_2 denote the voltages at the connection points. Since the coupling resistor r provides an energy dissipation with power $P_c = r i_c^2$, such interaction is also known as *dissipative* coupling.

Note that any type of cross-variable coupling (for example, if the term $(y_1 - y_2)$ appears in the equation for $\dot{x}_{1,2}$), as well as asymmetric coupling (for example, a unidirectional coupling) are not diffusive and are not considered in this chapter.

11.2 Effective Coupling Function

In the limit of weak coupling, the effect of mutual coupling between two slightly non-identical oscillators can be described by means of effective coupling introduced by Kuramoto [151]. Of course, the same approach is also valid for identical oscillators which we will be considering in this section.

11.2.1 Asymptotic Phase

Consider a single two-dimensional oscillator that demonstrates a limit cycle, and whose motion is characterized with the phase vector $\vec{X} = (x, y)$ being a state vector for the system (11.1). In order to simplify the description, the position $\vec{X}^0$ of a phase point on the limit cycle is replaced by a phase ϕ that has the following property: the rate of change of the phase along the limit cycle is the same at any point and is equal to 2π divided by the period T of the limit cycle, i.e.,

$$\frac{d\phi}{dt} = \frac{2\pi}{T}. \tag{11.6}$$

One can introduce ϕ with this property for a limit cycle of an arbitrary shape using, e.g., the return times to Poincaré section (see (8.10)). For any point Q located outside the limit cycle, the phase can be defined as follows: If the asymptotic state of Q converges to the asymptotic state of a point P on the limit cycle, then the phase of the point P is assigned to the phase of the point Q. The phase defined this way is called the *asymptotic phase*.

11.2.2 Effective Coupling Function

Consider a general form of two mutually coupled identical oscillators

$$\begin{aligned} \frac{d\vec{X}_1}{dt} &= \vec{F}(\vec{X}_1) + \gamma \vec{p}(\vec{X}_1, \vec{X}_2), \\ \frac{d\vec{X}_2}{dt} &= \vec{F}(\vec{X}_2) + \gamma \vec{p}(\vec{X}_2, \vec{X}_1), \end{aligned} \tag{11.7}$$

where $\vec{X}_i = (x_i, y_i)$, $i = 1, 2$, $\vec{p}$ is the perturbation function that describes the perturbation applied to each system as a result of the coupling between them, and γ is the strength of coupling. One can describe a single oscillator in terms of its phase, and (11.7) can be rewritten in terms of the phases ϕ_1 or ϕ_2 as follows:

$$\begin{aligned} \frac{d\phi_1}{dt} &= \frac{2\pi}{T} + \gamma \vec{Z}(\phi_1) \vec{p}(\phi_1, \phi_2), \\ \frac{d\phi_2}{dt} &= \frac{2\pi}{T} + \gamma \vec{Z}(\phi_2) \vec{p}(\phi_2, \phi_1), \end{aligned} \tag{11.8}$$

where T is the period of oscillations in oscillators when uncoupled, $\vec{Z}(\phi_i)$ is a sensitivity function defined as the change in $\dot{\phi}_i$ due to any perturbations from the position $\vec{X}_i^0(\phi_i)$ on the limit cycle of the *uncoupled* system,

$$\vec{Z}(\phi_i) = \text{grad}_{\vec{X}_i}(\phi_i(\vec{X}_i))|_{\vec{X}_i = \vec{X}_i^0(\phi_i)}, \quad i = 1, 2. \tag{11.9}$$

Now $\vec{p}(\phi_1, \phi_2)$ is a function that describes perturbation of the phases due to coupling. If coupling is weak, the changes in the phases induced by coupling over one period T of the unperturbed limit cycle are small, too. Therefore, one can average (11.8) over T,

$$\Gamma_i(\phi_i, \phi_{i'}) = \frac{1}{T} \int_0^T \vec{Z}(\phi_i) \vec{p}(\phi_i, \phi_{i'}) \, dt, \quad i = 1, 2, i \neq i'. \tag{11.10}$$

However, it is not the individual phases of interacting oscillators that are of interest to us, but rather the phase difference $\delta\phi = (\phi_1 - \phi_2)$ between them. Note that in (11.10) Γ_i can be represented as the functions of phase difference $\delta\phi$, and $\Gamma_2(\delta\phi) = \Gamma_1(-\delta\phi)$ [151]. Then (11.8) can be rewritten as

270   11 Synchronization of Anisochronous Oscillators

$$\frac{d\phi_1}{dt} = \frac{2\pi}{T} + \Gamma_1(\delta\phi), \qquad \frac{d\phi_2}{dt} = \frac{2\pi}{T} + \Gamma_1(-\delta\phi). \qquad (11.11)$$

The evolution equation for $\delta\phi$ then reads

$$\frac{d(\delta\phi)}{dt} = \Gamma_1(\delta\phi) - \Gamma_1(-\delta\phi) = \Gamma_a(\delta\phi), \qquad (11.12)$$

where $\Gamma_a(\delta\phi)$ is the effective coupling function expressed as

$$\Gamma_a(\delta\phi) = \frac{1}{T} \int_0^T \left(\vec{Z}(\phi_1)\vec{p}(\phi_1, \phi_2) - \vec{Z}(\phi_2)\vec{p}(\phi_2, \phi_1) \right) dt. \qquad (11.13)$$

How can we calculate the effective coupling in a real situation? In general, the evaluation of the quantity $\vec{Z}(\phi_i)$ in (11.9) $\mathrm{grad}_{\vec{X}_i}\, \phi(\vec{X}_i)$ for *any* point $\vec{X}_i$ with phase ϕ_i is very complicated. For weak coupling, however, it is approximately equal to the value calculated at a point $\vec{X}_i^0(\phi_i)$, where the point $\vec{X}_i^0$ is on the limit cycle and its phase is the same as that of point $\vec{X}_i$. For any point on the limit cycle, it is the gradient of the phase ϕ_i that is a function of the phase vector $\vec{X}_i$ and it can be obtained numerically.

Note that the effective coupling $\Gamma_a(\delta\phi)$ is an odd function of $\delta\phi$ because the coupling in (11.7) was introduced as symmetric. The equilibrium phase difference between the two oscillators is given by the condition $\Gamma_a(\delta\phi) = 0$ that corresponds to the phase locked regime of coupled oscillators. The stability of such regime is determined by the slope of the effective coupling $d\{\Gamma_a(\delta\phi)\}/d\{\delta\phi\}$ at the corresponding value of $\delta\phi$. The slope takes a negative or a positive value for the stable or unstable synchronous regimes, respectively.

11.3 Dephasing

The effective coupling function describes the system's response *averaged* over a period of the limit cycle. But the local inhomogeneity of the phase velocity field can in addition cause new important phenomena which we will introduce in this section.

There are three types of equilibrium states on a phase plane: stable, unstable and saddle. The asymptotic dynamics near either stable, or unstable points does not produce any non-trivial effects: all trajectories either converge to equilibrium or move away from it. However, the motion of the phase trajectory near a saddle point can lead to an interesting phenomenon known as *dephasing* [104, 264].

The mechanism of this phenomenon is illustrated in Fig. 11.3 where a superimposed phase plane for the two identical mutually coupled oscillators (11.4) is shown. S is a saddle equilibrium that existed in each of the oscillators before the coupling was introduced. Note that diffusive coupling does not change the positions of the equilibrium states, and therefore S remains at the same location as without coupling. W_s and W_u are the stable and the unstable manifolds of S, respectively.[3]

---

[3] For the properties of the manifolds see Sect. 5.1.

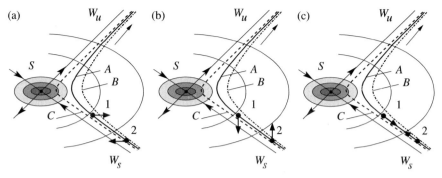

Fig. 11.3. Illustration of dephasing mechanism on the superimposed phase plane of two diffusively coupled identical oscillators (11.4). S is a saddle point with its stable W_s and unstable W_u manifolds. A (*solid line*) is the limit cycle that existed in the uncoupled oscillators. B (*dash-dotted line*) and C (*dashed line*) are the segments of the phase trajectory that go through the initial conditions set inside and outside A, respectively. *1* and *2* are the initial conditions of the coupled oscillators that are set on the cycle A at some distance from each other, and the *arrows* attached to them show the strength and direction of the forcing due to diffusive coupling: **a** along x-direction, **b** along y-direction and **c** along both variables. For more detail see text

Symbol A (solid line) denotes the limit cycle that existed in each of the identical systems when they were uncoupled (in the picture only a segment of the cycle is shown). We will refer to the area to the right of A as to the "inside of the limit cycle," and to the area to the left as to the "outside" of it.

The closed curves around S schematically represent the lines of constant phase velocity $|v|$: between the two successive lines there is the same increase in the value of $|v|$. Therefore, where these lines are denser, the gradient of $|v|$ is larger. At the saddle point itself, the phase velocity is equal to zero, near S the motion is slowed down, and generally the closer to the saddle, the slower the motion is. The cycle A crosses the lines of constant $|v|$ and thus the motion on A occurs with very different velocities: the portions that are closer to S are slower than those further away from it. Also, the inside of the limit cycle A is further away from S than the outside of it in the portion of the superimposed phase plane shown in Fig. 11.3. Therefore, inside A the evolution of the asymptotic phase is faster than outside of it. The curves B (dot-dashed line) and C (dashed line) schematically represent the phase trajectories that would go through the initial conditions set inside and outside the limit cycle A, respectively.

Phase points 1 with coordinates (x_1, y_1) and 2 with (x_2, y_2) represent the first and the second oscillators of (11.4), respectively. Note that the coordinates of the respective state of the full system (11.4) are then defined as (x_1, y_1, x_2, y_2). Suppose the initial conditions (x_1^*, y_1^*) for 1 and (x_2^*, y_2^*) for 2 are defined somewhere on the limit cycle A of the unperturbed system, but at slightly different positions: 1 is put closer to the saddle than 2. As seen from Fig. 11.3, the two points find themselves in the regions with substantially different phase velocities: 1 in a slower region,

272   11 Synchronization of Anisochronous Oscillators

and 2 in a faster one. Since the phase points move clockwise, the initial phase of the first oscillator appears larger than the phase of the second one, therefore we will call the first oscillator the "leading" subsystem and the second the "lagging" one. The arrows attached to points 1 and 2 in Fig. 11.3 show the direction of the force that is experienced by each point due to diffusive coupling. Three cases are illustrated: (a) coupling along x-direction only, $\gamma_x \neq 0$ and $\gamma_y = 0$, (b) coupling along y-direction only, $\gamma_y \neq 0$ and $\gamma_x = 0$, and (c) coupling along both variables, or "all-variable coupling" with $\gamma_x = \gamma_y$. Consider these cases separately.

For the x-coupling illustrated in Fig. 11.3(a) the phase points "attract" each other along the horizontal direction. With this, the leading subsystem is being pushed to the right, inside the limit cycle, where the phase flow is fast. On the contrary, the lagging subsystem is being pushed to the left, outside the limit cycle, where the phase flow is slow. As a result, a small phase difference between the two oscillators increases, and the phenomenon of dephasing occurs. Note that the mechanism described above is governed by the gradient of phase velocity field that is directed transversely to the phase trajectory, and therefore it does not disappear at vanishing coupling.

As the phase trajectories leave the vicinity of the saddle point, the "ordinary" mechanisms of interaction take control again. As a result, the phase difference settles down at some (non-zero) stable value. For coupling strong enough, the nonlinear mutual attraction takes over the dephasing effect. Hence, the phase difference between the two oscillators strongly depends on the strength of coupling.

If the coupling through y-variable is applied as illustrated in Fig. 11.3(b), the local behavior near S is opposite to the one considered above: the leading subsystem is slowed down while the lagging one is accelerated. Thus, a small phase difference rapidly decreases (the "inphasing" effect) and in-phase behavior becomes stable.

For the all-variable diffusive coupling $\gamma_x = \gamma_y$ illustrated in Fig. 11.3(c), points 1 and 2 always attract each other along the trajectory A and no dephasing or inphasing effects occur. The mechanism discussed above does not work any longer. This is why such coupling is sometimes referred to as "scalar."

The motion of a phase trajectory very close to a saddle point naturally takes place in the dynamical systems close to a homoclinic bifurcation. Thus, systems with a homoclinic transition appear to demonstrate the dephasing effect. For the two coupled Morris–Lecar (ML) models, Han et al. have shown [104] that the one-variable diffusive coupling leads to dephasing between two oscillations and is responsible for antiphase synchronization.

Figure 11.4 shows the effective coupling Γ_a for the diffusively coupled ML systems described by (11.21) that will be considered in the next section. Here the limit cycle appears through (a) Andronov–Hopf bifurcation and (b) homoclinic connection. For the Andronov–Hopf bifurcation (a) the slope of effective coupling for $\delta\phi = \pi$ is positive while for $\delta\phi = 0$ it is negative. Hence, the *antiphase* state is unstable and the *in-phase* state is stable. On the contrary, for the homoclinic bifurcation (b), the in-phase state is unstable and the antiphase state is stable. A small phase difference between the two oscillators gradually increases and settles down

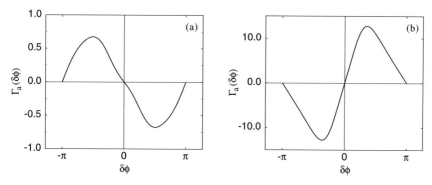

Fig. 11.4. Effective coupling $\Gamma_a(\delta\phi)$ as a function of the phase difference $\delta\phi$ between the two Morris–Lecar models (11.21): **a** for the limit cycle that appears via an Andronov–Hopf bifurcation ($J = 0.35$) and **b** for the limit cycle that appears via a homoclinic bifurcation ($J = 0.075$). A standard set of parameters is used: $u_c = 0.0$, $u_d = 0.3$, $\bar{g}_{Ca} = 1.1$, $f = 0.2$. In both cases, the effective coupling vanishes at $\delta\phi = 0, \pi$. In case **a**, the slope is positive at $\delta\phi = \pi$ and negative at $\delta\phi = 0$. Therefore, the in-phase solution is stable. In case **b**, the situation is opposite to the above and the antiphase solution is stable

as one achieves the antiphase state. Thus, the dephasing phenomena are the result of a strong deformation of the phase flow near the stable manifold of the saddle point [104].

11.4 Examples of 2D Anisochronous Oscillators

A two-dimensional oscillatory system in its most general form can be written as follows:

$$\varepsilon \ddot{x} + D(x, \dot{x})\dot{x} + \Omega(x) = 0, \qquad (11.14)$$

where functions $D(x, \dot{x})$ and $\Omega(x)$ can be chosen in various forms as long as they fulfill the conditions for the existence of self-oscillations. In (11.14) $D(x, \dot{x})$ is responsible for the energy dissipation and $\Omega(x)$ defines the force applied to the oscillator. The zeroes of $\Omega(x)$ determine the locations of equilibrium points, whose stability is determined by the signs of $d\Omega(x)/dx$ and of $D(x, \dot{x})$. Namely, $d\Omega(x)/dx$ is negative only for saddles, whereas nodes and foci are stable when $D(x, \dot{x})$ is positive. For more compact notations, thereafter we omit the arguments of D and Ω. The time separation parameter ε in (11.14) determines the difference in the time scales between x and $\dot{x}$.

Anisochronous features of a limit cycle oscillator can be introduced in two different ways. First, choosing $\varepsilon \gg 1$ or $\varepsilon \ll 1$ yields a strong difference in time scales between x and $\dot{x}$ that ensures the separation of fast and slow dynamics. Second, by variation of D and Ω one can change the features of the phase velocity field, since

$$|v(x, \dot{x})| = \sqrt{\dot{x}^2 + \frac{1}{\varepsilon^2}(-D\dot{x} - \Omega)^2}. \qquad (11.15)$$

274   11 Synchronization of Anisochronous Oscillators

Let us consider a number of representative examples of phase plane oscillators, most of which will be used throughout this chapter. Figure 11.5 displays isolines of phase velocity (left column) and the shapes of nonlinear functions D and Ω (right column) of oscillators (11.14).

van der Pol oscillator [292] is a well-known paradigmatic model for self-sustained oscillatory dynamics that was thoroughly considered in Chaps. 3, 4 and 6. It can be defined by setting

$$D = -\alpha(1 - x^2),$$
$$\Omega = x,$$

(11.16)

where α is a control parameter, and $\varepsilon = 1$ in (11.14). The phase velocity value is determined by

$$|v| = \sqrt{\dot{x}^2 + \left(\alpha(1 - x^2)\dot{x} - x\right)^2}.$$

(11.17)

At a vanishingly small α one can obtain

$$|v| \approx \sqrt{x^2 + \dot{x}^2} = \sqrt{r^2} = r,$$

(11.18)

with r being the radius of the limit cycle approximated by harmonic functions (as done in Chap. 3). Then the period of the limit cycle does not depend on its size and can be estimated as

$$T = \frac{2\pi r}{r} = 2\pi.$$

Thus, for small α, the van der Pol model provides an example of a perfectly *isochronous* limit cycle oscillator. However, for considerably large values of α, the trajectory goes through the regions of fast and slow motion, and oscillations become more complex. In Sect. 11.6 we will discuss how this affects synchronous behavior.

Bonhoeffer–van der Pol oscillator [54] is also referred to as a simplified form of the FitzHugh–Nagumo neuron model [80]:

$$\varepsilon \dot{x} = x - \frac{x^3}{3} - y,$$
$$\dot{y} = x + a.$$

(11.19)

This model corresponds to the following choice of the functions D and Ω:

$$D = -(1 - x^2), \qquad \Omega = x + a$$

in (11.14). Here, ε is the time separation parameter, which is assumed to be small, $\varepsilon \ll 1$, and a is the control parameter. By setting $a = 0$ and $\varepsilon \approx 1$, (11.19) is reduced to the van der Pol oscillator (11.16). However, we distinguish between these two cases since they come from different applications. For example, the FitzHugh–Nagumo model was derived from the four-dimensional Hodgkin–Huxley neural model [116] by means of simplifications and reduction of system's dimension. Figures 11.5(a) and (b) show that the phase velocity is slow in the vicinity of the

11.4 Examples of 2D Anisochronous Oscillators 275

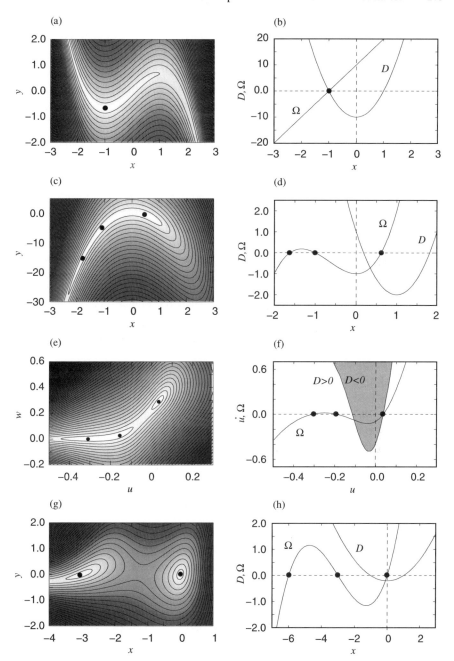

Fig. 11.5. The contour plot for the phase velocity field (**left column**) and the shape of functions D, Ω (**right column**) for the **a**, **b** Bonhoeffer–van der Pol oscillator; **c**, **d** Hindmarsh–Rose model; **e**, **f** Morris–Lecar model; and **g**, **h** modified van der Pol system

276   11 Synchronization of Anisochronous Oscillators

N-shaped nullcline defined by the condition $\dot{x} = x - x^3/3 - y = 0$. Here, the smaller the ε, the slower the motion of the phase point is. Function D is negative in the range of $x \in [-1, +1]$ and function Ω is a straight line. The intersection of D and Ω (Fig. 11.5(b)) corresponds to the equilibrium point that undergoes Andronov–Hopf bifurcation at $a = 1.0$.

The Hindmarsh–Rose model [115] was originally developed to describe a bursting behavior of a neuron, and has three equations in its original form. However, there is also a phase plane version of this model whose equations read

$$\begin{aligned} \dot{x} &= y - x^3 + 3x^2 + I, \\ \dot{y} &= 1 - 5x^2 - y. \end{aligned} \tag{11.20}$$

This is equivalent to the following choice of the corresponding functions D and Ω in (11.14):

$$D = -\left(3x^2 - 6x + 1\right), \qquad \Omega = x^3 + 2x^2 - 1 - I,$$

where I is the control parameter qualitatively describing the external applied current. In contrast to the FitzHugh–Nagumo model, the system (11.20) has three equilibrium points. Moreover, for negative x both x- and y-nullclines are located close to each other. As a result, a narrow "valley" of slow motion is formed on the phase plane, where two of three equilibrium points are located (Fig. 11.5(c)). Figure 11.5(d) shows that only one (right) zero of Ω function corresponds to the negative values of D, therefore the respective equilibrium is an unstable focus, whereas two other equilibrium points are a saddle (the middle one) and a stable (the left one) nodes.

The Morris–Lecar (ML) model [181] is a simplified model of a spiking neuron with a refractory period, which is similar to the Hodgkin–Huxley model [116]. Using two dynamical variables, this model takes into account most of the dynamical features of the real neurons, including a stimulus-dependent excitability and oscillatory behavior. The equations for a single ML model read

$$\begin{aligned} \frac{du}{dt} &= -J_{\text{ion}}(u, w) + J, \\ \frac{dw}{dt} &= f \frac{w_\infty(u) - w}{\tau_w(u)}. \end{aligned} \tag{11.21}$$

Here,

$$\begin{aligned} J_{\text{ion}}(u, w) &= \bar{g}_{\text{Ca}} m_\infty(u)(u - u_{\text{Ca}}) + \bar{g}_K w(u - u_{\text{K}}) + \bar{g}_{\text{L}}(u - u_{\text{L}}), \\ m_\infty(u) &= 0.5 \left[1 + \tanh\{(u - u_{\text{a}})/u_{\text{b}}\}\right], \\ w_\infty(u) &= 0.5 \left[1 + \tanh\{(u - u_{\text{c}})/u_{\text{d}}\}\right], \\ \tau_w(u) &= 1/\cosh\{(u - u_{\text{c}})/(2u_{\text{d}})\}, \end{aligned}$$

where the dynamical variables u and w represent the transmembrane voltage of a neuron and the activation of the potassium current, respectively. The driving forces for the membrane potential u are the external stimulus current J and the ionic channel current $J_{\text{ion}}(u, w)$. The ionic channel current consists of three terms: the calcium current $\bar{g}_{\text{Ca}} m_\infty(u)(u - u_{\text{Ca}})$ generating fast action potentials, the delayed rectifier potassium current $\bar{g}_{\text{K}} w(u - u_{\text{K}})$, and the leak current $\bar{g}_{\text{L}}(u - u_{\text{L}})$ maintaining a constant potential at the resting state.

The dynamical properties of this model at various sets of control parameters have been extensively analyzed by Rinzel and Ermentrout [244]. For very low or very high values of the excitation current J, it has a single stable equilibrium point. At intermediate values of J, a stable limit cycle appears either via the Andronov–Hopf bifurcation, or via the homoclinic connection, depending on the value of f. For the parameter values set as $u_a = -0.01$, $u_b = 0.15$, $u_c = 0.1$, $u_d = 0.145$, $\bar{g}_{\text{Ca}} = 1.0$, $\bar{g}_{\text{K}} = 2.0$, $\bar{g}_{\text{L}} = 0.5$, $u_{\text{Ca}} = 1.0$, $u_{\text{K}} = -0.7$, $u_{\text{L}} = -0.5$ and $f = 1.15$, a limit cycle arises at $J = 0.0730$ via a homoclinic connection. A one-parameter bifurcation diagram and the corresponding phase portraits are shown in Fig. 11.6.

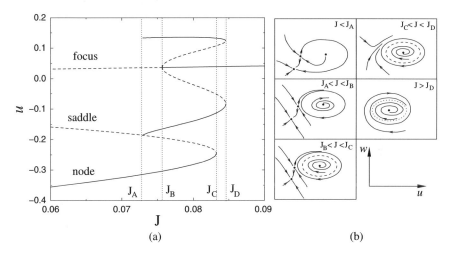

Fig. 11.6. The dynamics of a single Morris–Lecar model (11.21). **a** One-parameter bifurcation diagram. *Solid lines* denote the stable solutions while *dashed lines* correspond to nonstable (saddle and unstable) solutions. **b** Evolution of the phase portraits. For the external stimulus $J < J_A$, a stable fixed point, a saddle point, and an unstable fixed point coexist. At $J = J_A$, a stable limit cycle appears through a homoclinic connection at the saddle point. At $J = J_B$, the unstable fixed point becomes stable, and an unstable limit cycle is born in its vicinity. Two limit cycles, stable and unstable ones, coalesce to disappear at $J = J_D$ via a saddle-node bifurcation. At $J = J_C$, the stable and the saddle equilibria disappear via a saddle-node bifurcation. Here, $J_A = 0.0730$, $J_B = 0.0756$, $J_C = 0.0833$ and $J_D = 0.0845$. In **b** we show schematically how phase portraits are evolving for each range of J indicated in **a**

278   11 Synchronization of Anisochronous Oscillators

For the case of the Morris–Lecar model, functions D and Ω are given by

$$
\begin{aligned}
D(u, \dot{u}) &= \bar{g}_{\text{Ca}} \frac{\partial m_\infty(u)}{\partial u}(u - 1) + \frac{f}{\tau_w(u)} \\
&\quad + \frac{\bar{g}_{\text{Ca}} m_\infty(1 - u_{\text{K}}) + \bar{g}_{\text{L}}(u_{\text{L}} - u_{\text{K}}) + I_{dc} - \dot{u}}{u - u_{\text{K}}}, \\
\Omega(u) &= \frac{f}{\tau_w(u)}\Big\{\bar{g}_{\text{K}}(u - u_{\text{K}})w_\infty(u) \\
&\quad + \bar{g}_{\text{Ca}} m_\infty(u)(u - 1) + \bar{g}_{\text{L}}(u - u_{\text{L}}) - J\Big\}.
\end{aligned}
\tag{11.22}
$$

Note that unlike in the models considered above, function D here depends on both u and $\dot{u}$, i.e., is a function of two variables which is illustrated in Fig. 11.5(f) as contour plot. Within a typical range of the parameters, Ω has three zeroes that correspond to a stable node, a saddle, and an unstable focus, respectively. When the value of J is small, the stable node is the only attractor in the phase space. At larger J, the system becomes excitable: a small stimulus is unable to induce firing, producing only insignificant fluctuations of the phase flow near the stable node. However a sufficiently large stimulus might lead to the firing of the neuron that corresponds to a long excursion of the phase trajectory along the separatrix formed by the stable manifolds of the saddle. Thus, in this regime firing is not a self-sustained process, but is realized due to applied stimuli.

As J increases, a homoclinic bifurcation[4] occurs at $J \approx 0.0729$, and the stable and unstable manifolds of the saddle are connected to form a homoclinic loop to the saddle. Further increase of J leads to the self-sustained firing that is represented in the phase space of the system by a stable limit cycle. Consequently, now the phase space contains three equilibria together with a limit cycle. Such structure of the phase space can be predicted by the shape of the functions D and Ω as shown in Fig. 11.5(f). Three equilibria are located at the zeroes of Ω, and the type of each equilibrium is determined by the signs of D and $d\Omega/du$. In the figure, D is given as a contour plot on the $(u, \dot{u})$ plane, where the dark area corresponds to the negative dissipation (the energy generation). Figure 11.5(e) shows that the decrease of the phase velocity is observed around the line connecting the equilibrium points.

The modified van der Pol model [219] is a generic oscillator with a homoclinic bifurcation.

The examples of anisochronous oscillators discussed above show that one can single out two patterns of the inhomogeneous vector fields. The first pattern is related to the existence of a nullcline for the fast variable. In the vicinity of this nullcline the motion is slow. The second pattern is related to the slowing down of the trajectory near the singular points that can be either stable or unstable. In order to understand the general aspects of interaction between such oscillators, it is useful to develop a simple model that would mimic the main features of neuronal oscillators and such that the above features could be easily adjusted by choosing the appropriate values of control parameters.

---

[4] Homoclinic bifurcation was also discussed in Sect. 5.1.

11.5 Synchronization near the Homoclinic Bifurcation

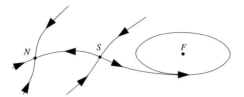

Fig. 11.7. A typical phase portrait of neuronal oscillators

The inhomogeneity of the vector field can be induced by a singular point. For example, this takes place when some segments of a limit cycle are located nearby a saddle point. Such situation usually precedes a homoclinic bifurcation and is typical, for example, for Morris–Lecar and Hindmarsh–Rose models. A schematic structure of the phase space for this case is sketched in Fig. 11.7.

As a model featuring such structure of the phase space we propose a modification of the van der Pol oscillator [219], which we refer to below as modified van der Pol (MVP) model. We choose function D and Ω in the following form:

$$D = -\alpha(\mu - x^2), \qquad \Omega = \frac{x(x+d)(x+2d)}{d^2}, \qquad (11.23)$$

where α, μ and d are the control parameters that are assumed to be positive. The chosen cubic form of $\Omega(x)$ provides three equilibria in the system. A canonical form for the MVP model is provided by (11.14), and taking into account (11.23), we can rewrite it as

$$\begin{aligned}\dot{x} &= y, \\ \dot{y} &= \alpha(\mu - x^2)y - \frac{x(x+d)(x+2d)}{d^2}.\end{aligned} \qquad (11.24)$$

Three equilibrium points are located at $y_{F,S,N} = 0$ and $x_{F,S,N} = 0, -d, -2d$ for the focus (index F), the saddle (S), and the stable node (N), respectively. The phase velocity field as well as the plots for D and Ω are shown in Fig. 11.5(g), (h). The slopes of Ω at the equilibria are $d\Omega/dx = 2, -1, 2$, respectively. The focus is unstable since D is negative at x_F, and the limit cycle is located between the unstable focus and the saddle. At the fixed values of d and α, the parameter distance from the homoclinic bifurcation is controlled by μ. At $d = 3$ and $\alpha = 0.2$, the limit cycle approaches the saddle as μ is increased and the homoclinic connection occurs at $\mu \approx 1.255$. At small μ, the limit cycle is located close to the unstable focus, and the behavior of MVP model is similar to the dynamics of van der Pol oscillator.

11.5 Synchronization near the Homoclinic Bifurcation

In this section we consider synchronization between the limit cycle oscillators near the homoclinic bifurcation with account of the dephasing mechanism near the saddle point. The coupled MVP models are taken as a toy system.

Let us first study the simplest case of the one-variable coupling. Two diffusively coupled MVP models read

$$\begin{aligned}\dot{x}_{1,2} &= y_{1,2} + K_x(x_{2,1} - x_{1,2}),\\ \dot{y}_{1,2} &= -\alpha\left(\mu - x_{1,2}^2\right)y_{1,2} - \left(x_{1,2}(x_{1,2}+d)(x_{1,2}+2d)\right)/d^2\\ &\quad + K_y(y_{2,1} - y_{1,2}),\end{aligned} \quad (11.25)$$

where the coupling strength K_x (or K_y) tends to 0, and K_y (or K_x) are assumed to be sufficiently small so that perturbations of each subsystem are negligibly small. In these equations, the variable x can be expounded as a coordinate of the system, and y is its velocity. Therefore, the particular case of $K_y = 0$ is sometimes called *position-coupling*, and the case of $K_x = 0$ is referred to as *velocity-coupling*.

Consider a single oscillator in the absence of coupling. Figure 11.8 shows a contour plot for the magnitude of the phase velocity $|v|$. The phase velocity vanishes at the equilibria (S and F). At small values of μ, the limit cycle X_1 is located far away from the saddle, an example being given in Fig. 11.8 for $\mu = 0.2$. In this case, the phase space structure in terms of the $|v|$-surface along the limit cycle is qualitatively equivalent to that of the van der Pol oscillator, for which the diffusive coupling typically leads to the stable in-phase solution.

However, as μ is increased, the limit cycle gradually approaches the saddle, and the phase trajectories visit the regions with different phase velocities. This case is qualitatively different as compared to the case of small μ. The limit cycle X_2 for $\mu = 1.0$ is shown in Fig. 11.8. The trajectory lying on X_2 spends most of the time near the saddle, and therefore the interaction due to coupling in this region becomes important. The trajectory is obviously subjected to dephasing mechanism as discussed in Sect. 11.3. Note that since $|v|$ and the vector field change with the

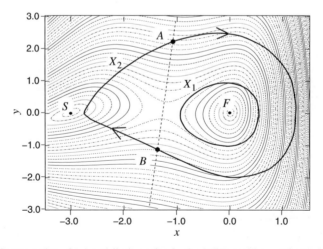

Fig. 11.8. Contour plot of $|v|$ and limit cycles in the MVP oscillator. The limit cycles are shown for $\mu = 0.2$ (X_1) and $\mu = 1.0$ (X_2). The phase velocity field is given for $\mu = 1.0$

11.5 Synchronization near the Homoclinic Bifurcation

change in μ, it is impossible to plot correctly both limit cycles X_1 and X_2 in the same plot. Therefore, in Fig. 11.8, the phase velocity field is given for $\mu = 1.0$.

Now, consider two mutually coupled identical oscillators (11.25), for which the plane (x, y) in Fig. 11.8 will be a superimposed phase plane. In order to obtain a *quantitative* estimation for the dephasing effect between them, let us divide the limit cycle into two parts by a line AB as shown in Fig. 11.8. It is expected that the part located near the saddle (to the left of AB) is relevant to the dephasing and, thus, to the antiphase synchronization. One can introduce measures P and Q for the linear rate of dephasing as follows. Assume that the motion occurs clockwise and the initial conditions for both oscillators are exactly on the limit cycle X_2, but are slightly separated from each other. Then the distance between them along the cycle can be associated with a certain time lag Δt. Assume that when the leading oscillator at point A this lag was Δt_{A0} and at point B the lag was Δt_{B0}. Let us follow the motion of the phase points in two systems along the cycle X_2 following the route ABA clockwise. Δt_B and Δt_A will be the time lags between the two systems at the endpoints of these routes. Then one can introduce P and Q numerically as

$$\Delta t_B = P \Delta t_{A0},$$
$$\Delta t_A = Q \Delta t_{B0}. \qquad (11.26)$$

The values of P and Q should be determined in the limit of a small initial time lag.

Numerical analysis of the coupled MVP model shows that while P is insensitive to variations in μ ($P \approx 1$), Q strongly depends on μ. Figure 11.9 illustrates the variations of Q versus μ for different coupling strengths and different coupling variables. The position-coupling ($K_y = 0$) leads to stronger dephasing as μ approaches the value of the homoclinic bifurcation ($\mu \approx 1.255$). Curves 1 and 2 correspond

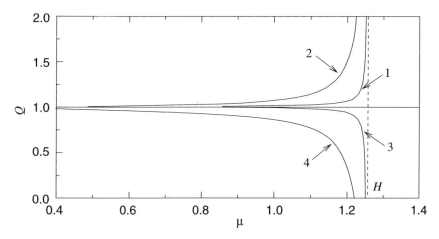

Fig. 11.9. The linear rate of dephasing Q versus parameter μ. The *dotted line* H denotes the homoclinic bifurcation point. Curves *1, 2* and *3, 4* correspond to different types of coupling (position- and velocity-coupling, respectively). Coupling strength is changed from 0.001 to 0.01 in each pair of curves

to the coupling strength $K_x = 0.001$ and $K_x = 0.01$, respectively. The case of the velocity-coupling ($K_x = 0$) leads to in-phasing (curves 3 and 4). Curves 3 and 4 correspond to the coupling strength $K_y = 0.001$ and $K_y = 0.01$, respectively. Note that the effect of both in-phasing and dephasing becomes stronger with larger coupling strength (compare curves 2 and 4 with curves 1 and 3).

Now consider the general case of the two-variable coupling as discussed in Sect. 11.1. We introduce a vector of coupling using the polar coordinates:

$$\begin{aligned} K_x &= K \cos \Psi, \\ K_y &= K \sin \Psi. \end{aligned} \tag{11.27}$$

Here, K denotes the coupling strength and the angle Ψ reflects the relative weight of coupling between two variables x and y. Ψ can be also regarded as the orientation angle of the coupling force in the two-dimensional subspace of each oscillator. A special case includes the scalar coupling when $K_x = K_y$. With this, the coupling forces become attractive when $\Psi = \pi/4$ and repulsive when $\Psi = 5\pi/4$. The one-variable coupling is achieved when $\Psi = 0, \pi$ (position-coupling) and $\Psi = \pm\pi/2$ (velocity-coupling), respectively. The diffusive coupling refers to the case where neither K_x nor K_y is negative, i.e., $0 \leq \Psi \leq \pi/2$.

11.5.1 Weak Coupling Limit

In this section we consider the case when the coupling is weak, so that an analytical method based on a phase reduction model (effective coupling function) explained in Sect. 11.2 can be applied. The calculation of the effective coupling function $\Gamma_a(\delta\phi)$ at different values of coupling angle Ψ (while K is assumed to be vanishingly small) reveals typical regimes. Figure 11.10 summarizes the main dynamical patterns. The

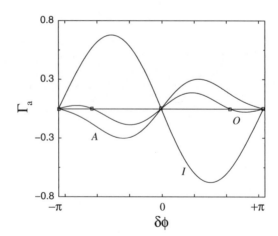

Fig. 11.10. Function of effective coupling $\Gamma_a(\delta\phi)$ for three synchronous states (in-phase I, antiphase A, and out-of-phase O). The phase difference for each state is marked by a square

11.5 Synchronization near the Homoclinic Bifurcation

three curves in this figure correspond to the three main synchronous states: the in-phase (I), antiphase (A), and out-of-phase synchronization (O). The existence of the in-phase state is expected due to the symmetry of (11.25). The diffusive coupling term vanishes when the state variables of coupled oscillators are equal. The appearance of an antiphase state is also guaranteed by the periodicity of $\Gamma_a(\delta\phi)$. The out-of-phase state corresponds to the phase-locked state with the phase difference $\delta\phi \in [0; \pi]$. The symmetry of the solution of (11.25) is broken for out-of-phase states, but they occur in pairs, and the two solutions demonstrate reflection symmetry with respect to each other.

By using the effective coupling approach, one can learn how the formation of the synchronous states depends on the coupling vector. The results are presented in Fig. 11.11 in the form of a diagram in polar coordinates (Ψ, μ). The angle Ψ

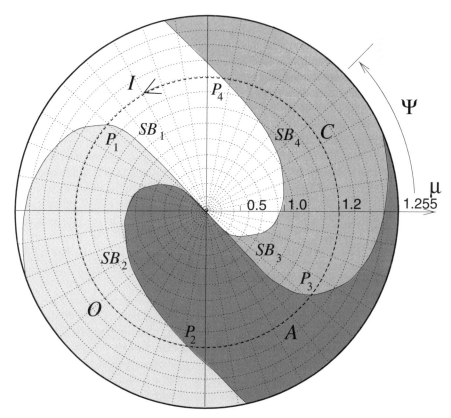

Fig. 11.11. (Color online) Phase diagram for the coupled MVP models (11.25) in the weak coupling limit on the plane of the polar coordinates "coupling angle Ψ–control parameter μ." Areas of in-phase, antiphase, and out-of-phase states are labelled as I, A, and O, respectively, and shaded by *white*, *dark-grey*, and *hatched pattern*. In the *grey area* C both in-phase and antiphase regimes are stable and coexist. Curves of symmetry breaking bifurcations are denoted SB. *Dashed line with arrow* is the selected circular path P discussed in text

284   11 Synchronization of Anisochronous Oscillators

varies from 0 to 2π, and the radius is limited by the value of μ corresponding to the homoclinic bifurcation.

Four prominent areas of the parameters denoted by different colors can be distinguished in the diagram. The white area corresponds to the stable in-phase synchronization (I), and the antiphase state is unstable there. The dark grey area corresponds to the stable antiphase (A) and the unstable in-phase states. In the hatched area, a pair of out-of-phase solutions (O) are stable, and they are separated by unstable in-phase and antiphase states. Overlapping of areas I and A is denoted by the light grey area (C). For these values of control parameters, the in-phase and the antiphase states coexist and are separated by a pair of unstable states with reflection symmetry. Note that the scale of the radial axis has been non-linearly transformed in order to magnify the part of the diagram for the larger values of μ.

As one can see from the diagram, at small values of μ ($\mu \approx 0.1$) the system behaves like two coupled van der Pol oscillators. Namely, the in-phase synchronization is the only stable state in the case of diffusive coupling ($\Psi \in [0, \pi/2]$). Generally, depending on Ψ, i.e., on the combination of signs of K_x and K_y in (11.27), the system demonstrates either in-phase or antiphase behavior. The area where the pair of out-of-phase states exist degenerates into a line.

At larger values of μ, however, synchronization states essentially depend on both Ψ and μ. Let us fix $\mu = 1.2$ and change Ψ along the circular path P denoted in Fig. 11.11 by a dashed circle with arrows. The points at which this path crosses the boundaries between different areas labelled as P_1–P_4. Along this way, three different synchronization states change their stability via symmetry-breaking bifurcations. The sequence of the symmetry-breaking bifurcations is schematically illustrated in Fig. 11.12. Large circles represent the variation of the phase difference $\delta\phi$, whereas small circles on them denote synchronization states. Filled circles mark stable states and the open circles represent the unstable states. In the insets of the diagram the branches of in-phase and antiphase states are denoted as straight lines, and the emerging pairs of branches for symmetry-breaking O states are denoted as parabolic curves. A solid line denotes a stable branch and a dotted line corresponds to an unstable branch.

The in-phase state is the only stable state of the system until it reaches P_1 where the largest Floquet multiplier becomes equal to $+1$. At this point, the in-phase state loses its stability, and two other stable states with the broken symmetry (O states) are born. The curve of the symmetry-breaking bifurcations and the corresponding branch in Fig. 11.12 are denoted as SB_1. As Ψ is increased, the out-of-phase states collide and disappear at P_2 where the inverse symmetry-breaking bifurcation (SB_2 in Fig. 11.12) occurs, and this gives rise to a stable antiphase state (A).

With the further increase of Ψ, the in-phase state becomes stable at P_3 but the antiphase state remains stable as well. Thus, there is a range of the parameter values where both in-phase and antiphase states are stable (as denoted by C in Figs. 11.11 and 11.12). These regimes coexist until the antiphase state loses its stability at P_4 via the symmetry-breaking bifurcation with increasing Ψ. The bifurcation curves passing through points P_3 and P_4 are denoted as SB_3 and SB_4, respec-

11.5 Synchronization near the Homoclinic Bifurcation

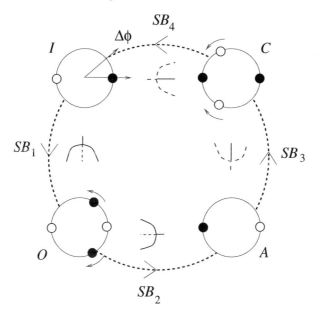

Fig. 11.12. Stability and bifurcations of the synchronization states along the circular path P in Fig. 11.11. The *insets* illustrate the corresponding symmetry-breaking bifurcations. Stable and unstable regimes are denoted by *small filled* and *empty circles*. They are connected with large circles that schematically represent the possible phase difference $\delta\phi$

tively. The bifurcations at SB_3 and SB_4 are subcritical since they are accompanied by two unstable (out-of-phase) states shown as a pair of open circles in Fig. 11.12.

To summarize the described behaviors, the phase diagram in Fig. 11.11 shows that the in-phase synchronization is the only stable state for weak diffusive coupling that is similar to the behavior of the coupled van der Pol oscillators. The area of stability of the in-phase state is particularly large for $0 \leq \Psi \leq \pi/2$.

However, with increasing μ the limit cycle approaches the homoclinic bifurcation, and the situation changes drastically. Actually, at sufficiently large μ even for purely diffusive coupling the coexistence of several synchronization states is possible. This tendency seems to be more pronounced in the case of the position coupling.

11.5.2 Finite Coupling Strength

At the finite coupling strength the perturbation of the limit cycle can be significant and, consequently, the phase model reduction considered above might not be appropriate for the prediction of the behavior of the coupled systems. Thus, direct numerical methods have to be applied. The question is: To what extent the results for the weak coupling limit can be generalized to the case of finite coupling?

Since our primary interest is limit cycle oscillations near a homoclinic bifurcation, we fix $\mu = 1.2$ (close to the bifurcation point) for both oscillators and vary the

11 Synchronization of Anisochronous Oscillators

two coupling parameters, K and Ψ. Since each of the interacting oscillators demonstrates multistability, i.e., in our case the coexistence of a stable fixed point and a limit cycle for the same parameters values, the general structure of the phase space of the coupled system is very complicated. Therefore, for simplicity let us focus on the region in the phase space where each system has an attractor corresponding to the oscillatory behavior. If for some parameters a trajectory leaves this region, e.g., due to a boundary crisis, we assume that there are no stable attractors in the phase space of the system.

By analogy with Fig. 11.11, Fig. 11.13 presents the resulting phase diagram in polar coordinates (K, Ψ) with K varying within the interval [0; 0.013]. The curve BC denotes the line of the boundary crisis, and the region of the parameter space without attractors is colored with black.

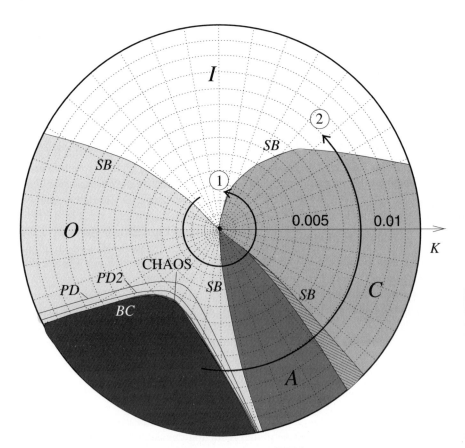

Fig. 11.13. (Color online) Phase diagram for the coupled MVP model (11.25) at $\mu = 1.2$ at the finite coupling strength. The polar coordinates are the coupling angle Ψ and the coupling strength K. Other notations are described throughout the text

11.5 Synchronization near the Homoclinic Bifurcation

Similarly to the case of the weak coupling limit, the system demonstrates three synchronization states: the in-phase (I), antiphase (A), and out-of-phase states (O). The parameter area of coexisting A and I states is labelled as C. As K becomes larger, the region of each state is deformed with the deflection of the bifurcation curves depending on K. Besides the symmetry-breaking bifurcation similar to the one described above, the states can also undergo other bifurcations such as period doubling, that cannot be predicted by the phase model with effective coupling function. Typical bifurcational transitions are illustrated in Fig. 11.14. They correspond to the variation of Ψ at two different values of K along paths labeled 1 and 2 in Fig. 11.13. The upper horizontal line denotes the in-phase state branch (I), and the lower line denotes the antiphase state branch (A) with solid and dashed curves corresponding to stable and unstable solutions, respectively. Note that the diagram related to path 1 (smaller K) demonstrates the behavior which is qualitatively the same as the one observed in the case of the weak coupling limit (Fig. 11.12).

The path 2 for larger K is more complicated. In particular, it includes the period-doubling routes to chaos. The transition between regions A and I also becomes more complex, since it involves an additional pair of the stable out-of-phase limit cycles, whose area of existence is shown in Fig. 11.13 as a hatched region between areas A and C.

If we start from area A, then with decreasing Ψ the antiphase state loses its stability via the symmetry-breaking bifurcation (curve SB in Fig. 11.13) that gives

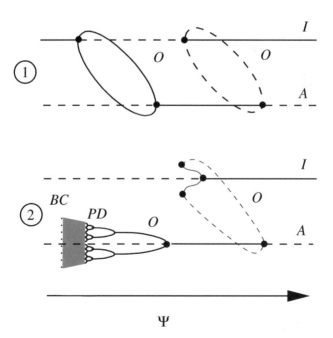

Fig. 11.14. Schematic bifurcation diagrams along paths *1* and *2* in Fig. 11.13, which correspond to $K = 0.0025$ and $K = 0.009$, respectively

288   11 Synchronization of Anisochronous Oscillators

birth to a pair of stable O solutions. For even smaller Ψ, the O states of each branch undergo a cascade of period-doubling bifurcations that lead to the appearance of two chaotic attractors that are symmetrical with respect to each other. As Ψ is decreased further, these two chaotic attractors merge to form a single chaotic attractor and thus to restore the symmetry in the system. The merged chaotic attractor eventually disappears via a boundary crisis, and the trajectory leaves the region of stable oscillations after crossing the line BC in Fig. 11.13. Note that the period-doubling cascade of O solutions also occurs on the other side of the BC region.

Let us consider the transition between regions A and I. As one can see from Fig. 11.14, with decreasing Ψ, I state loses its stability via the symmetry-breaking bifurcation, and a pair of stable out-of-phase cycles appear. With this, at the moment when the A state becomes stable, a pair of unstable O solutions are born. These two pairs of O branches are linked via saddle-node bifurcations. Remarkably, such saddle-node bifurcations are not observed in the weak coupling limit (Fig. 11.11), although they can occur in the phase model when Γ_a in Fig. 11.10 touches abscissa.

11.5.3 Strong Coupling with Moderate μ

In the previous subsection we considered two limit cases: interaction of weakly non-linear (van der Pol like) oscillators and interaction of strongly non-linear oscillators, when the coupled units operate close to the homoclinic bifurcation. In the latter case, the increase of coupling parameter K provides the stronger perturbation of each oscillator and the systems are pushed out of the self-sustained regime via the boundary crisis.

Here we examine the behavior of the coupled MVP model (11.25) in the regime with moderate $\mu = 1.0$, but with strong coupling $K = 0.23$. It is interesting to find out if there are any new regimes and transitions as compared to the cases discussed above.

Figure 11.15 represents the phase diagram on the (K, Ψ) parameter plane. The structure of the bifurcation diagram becomes more complicated as compared to Figs. 11.11 and 11.13. The in-phase synchronous states, in addition to the symmetry-breaking bifurcations, demonstrate supercritical (PD) and subcritical (PDS) period-doubling bifurcations, and torus birth, a Neimark–Sacker (T), bifurcations.

The supercritical period-doubling bifurcation gives birth to a stable period-doubled in-phase state. With variation of the parameters, this period-doubled in-phase solution undergoes the symmetry-breaking bifurcation that gives rise to a pair of the out-of-phase regimes. The out-of-phase states are involved in the cascade of period-doubling bifurcations that leads to chaos in a way similar that observed for O states in Fig. 11.13.

The subcritical period-doubling bifurcation entails no attractors, but is related to the boundary crisis (lines PDS outlining the black area in the diagram). After this crisis no attractors exist in the region of the phase space considered.

The torus birth bifurcation of the in-phase solutions occurs on the curve T when a pair of complex-conjugate Floquet multipliers leave the unit circle. Below this curve there exist Arnold tongues corresponding to the frequency-locked states

11.5 Synchronization near the Homoclinic Bifurcation

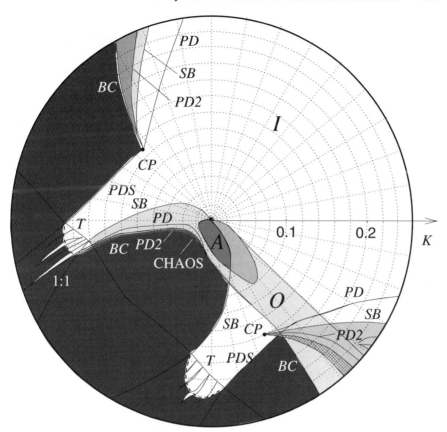

Fig. 11.15. (Color online) Phase diagram for the coupled MVP model (11.25) at $\mu = 1.0$ for strong coupling. Notations for bifurcation lines and points are as follows: *SB*—symmetry-breaking bifurcation; *PD*—supercritical period-doubling bifurcation; *PD2*—second supercritical period-doubling bifurcation; *PDS*—subcritical period-doubling bifurcation; *T*—torus birth, a Neimark–Sacker, bifurcation; *CP*—cusp point; and *BC*—boundary crisis

with rational rotation numbers (Fig. 11.15). The most prominent among them is the tongue of the 1 : 1 locking. Note that the tongues can persist even in the absence of the stable torus nearby in the parameter space, after the torus has disappeared via a boundary crisis.

11.5.4 Summary on Synchronization near Homoclinic Bifurcation

Synchronization between the coupled oscillations acquires special features as the limit cycle in any of the subunits involved approaches the homoclinic bifurcation. The homoclinic bifurcation implies the presence of a saddle point nearby the limit cycle, and the latter causes dephasing. Namely, the dephasing rate calculated as the linear rate Q in (11.26) is shown to increase dramatically as the limit cycle

290   11 Synchronization of Anisochronous Oscillators

approaches the homoclinic bifurcation. Although dephasing is an effect which is localized in the phase space, it affects the behavior of coupled oscillators in a wide range of their control parameters.

In the weak coupling limit the use of the effective coupling function allows one to reveal the existence of three main synchronization states and identifies the transitions between them through the symmetry-breaking bifurcations. At the finite coupling strength the effective coupling approach fails, and we have resorted to the direct numerical calculations. It has been shown that the in-phase synchronization might not be the only stable state in a system of diffusively coupled oscillators when a limit cycle approaches the homoclinic bifurcation. A variety of complex transitions, including the period-doubling and the torus birth bifurcations, the mode-locking regimes and chaos arise as the coupling strength becomes larger, even if the control parameter is selected below the homoclinic bifurcation.

The modified van der Pol model introduced as a simple modification of the generic van der Pol oscillator mimics the dynamical patterns typical of neuron models. Another more realistic example of a system with similar features will be considered in Sect. 11.7 below.

11.6 Phase Locking Patterns of Coupled Fast-and-Slow Oscillators

In the previous section we discussed how synchronous dynamics changes when a limit cycle approaches the vicinity of a saddle point. The mechanism of dephasing described in Sect. 11.5 requires the presence of a singular point (equilibrium). With this, the difference between the time scales of different variables of the same oscillator might not be pronounced. Below we consider another situation that occurs in systems demonstrating motion involving fast and slow time scales which will be referred to as fast-and-slow motion.

11.6.1 Antiphase Locking in Coupled FitzHugh–Nagumo Models

At small values of time separation parameter ε in the model (11.19), the trajectory of the limit cycle is split into intervals of fast horizontal jumps and slow drifts up and down along the right and left branches of the cubic nullcline, respectively, as schematically shown in Fig. 11.16. The smaller the values of ε, the more pronounced such fast-and-slow dynamics is.

When two models (11.19) are diffusively coupled, the fast-and-slow dynamics of the individual units underlies their mutual adjustment. At $\varepsilon \to 0$, segments of fast motion turn into instantaneous jumps and the first equation in (11.19) converges to the nullcline equation $x - x^3/3 - y = 0$. Theoretical results on the phase equation for the weakly coupled relaxation oscillators for such relaxation limit have been obtained in [127, 149]. Application of this approach to (11.19) shows that (i) the rate of convergence to the in-phase state is relatively fast compared to the case of

11.6 Phase Locking Patterns of Coupled Fast-and-Slow Oscillators

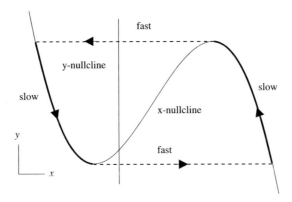

Fig. 11.16. Nullclines and the periodic solution of the model (11.19). Motions along the *branches* can be fast or slow

smooth oscillators, and (ii) the antiphase synchronous pattern can be stable. The latter is realized when the limit cycle spends more time on one of the branches of the fast nullcline and when the motion on that branch slows down before the jump point [127]. Note that the above results are valid in the relaxation limit $\varepsilon \to 0$ but should be used carefully for finite values of ε.

The results of numerical calculation of effective coupling function within certain ranges of ε and a parameter values are summarized in Fig. 11.17. The vertical line at $a = 1.0$ corresponds to the supercritical Andronov–Hopf bifurcation. At $a > 1.0$ the individual systems are in excitable regime, thus there are no oscillations to be synchronized. At $a < 1.0$ there is an area where the antiphase locked regime is stable both for x-coupling (shaded light grey) and for y-coupling (shaded dark grey).

Let us first consider x-coupling. Figure 11.17 illustrates that for small a the stable antiphase regime is observed at extremely small values of ε. However, for moderate values of a there is a wide range of ε where antiphase locked regime is stable. Close to the Andronov–Hopf bifurcation line the area of antiphase solution sharply shrinks. For the y-coupling, the stability area for the antiphase locked regime is much smaller and located at $a \in [0.8; 1.0]$.

Thus, when $a \leq 1.0$ both x- and y-coupling lead to antiphase synchronization in addition to the in-phase locked regime. Two insets in Fig. 11.17 show the representative examples of the effective coupling function Γ_a. It is clearly seen that in the light-grey area of the diagram, the curve Γ_a corresponding to x-coupling (solid line) has four zeroes with negative slope at $\delta\phi = 0$ and at $\delta\phi = \pi$. With this, there are two unstable out-of-phase solutions. The curve for y-coupling (dashed line) related to the dark grey part of the diagram has a similar form.

The qualitative explanation of the observed effects can be given in terms of superimposed phase plane and of geometrical interpretation of coupling. Let us calculate how much time the phase point spends on the particular segments of the limit cycle. The left panel in Fig. 11.18 represents the probability density distribution P of the phase points at $a = 0.9$ and $\varepsilon = 0.05$ along the limit cycle discretized in

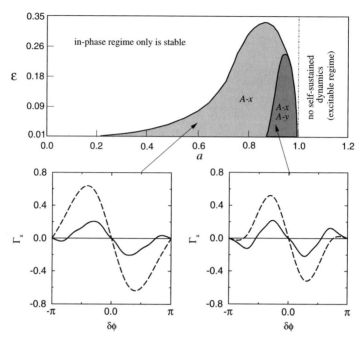

Fig. 11.17. Antiphase synchronization in two FitzHugh–Nagumo models (11.19) with one-variable coupling. The corresponding parameter area for x-coupling is shaded by *light grey* and labeled $A - x$. The same parameter area for the y-coupling is shaded by *dark grey* and labeled $A - y$. *Inserts* show the antisymmetric part of the effective coupling function (11.13) with *solid* and *dashed lines* for the x- and y-coupling, respectively

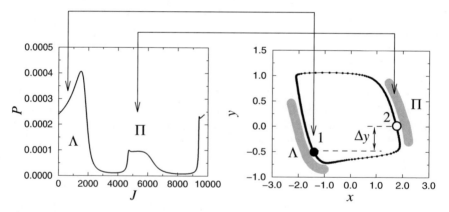

Fig. 11.18. (Color online) **Left panel**: Probability density distribution along the trajectory of a limit cycle. Two areas L and M where the phase point spends most time are clearly distinguished. **Right panel**: Position of both subsystems and interacting force are shown schematically in the superimposed phase plane. Areas of slow motion are shaded *grey*

11.6 Phase Locking Patterns of Coupled Fast-and-Slow Oscillators   293

10 000 points. Λ and Π denote the groups of points that correspond to the left and right branches of slow motion in Fig. 11.16, respectively. The right panel shows the superimposed phase plane. The filled and open circles denote the first and the second oscillator, respectively.

Let us first consider y-coupling. It vanishes if at a certain time moment vertical levels of points 1 and 2 are the same. If the phase point of the second subsystem is located higher, it is being pulled down by the first subsystem. Due to symmetry of the coupling term, the phase point of the first subsystem is pulled upwards by the second one. Thus, both systems are slowing down. However, the area Λ where the phase points of the first subsystem are accumulated, is much slower than the area Π, therefore the coupling-induced perturbation of the first subsystem is less than of the second one. As a limit case, we can assume that the first subsystem remains in point 1. Hence, coupling will try to hold point 2 at the same y level, by slowing it down when its y-position is too high, and by accelerating if it is too low. Note that both clouds of phase points 1 and 2 move and the process described repeats in time. As a result, the phase lag of the synchronized regime is determined by the reciprocal arrangement of the phase points in clouds 1 and 2 as illustrated in Fig. 11.18. For limited cases of identical coupled systems this corresponds to antiphase locking.

For x-coupling the similar mechanism takes place. Since the second subsystem is on the fast upper branch (see Fig. 11.16), the effect is even more pronounced. This explains why the x-coupling provides wider region of the antiphase locking in the diagram in Fig. 11.17.

11.6.2 Out-of-phase Synchronization via Slow Channels

The key point of the mechanism of the antiphase locking described above is *asymmetric* anisochronous motion on a periodic orbit. However, there is another anisochronous mechanism of the phase adjustment based on symmetric geometry of the limit cycle.

Let us consider how the phase velocity field changes in the van der Pol oscillator with increasing parameter α in (11.16). The representative snapshots of phase dynamics are shown in the left panels of Fig. 11.19 where a phase portrait of periodic oscillations is shown as a solid line, and phase velocity magnitude is coded by color gradient. The top panel illustrates the regime of smooth oscillations at $\alpha = 0.2$. In this case, the phase velocity field on the limit cycle does not demonstrate essential variations, thus, the motion is close to isochronous.

For larger $\alpha = 1.5$ (middle panel in Fig. 11.19), there is considerable inhomogeneity in the velocity along the trajectory on the limit cycle. One can distinguish two narrow symmetrically located "channels" of slow motion that approach the limit cycle from the left and from the right. They are situated in the area where both x- and y-nullclines defined as

$$y = \frac{x}{\alpha(1 - x^2)}, \qquad y = 0 \qquad (11.28)$$

come close to each other, but do not intersect. For brevity, thereafter we call them *slow channels*.

294   11 Synchronization of Anisochronous Oscillators

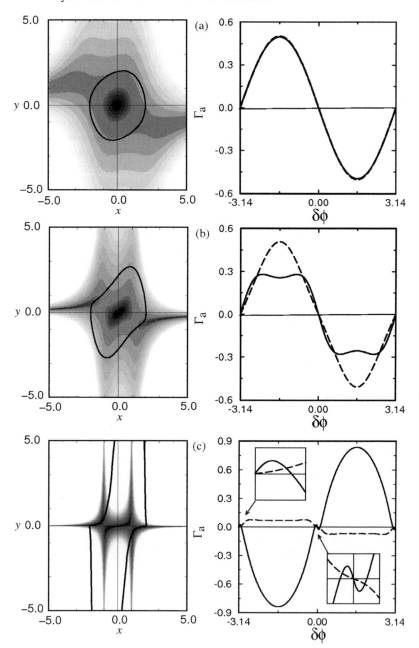

Fig. 11.19. Formation of slow channels in two coupled van der Pol oscillators (11.14), (11.16). **Left panel**: the phase velocity field (*gradient plot*) with superimposed limit cycle trajectory (*solid line*) for **a** $\alpha = 0.2$, **b** $\alpha = 1.5$, and **c** $\alpha = 10.0$, respectively. Areas of slow motion are shaded *dark grey*. **Right panel**: The corresponding evolution of effective coupling function Γ_a for x- and y-coupling (*solid* and *dashed lines*, respectively)

11.6 Phase Locking Patterns of Coupled Fast-and-Slow Oscillators 295

At $\alpha = 10.0$ the slow channels are fully developed (the bottom left panel in Fig. 11.19). For such high non-linearity the trajectory on the limit cycle runs very close to nullclines and inevitably passes through both slow channels spending there most of the time within a period of oscillations.

Let us now investigate synchronization features of coupled oscillators using the effective coupling Γ_a approach (right panels in Fig. 11.19). For small non-linearity $\alpha = 0.2$ (top panel), and correspondingly weak anisochronicity, the curves of Γ_a calculated for x-coupling (solid line) and y-coupling (dashed line) almost coincide. Any one-variable coupling, as well as any of its combinations, provide only in-phase synchronization: the effective coupling curve has one zero with negative slope at $\delta\phi = 0$. For $\alpha = 1.5$ (middle panel), the curves for x- and y-coupling have different shapes, but there is a still single stable synchronous regime with zero phase lag $\delta\phi$. However, at $\alpha = 10$ (bottom panel), one can observe significant changes in synchronization patterns. The x-coupling curve has more zeroes located nearby the in-phase and antiphase states. Inspection of the plot shows that (i) the in-phase regime is still stable, but its basin of attraction is determined by two unstable out-of-phase solutions and is thus narrow (right inset), and (ii) the antiphase regime is still unstable but there are two stable out-of-phase states nearby. In this way, no qualitative changes are observed for the y-coupling.

To explain the observed phenomena we use a superimposed phase plane and the geometrical interpretation of coupling. Note that:

- Both interacting subsystems spend most of the time in the slow channels, thus, we can focus on this state assuming that the contribution from the other segments of the limit cycle is small
- The antiphase locking is geometrically represented by the symmetric location of phase points in the left and right slow channels

Suppose that the first (leading) subsystem reaches the slow channel, while the second (lagging) subsystem is sufficiently delayed (Fig. 11.20(a)). We consider the leading subsystem being in almost resting state since its position changes very slowly. Without y-coupling, the phase point of lagging subsystem should follow the unperturbed limit cycle trajectory (solid line). Once coupling is introduced, the phase trajectory of the lagging subsystem is "pulled" up by the leading subsystem by means of a coupling force. The perturbed trajectory is located closer to the center of the limit cycle. The motion on this trajectory segment is fast, so the interaction time and, hence, the resulting perturbation of trajectory is quite small. However, it results in skipping some piece of the slow channel, that will save time considerably. Thus, the described mechanism accelerates the lagging subsystem to diminish the initial phase lag.

Let us consider small perturbations of antiphase locking (Fig. 11.20(b)). The exactly symmetrical location of the leading subsystem with respect to the lagging subsystem is marked by a cross. Assume that a small perturbation brings the leading subsystem to the position indicated by filled circle. Two effects can be distinguished: (i) the phase velocity rises since the trajectory now is close to the exit from the slow channel, and (ii) the coupling force becomes weaker because the y-distance is

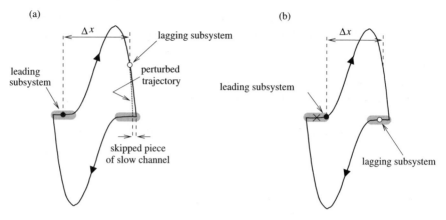

Fig. 11.20. On the mechanism of formation of out-of-phase stable regimes **a** when phase lag is considerable and **b** when it is small

reduced. Thus, the lagging subsystem is not pulled or pushed to diminish the time lag. With this, anti-phase regime might be unstable.

In summary, the antiphase regime is found to be locally unstable, but a close-to-antiphase state could be attractive when a time lag between systems is considerable. This statement is in a good agreement with the prediction provided by the effective coupling function (Fig. 11.19). Note that the in-phase regime is stable as long as both subsystems are in the same slow channel. As one of subsystems escapes from it, the situation depicted in Fig. 11.20(a) occurs. This explains why the attraction basin of the in-phase state is quite narrow.

11.7 Synchronous Patterns in Coupled Morris–Lecar Models

In the previous sections we discussed the main mechanisms for the formation of antiphase and out-of-phase synchronous patterns in coupled anisochronous oscillators. So far we have used simplest models. In this section, we investigate in detail the synchronization patterns that can be observed in a more realistic model with an arbitrary coupling strength.

In this respect, several important questions arise: What are the bifurcational transitions to and between the coexisting synchronous states? What are the scenarios leading to the breakdown of quasiperiodic motion and to chaotic bursting? To what extent can we understand the complexity of the cooperative dynamics of general anisochronous oscillators, if interaction of even simple periodic oscillators leads to quite complicated behavior?

11.7.1 Model

The corresponding model equations (11.21) are described in Sect. 11.4. According to the bifurcation diagram in Fig. 11.6, the homoclinic bifurcation takes place at

11.7 Synchronous Patterns in Coupled Morris–Lecar Models 297

$J \approx 0.0729$. We fix $J = 0.0750$ which corresponds to the regime of periodic oscillations close to the homoclinic bifurcation. In this regime an individual Morris–Lecar model demonstrates a pacemaker activity, but an isolated resting state (stable equilibrium) also exists, being separated from the limit cycle by the stable manifold of a saddle point.

Consider two mutually and diffusively coupled Morris–Lecar (ML) models

$$\frac{du_{1,2}}{dt} = -J_{\text{ion}}(u_{1,2}, w_{1,2}) + J_{1,2} + \gamma \cos \Psi (u_{2,1} - u_{1,2}),$$
$$\frac{dw_{1,2}}{dt} = f\frac{w_\infty(u_{1,2}) - w_{1,2}}{\tau_w(u_{1,2})} + \gamma \sin \Psi (w_{2,1} - w_{1,2}). \tag{11.29}$$

In order to compare their dynamics with the behavior of the coupled modified van der Pol (MVP) oscillators considered above, we plot a bifurcation diagram in the parameter plane of coupling angle Ψ and coupling strength γ (Fig. 11.21) in a similar way as in Fig. 11.15 of Sect. 11.5.3.

Note that as in the case of coupled MVP oscillators, the interacting ML models demonstrate three distinctive synchronous states: in-phase, antiphase, and out-of-phase. The main dissimilarity between the Morris–Lecar models and the modified van der Pol oscillators is that areas I and A are located in a different range of the coupling angle (compare with Figs. 11.15 and 11.21). Thus, coupling via w-variables for the ML models leads to the synchronous behavior that is qualitatively similar to the behavior of MVP oscillators coupled via x-variables. Besides the above mentioned discrepancy, both models demonstrate similar bifurcation transitions and coexisting synchronous patterns. The comparison of the diagrams shows that both systems demonstrate stable antiphase synchronization in a wide range of the parameters of the diffusive coupling and similar transitions between the states under the variation of coupling angle Ψ. Moreover, both Figs. 11.15 and 11.21 demonstrate the existence of a cusp point (CP) at which several regions with different states merge. Thus, coupled Morris–Lecar models indeed show cooperative dynamics that is typical of simplified (MVP) models of oscillators near the homoclinic bifurcation considered in the previous sections.

When studying neuron models, it is important to bear in mind that not every type of coupling is biologically approved. With this, we focus on the particular type of one-variable coupling which is realized via variable u representing a transmembrane voltage. Then we come to the following form of the model equations:

$$\frac{du_{1,2}}{dt} = -J_{\text{ion}}(u_{1,2}, w_{1,2}) + J_{1,2} + \gamma (u_{2,1} - u_{1,2}),$$
$$\frac{dw_{1,2}}{dt} = f\frac{w_\infty(u_{1,2}) - w_{1,2}}{\tau_w(u_{1,2})}, \tag{11.30}$$

where γ is the coupling strength and the subscripts 1 and 2 denote the first and the second neuron, respectively. The above-introduced coupling represents the so-called gap junction between the neurons, when two intracellular volumes are connected by means of ion-permittable channels that provide diffusive processes between the

298   11 Synchronization of Anisochronous Oscillators

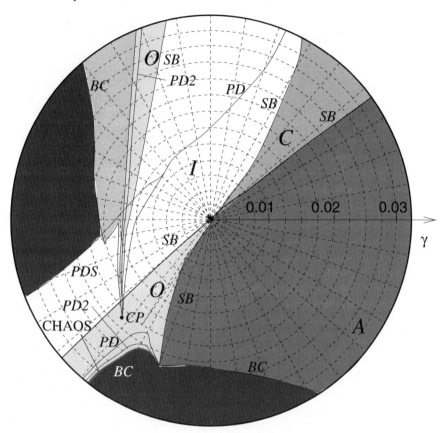

Fig. 11.21. (Color online) Phase diagram for the coupled Morris–Lecar models (11.29) at $J = 0.0750$. In-phase, antiphase, and out-of-phase synchronization patterns are labelled as I, A, and O, respectively. Overlapping of I and A is labelled as C. As in Sect. 11.5, *SB* denotes the symmetry breaking bifurcation, *BC* denotes the boundary crisis, and *CP* is a cusp point. Lines of the first and the second supercritical period-doubling bifurcations are labelled as *PD* and *PD2*, respectively. Subcritical period-doubling bifurcation is labeled as *PDS*. The region of chaos following the period-doubling cascades is denoted by the *hatched area*

cells. In terms of the diagram Fig. 11.21, we deal with the case of $\Psi = 0$, i.e., we require that only the strength of coupling γ is changed.

11.7.2 Overview of the Dynamics

Figure 11.22(a) presents an overview of the dynamical regimes of the coupled ML system in two-dimensional parameter space (J_2, γ). We have to note that the parameters $J_{1,2}$ affect the frequencies of oscillations in the interacting subsystems. Since in our study we fix $J_1 = 0.075$, J_2 determines the frequency mismatch, or detun-

11.7 Synchronous Patterns in Coupled Morris–Lecar Models

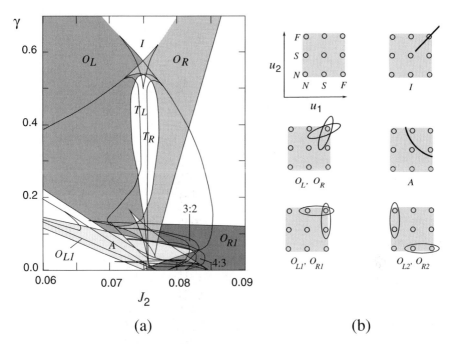

Fig. 11.22. (Color online) **a** Overview of dynamical regimes for the diffusively coupled Morris–Lecar models (11.30). Different synchronous regions are denoted by different symbols and shadings. **b** Qualitatively different dynamical states are characterized in terms of the dynamical features of the single oscillator, that is, nine different combinations of the equilibrium points

ing. Thus, we have a bifurcation diagram in the "classical" parameter space used in synchronization problems: "frequency detuning–coupling strength."

To mark qualitatively different dynamical states, we use different shadings in grey scale. It is convenient to characterize new regimes induced by coupling in terms of states typical of an unperturbed single oscillator. Each of the subsystems has three coexisting equilibrium points (Fig. 11.6): a stable node N, a saddle S, and an unstable focus F. Thus, for coupled systems there are nine different equilibrium states (Fig. 11.22(b)). Possible synchronous regimes can therefore be treated as oscillations in the vicinity of one of nine equilibrium points:

- The in-phase regime I: When $J_1 = J_2$, the time evolution of two oscillators coincides completely, and the in-phase attractor of the coupled system belongs to the symmetric subspace $v_1 = v_2$, $w_1 = w_2$. When $J_1 \neq J_2$, the in-phase attractors slightly deviate from the perfect symmetric state. For this type of the symmetric solution, the trajectories projected in the subspace $(u_{1,2}, w_{1,2})$ are similar to those of uncoupled system. In Fig. 11.22(b), I is represented by a diagonal line near the equilibrium point.
- The antiphase regime A: When $J_1 = J_2$, the phases of the two oscillators are shifted by π. When $J_1 \neq J_2$, the phase shift is not exactly π, but the regime

300   11 Synchronization of Anisochronous Oscillators

keeps the main features. In (b), it is depicted as a curve that is perpendicular to the diagonal direction.

- The out-of-phase regime O: When $J_1 = J_2$, there exists a pair of solutions, O_L and O_R, with a reflection symmetry with respect to the change of coordinates $(u_1, w_1) \longleftrightarrow (u_2, w_2)$. They are a mirror image of each other. The subscripts L and R are used for states $J_2 < J_1$ and $J_2 > J_1$, respectively. For $J_1 \neq J_2$, the phase relation between the two oscillators changes continuously, therefore, labelling O_L and O_R is preserved. These two solutions are schematically drawn in the left panel of the middle row of Fig. 11.22(b). With varying parameters, O_L and O_R are transformed into the quasiperiodic solutions T_L and T_R via a torus birth (Neimark–Sacker) bifurcation. When one of the two oscillators is winding around the fixed point F or N, and the other oscillates around F, the joint dynamical state is denoted by $O_{L1,R1}$, or $O_{L2,R2}$, respectively (bottom row in the figure).

Besides the in-phase (I), the antiphase (A), and the out-of-phase (O) states, a variety of synchronous periodic solutions appear in the phase space. These solutions lying on the resonant tori exist in the horn-like regions of the diagram. Inside each of those regions, the frequency ratio between the two oscillators is locked to $p:q$, where p and q are integers [30].

As one can see from Fig. 11.22, the multistability phenomena, i.e., the coexistence of several stable solutions, is one of the most prominent dynamical features of the diffusively coupled ML oscillators. For a weak coupling ($\gamma < 0.1$), the antiphase synchronization A and higher-order resonant solutions are typical states. For intermediate coupling, the out-of-phase solutions undergo a sequence of bifurcations leading to the onset of the quasiperiodic behavior and chaos. For strong coupling ($\gamma > 0.4$), the stable in-phase synchronous regime dominates. Below we consider the most important bifurcation scenarios between the described dynamical states that are related to anisochronous properties of the individual subsystems that are close to a homoclinic bifurcation.

11.7.3 Structure of Arnold Tongue for Antiphase Solution

Most studies on the interacting nonlinear oscillators were focused on the identification of the generic bifurcations leading to synchronization [34, 35, 240]. As we already know from Part I of this book, in the case of weak coupling, one of the three general mechanisms of synchronization is the formation of a resonant torus via the saddle-node (SN) bifurcation of a pair of cycles on ergodic torus. With this, the mechanisms leading to the breakdown of resonant torus remain one of the interesting research topics. Several different schemes of resonant torus breakdown were reported, depending both on the type of the non-linear oscillators and on the configuration of coupling [138, 285]. Most studies, however, were focused on the breakdown of the in-phase resonant torus, where the stable solution is an in-phase regime. We now analyze the formation and the destruction of the antiphase resonant solution.

11.7 Synchronous Patterns in Coupled Morris–Lecar Models

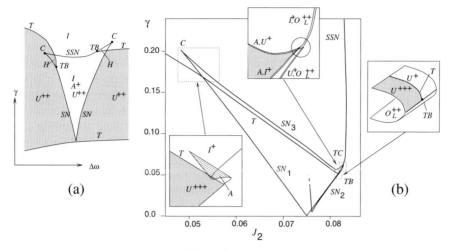

Fig. 11.23. (Color online) **a** A typical bifurcation structure in the vicinity of the 1 : 1 in-phase synchronization region at weak coupling. **b** Details of the bifurcation diagram in Fig. 11.22(a) in the vicinity of the 1 : 1 antiphase synchronization regions. Here, I, A^+, and U^{++} are stable, saddle, and twice saddle solutions, respectively. SN is a saddle-node bifurcation curve for a stable and saddle cycles, SSN is a saddle-node bifurcation curve for a saddle and a twice saddle cycles, T is a torus birth bifurcation, and H is a homoclinic bifurcation curve. The Takens–Bogdanov, the transcritical, and the cusp points of co-dimension-two are denoted as TB, TC, and C, respectively

To compare the cases of in-phase and of antiphase synchronization, let us briefly summarize the structural features of synchronization region for the in-phase resonant solutions. Figure 11.23(a) illustrates the bifurcation curves typically observed in the case of in-phase synchronization that were discussed in Chaps. 4 and 5. For weak coupling, two resonant periodic solutions, one being in-phase solution denoted as I, and the other antiphase solution A^+, are generated via a SN bifurcation. The resonant region is bounded by two curves of saddle-node SN bifurcations. To characterize the stability of the periodic solutions mentioned above, we use superscript "+" for each direction of the instability. For example, I, I^+, and I^{++} denote a stable, a saddle with one direction of instability, and a twice saddle (with two directions of instability) solution with two directions of instability, respectively. Within the synchronization region, two oscillators are synchronized through the frequency/phase locking. Note that the periodic solution U^{++} is a twice saddle limit cycle which is located inside the phase-locked region. For a large coupling strength, the resonant torus apparently disappears. At co-dimension-two Takens–Bogdanov (TB) bifurcation point [53, 282], the saddle-node SN bifurcation between a stable and a saddle cycle changes to SSN bifurcation, which is the saddle-node bifurcation between a saddle and a twice saddle cycle. A torus birth bifurcation curve (T) emanates from the TB point. This bifurcation corresponds to the suppression of natural dynamics in one of the systems [15, 205, 285]. The upper boundary of the resonant

302   11 Synchronization of Anisochronous Oscillators

region is the bifurcation curve denoted as *SSN*. On this curve, the saddle cycle A^+ and the twice saddle solution U^{++} merge together to disappear. Above this curve, the resonant torus does not exist any longer. However, the stable periodic solution I is not involved in the process of the torus destruction. Hence, the in-phase synchronization region extends up to high values of coupling strength.

An important question arises: What will happen with the structure of synchronization region if the in-phase solution is unstable while the antiphase solution is stable? This situation can arise, for instance, in weakly interacting ML neural oscillators with diffusive coupling, when the individual oscillators are close to the homoclinic bifurcation. Taking into account the dynamical regimes presented in Fig. 11.22, we focus on the bifurcations leading to the 1 : 1 resonant solution. In Fig. 11.23(b) the area around 1 : 1 antiphase synchronization tongue in two coupled Morris–Lecar systems (11.30) is shown (compare with Fig. 11.22(a)).

Within the main 1 : 1 synchronization region in Fig. 11.23(b), there are three periodic solutions: the stable antiphase solution A, the saddle in-phase solution I^+, and the unstable solution U^{+++} (the latter means that the solution has no stable manifolds). The unstable solution U^{+++} corresponds to a topological product of an unstable fixed point with an unstable periodic solution which appears via a subcritical Andronov–Hopf bifurcation at $J = J_B = 0.0756$ (for the bifurcation diagram of a single ML system, see Fig. 11.6).

If the coupling strength is small enough, the synchronization tongue is bounded by two curves of saddle-node bifurcation, SN_1 and SN_2, where the stable cycle A and the saddle cycle I^+ are born. As the coupling strength increases, the unstable cycle U^{+++} undergoes an inverse torus birth bifurcation (the curve T) and becomes a saddle cycle U^+. Then it collides with the stable cycle A, and they disappear via a saddle-node bifurcation (the curve SN_3) at the top of the synchronization region. This transition is also schematically illustrated in Fig. 11.27, bottom panel. Above the curve SN_3 the stable solution no longer exists, and the phase trajectory escapes to one of the four coexisting out-of-phase solutions. The saddle in-phase cycle I^+ persists above the saddle-node bifurcation curve SN_3 and below the saddle-node bifurcation curve *SSN* for a saddle and a twice saddle cycles, where I^+ collides with one of the out-of-phase solutions, either O_R^{++} or O_L^{++}.

As shown in the insets of Fig. 11.23(b), there exist several co-dimension-two bifurcation points. The point TC shown in the upper inset of Fig. 11.23(b) corresponds to a co-dimension-two transcritical bifurcation point. Two pairs of periodic solutions involved in the bifurcations SN_2 and SN_3 merge at this point to change their stability. At a co-dimension-two Takens–Bogdanov TB point (the right inset in Fig. 11.23(b)) the bifurcation curves T and SN_2 merge and two real Floquet multipliers become equal to one. At the cusp point C located in the left upper part of 1 : 1 resonant region, the stable solution A and the two saddle solutions I^+ join together to give rise to cusp structure [32, 235].

Next, we investigate how the phase space structure consisting of the three periodic solutions and a 1 : 1 resonant torus evolves with increasing coupling strength. In Fig. 11.24(a), we plot the Poincaré sections of the three periodic solutions A,

11.7 Synchronous Patterns in Coupled Morris–Lecar Models   303

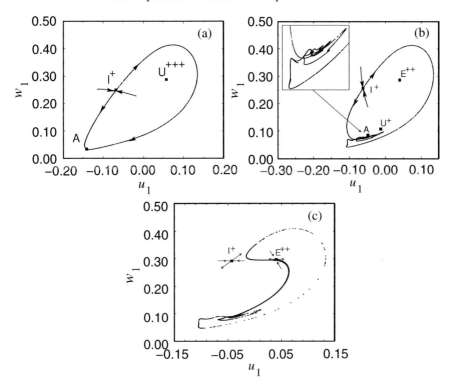

Fig. 11.24. The destruction of resonant torus in 1:1 antiphase synchronization region shown in Fig. 11.23(b) for $J_1 = J_2 = 0.075$ and with increasing γ. The Poincaré sections are shown, namely **a** a smooth torus surface at $\gamma = 0.025$; **b** the wrinkling of the torus in the vicinity of the saddle cycle U^+ at $\gamma = 0.080$; **c** for $\gamma = 0.084$, A and U^+ have disappeared via a saddle-node bifurcation, and the trajectory follows to the saddle-focus fixed point E^{++} ($J_1 = J_2 = 0.075$)

I^+, and U^{+++}, together with the invariant curve representing the projection of the invariant torus manifold. Since two resonant solutions A and I^+ lie on the torus, A and I^+ are located on the invariant curve. Since the solution U^{+++} does not lie on the resonant torus, it is located inside the invariant curve. We calculate the invariant manifold numerically using a modification of the technique suggested by Kevrekidis et al. [138]. Namely, we follow a large number of phase trajectories launched from the initial conditions distributed around the saddle solution I^+. As time passes by, the phase points scatter along the unstable manifold of the I^+ and tend to approach the stable solution A, thus revealing the manifold sought. At any point located deep inside the 1:1 resonance region (Fig. 11.24(a)) this closure is smooth, i.e., the derivative at a node does not suffer any discontinuity.

As shown in Fig. 11.24(b), with increasing coupling strength γ, the smooth resonant torus is destroyed by the folding of the torus surface. The folding occurs nearby the stable antiphase solution A. Note that instead of the saddle solution I^+,

A merges with the saddle cycle U^+ which does not belong to the torus surface. With the further increase of the coupling strength γ, the trajectory leaves the torus because of the crisis that occurs at this point (Fig. 11.24(c)). Thus, the way in which the resonant torus is broken with the increasing coupling is quite different from the way typical for the case of in-phase locking.

Chaotic Bursting

What happens with a trajectory launched in the vicinity of the limit cycle A that has just disappeared (see Fig. 11.24(c))? Can it be reinjected again to this vicinity? If the answer "yes," it belongs to an attracting set. To understand this, we have to take into account that the unstable manifold of the limit cycle U^+ is connected with the stable manifold of the saddle-focus equilibrium point E^{++}, which possesses two-dimensional stable and unstable manifolds (Fig. 11.24(c)). In turn, the unstable manifold of the E^{++} is connected with the stable manifold of I^+. Along the unstable manifold of I^+, the trajectory is reinjected back to the vicinity of the vanished cycle A. This means that there is a closed loop connecting several unstable solutions. Since there are no stable periodic orbits on its way, the trajectory returns again and again to the same part of the phase space. Thus, the connection of the stable and unstable manifolds of several saddle solutions provides a possibility for a new attractor. Since the trajectories on the attractor have to pass through the folding structure, it becomes chaotic.

The temporal behavior of the resulting trajectory is quite remarkable. As illustrated in Fig. 11.25(a), it is formed by two dynamical components: spiking trains, and non-spiking silent zones alternating each other.

Note that the time intervals between two subsequent spikes are of the order 10 (in arbitrary units), while the time distance between the successive spiking trains is of the order 1 000. With this, the interspike interval almost does not change within one spiking train, while the distance between the successive spiking trains depends on the coupling strength. Note that the oscillations within one spiking train also have two components: high-amplitude and medium-amplitude oscillations. This means

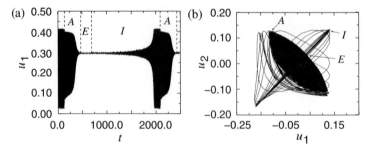

Fig. 11.25. **a** The realization and **b** the phase projection of the chaotic bursting. A, I, and E denote oscillations near the antiphase regime, in-phase regime, and equilibrium state, respectively ($J_1 = J_2 = 0.075$ and $\gamma = 0.084$).

that the resulting dynamics is in fact composed of three different kinds of behavior: a high-amplitude regular spiking, a medium-amplitude regular spiking, and a non-spiking small-amplitude oscillation (a *silent* zone). Projection of the phase portrait onto a phase plane (u_1, u_2) shown in Fig. 11.25(b) clarifies the properties of those three different regimes. As we can see, high-amplitude oscillations correspond to the trajectory nearby the in-phase synchronous state, the medium-amplitude spiking is related to the walking around the antiphase synchronous state, and non-spiking *silent* state reflects small oscillations around the equilibrium point. Thus, the temporal behavior is defined by itinerant phase trajectories that trace from the in-phase state to antiphase oscillations, then to the silent state and back again to the in-phase oscillations.

Such behavior is known as a chaotic bursting [104]. The bursting dynamics has been observed in neuronal systems [9] and in some biological cells [69]. Typically, the bursting behavior occurs in a spike generating system, when an additional slow variable is introduced [107, 126, 244]. This slow component makes the system oscillate between the equilibrium and spiking states, thus producing bursting behavior. However, the bursting dynamics observed in the coupled ML models does not require an additional slow variable. It is caused not by the presence of a slow variable, but by the mutual coupling between interacting subsystems. Thus, the provision of additional (unstable) synchronous states in the dynamics of coupled spiking systems provides an alternative way to bursting activity [103, 104].

In two-parameter bifurcation diagram (Fig. 11.26(a)) the regime of chaotic bursting occupies a triangular white area in the center of the figure. This area is bounded by the line of saddle-node bifurcation *SN* of the limit cycles and the lines of the boundary crisis *BC*. Below the *SN* line the antiphase synchronous solution is stable, and above the line the out-of-phase solutions become stable. On the *BC* line, the chaotic burst attractor undergoes boundary crisis [98, 158], colliding with other limit cycles or with their manifolds. Therefore, within the triangular region, neither the antiphase solution, nor the out-of-phase solutions are stable. In Fig. 11.26(b) the

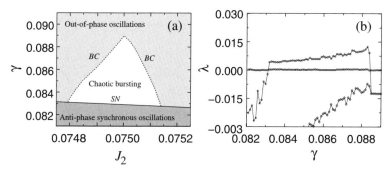

Fig. 11.26. (Color online) **a** Two-parameter bifurcation diagram representing the transition to chaotic burst attractor. *SN* and *BC* denote saddle-node bifurcation and the boundary crisis, respectively. Parameter J_1 is fixed at 0.075. **b** Three largest Lyapunov exponents vs coupling strength γ at $J_1 = J_2 = 0.075$. Chaotic behavior is observed for $\gamma \in [0.08315; 0.08841]$

three largest Lyapunov exponents of the attractor are plotted as functions of coupling strength γ for $J_1 = J_2 = 0.075$. The largest Lyapunov exponent λ_1 has a distinctive positive value within the limited interval of $\gamma \in [0.08315; 0.08841]$ that indicates chaotic dynamics of bursting activity. An abrupt change of λ_1 from negative to positive values is associated with the boundary crisis.

Chaotic Bursting and Torus Breakdown

It is important to emphasize that the formation of chaotic bursting discussed above, represents a specific scenario of resonant torus destruction. The latter is an important topic in the theory of dynamical systems, since it is related to the generic mechanisms of development of deterministic chaos.

Let us compare our findings with the known theoretical results. According to the mathematical theorem on the mechanism of resonant torus breakdown [15, 33] known as the Afraimovich–Shilnikov theorem, the smooth invariant torus can be destroyed in one of the three following ways: (i) near the stable resonant solution, a torus can lose its smoothness via discontinuous folding of the invariant curve in its Poincaré section (see Fig. 11.27, top panel); (ii) breakdown of a torus can be caused by formation of homoclinic structure involving both the stable and unstable manifolds of the saddle resonant solution, and (iii) a torus can be destroyed by period-doubling (or some other) bifurcations of the stable resonant solution. The transition to chaotic bursting described above reminds pretty much the first scenario, but there are some differences. To understand these differences let us take a look at Fig. 11.27 where the upper panels (1a)–(1d) schematically illustrate a "classical" scenario of

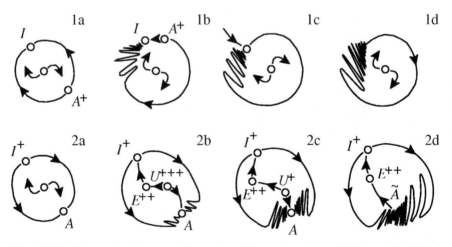

Fig. 11.27. Schematic diagrams illustrating deformation of manifold structure. **Top panel**: Torus destruction according to Afraimovich–Shilnikov theorem (**1a–1d**). **Bottom panel**: The mechanism of torus breakdown leading to chaotic bursting behavior in two diffusively coupled ML models (**2a–2d**)

11.7 Synchronous Patterns in Coupled Morris–Lecar Models   307

torus breakdown, and the lower panels (2a)–(2d) correspond to the bursting transition. The notations in the figure are similar to the ones used in other diagrams. Figures 11.27(1a) and (2a) show that the structures of the manifolds at the tip of the resonant tongue are qualitatively the same in both cases. Also, in both cases the invariant closed curves are formed by the smooth closure of the manifolds of saddle cycles [138, 139]. The phase space structure includes saddle cycle A^+ or I^+ and stable cycle I or A, an unstable equilibrium point E, and two additional saddle limit cycles. The latter saddle cycles are not involved in the transition, and therefore are not shown in the figures. At this stage the only difference between the classical mechanism of torus destruction and the discussed transition to bursting behavior is the stability of the cycles on the resonant torus. So, in the coupled ML models the antiphase solution A is stable, whereas in the classical scenario the stable solution is the in-phase orbit I.

The increase of the coupling strength leads to the formation of the folded structure near the stable limit cycle (Figs. 11.27(1b) and (2b)). This formation is, however, different in these two cases. Namely, the folding in ML model requires an additional bifurcation as a result of which the unstable limit cycle U^{+++} appears from an equilibrium point E^{++++} via an inverse subcritical Andronov–Hopf bifurcation $E^{++++} \longrightarrow E^{++} + U^{+++}$. Thus, the unstable fixed point E^{++++} is transformed into a saddle-focus point E^{++}, whose stable manifold comes from U^{+++}, and the unstable manifold connects to the stable manifold of the saddle cycle I^+.

Between Figs. 11.27(1a) and (1b), the control parameter is changed in such a way that the systems move towards the boundary of the resonant tongue, on which the stable and saddle cycles collide in the saddle-node bifurcation as shown in (1c). Between Figs. 11.27(2a) and (2b), the increase of the coupling strength along the line $J_2 = 0.075$ also leads to a saddle-node bifurcation that occurs at $\gamma = 0.08315$ (2c). In between (2b) and (2c) the unstable cycle U^{+++} undergoes an inverse torus birth bifurcation and becomes a saddle U^+, which then collides with the stable resonant cycle A. Thus, only in the classical scenario (1c) the stable and the saddle limit cycles are both lying on the same torus at the moment of saddle-node bifurcation.

Although in both cases Figs. 11.27(1c) and (2d) periodic attractors disappear as a result of the saddle-node bifurcation, they leave a "ghost" formed by the folds of the invariant curve, where the phase trajectory spends quite a long time before it finally escapes along the unstable manifold (see stages (1d) and (2d)). However, the way to escape from this region depends on the scenario. In the classical case (1d), the escaped trajectory moves along the invariant closed curve and is then reinjected back to the area of the ghost. In the ML system (2d), the escaped trajectory does not follow the former torus surface because the saddle cycle I^+ still exists. After a long wandering across $\tilde{A}$, the trajectory reaches the vicinity of the point where the stable manifold of equilibrium point E^{++} is connected with the folded invariant curve $\tilde{A}$. Here, the trajectory leaves $\tilde{A}$ and first approaches E^{++}, then, following its unstable manifold, finally arrives at the vicinity of I^+. After spending some time in the vicinity of I^+ the phase trajectory is reinjected back into $\tilde{A}$. In general, the phase

308   11 Synchronization of Anisochronous Oscillators

state of the system sequentially passes the stages of antiphase, small-amplitude, and in-phase motion, producing chaotic bursting dynamics.

Thus, the torus breakdown in coupled ML models differs substantially from the classical scenario. This seems to be a typical mechanism of torus destruction for high-dimensional systems that demonstrate similar "itinerant" dynamics associated with chaotic bursting [109, 147].

11.7.4 Crises at the Boundary of Quasiperiodic Regions

In the previous subsection, we investigated the structure of the $1:1$ resonance. However, the region of the existence of the invariant (resonant or non-resonant) torus is not limited by the main $1:1$ synchronization region. For $J_2 \in [J_A = 0.0730; J_D = 0.0845]$ and $J_1 = 0.075$, there are regions of higher order resonant tori (Fig. 11.28). We observe alternating of non-resonant and resonant regions with different winding numbers $p:q$. We would like to illustrate three different characteristic routes leading to the destruction of the quasiperiodic solution.

Route A

As J_2 approaches the homoclinic bifurcation point $J_A = 0.0730$, several resonant tori with the winding numbers $p:q$ less than 1 change each other in the phase space. In the inset of Fig. 11.28, a cascade of period-doubling bifurcations for several resonant solutions is illustrated, see lines on the top of each tongue. This bifurcation sequence results in the appearance of chaos, whose area of existence is shaded by dark grey color. This scenario of resonant torus breakdown is in a full agreement with the scenario suggested by Afraimovich and Shilnikov [15, 33]. At some critical values of J_2, the chaotic states disappears via crisis, and the system goes out of self-sustained regime (light grey area). Remarkably, with decreasing J_2, the curves of period-doubling bifurcations and the critical curve of the transition to chaotic behavior are shifted downwards along the axis γ.

Route B

For $J_2 \in [0.081; 0.083]$, we observe the resonant tori with winding numbers $9:7$, $4:3$, and $3:2$, etc. This region is characterized by hysteresis phenomenon that was shown to be prominent in strong resonances.[5] In this case, the mechanism of the transition from non-synchronous to synchronous behavior involves a homoclinic bifurcation [220].[6] Route B assumes two main ways of the evolution of a synchronous state on a resonant torus:

---

[5] Following the widely accepted definition, we refer to strong resonances the cases of the synchronization $p:q$, for which $q < 4$, see [30] for details.

[6] See also Chap. 5.

11.7 Synchronous Patterns in Coupled Morris–Lecar Models

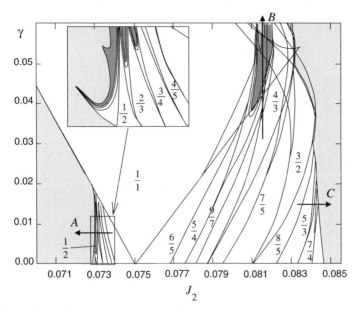

Fig. 11.28. (Color online) Two-parameter bifurcation diagram illustrates the higher order resonance $p:q$ regions. The three different routes leading to the destruction of quasiperiodic regimes are marked as A, B, C

(i) Similarly to route A, the synchronous state demonstrates a cascade of period-doubling bifurcations destroying an invariant torus and leading to chaos. For the example considered, this scenario arises in the weak $9:7$ resonance. As γ increases, the chaotic attractor touches the boundary of its basin of attraction. Since it takes relatively large time before the trajectory arrives at the boundary crisis area, the so-called "chaotic transient" is observed within the narrow range of the coupling parameter above the boundary crisis.

(ii) A period-doubling cascade is not developed inside the resonant region. This case is illustrated in more details, in Fig. 11.29 where the enlargement of strong $3:2$ and $4:3$ resonance tongues of the bifurcation diagram in Fig. 11.28 is presented. Figure 11.29 shows that in both synchronization regions the stable resonant solution loses its stability via a torus birth bifurcation (line T) with the increasing coupling strength. Since the newly born torus is not stable, the trajectory leaves the vicinity of the unstable resonant cycle for another attractor. Note that a pair of resonant limit cycles, namely a saddle and a twice saddle ones, disappear at the two SSN curves via saddle-node bifurcations. The two SSN curves meet at one point to make a cusp structure C.

Route C

When the systems are uncoupled, a saddle-node bifurcation occurs at $J_2 = J_D$ (see Fig. 11.6), where the stable and the unstable periodic solutions collide and disap-

310   11 Synchronization of Anisochronous Oscillators

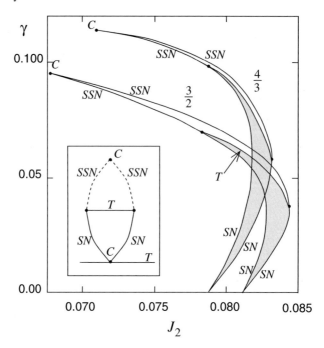

Fig. 11.29. (Color online) Enlargement of the bifurcation diagram in Fig. 11.28 showing 3 : 2 and 4 : 3 resonance tongues. The *inset* schematically illustrates the bifurcation structure in these regions

pear. A non-zero coupling transforms this bifurcation into a crisis occurring on the route C in Fig. 11.28. As we show below, this crisis is different from the one observed along the route A.

For weak coupling the cooperative dynamics of the interacting systems can be considered as a direct product of their sub-spaces. In the range of $J_2 \in [0.0833; 0.0845]$ and $J_1 = 0.075$, the phase space of the coupled systems contains twelve asymptotically stable solutions. They appear from the direct product of four states for the first subsystem (three equilibrium points and one periodic solution) and three states for the second subsystem (one equilibrium point and two periodic solutions).

Figure 11.30(a) illustrates three states responsible for the crisis of the quasiperiodic regime: a stable torus, a saddle torus,[7] and a stable limit cycle. These objects are the direct product of the space of the stable limit cycle for the first subsystem and the three states for the second subsystem, which are one stable fixed point, one stable limit cycle, and one unstable limit cycle. Thus, a saddle-node bifurcation for the limit cycles in the individual oscillator will evoke a saddle-node bifurcation for the tori in the joint coupled system that happens at $J_2 = 0.0845$ along the route C.

If we deal with resonant tori, each having a pair of cycles lying on their surfaces, then the saddle-node bifurcation for tori involves those cycles. Namely, a stable

---

[7] A resonant saddle torus was described in Sect. 4.5 and an ergodic saddle torus in Sect. 8.4.

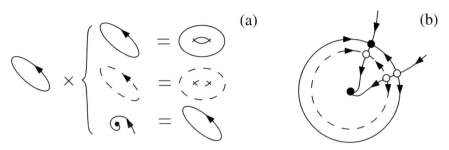

Fig. 11.30. a Schematic illustration of the solutions involved in crises. **b** Merging of a stable and a saddle tori (*solid* and *dashed curves* in Poincaré section, respectively) is described as bifurcations for a pair of cycles on the surface of tori

cycle from the stable torus collides with a saddle cycle from the saddle torus, and a saddle cycle from the stable torus collides with a twice saddle cycle from the saddle torus (Fig. 11.30(b)). In this transition, an additional Floquet multiplier of each of the cycles becomes equal to unity.

For a non-resonant torus, the crisis can be identified through the spectrum of Lyapunov exponents which at the moment of the crisis is characterized by three zero values.

11.7.5 Transition to In-phase Synchronization

Complexity of the bifurcation diagrams that we observed in interacting Morris–Lecar systems (11.30) for weak and intermediate coupling is reduced significantly when the interaction between the oscillators becomes stronger.

Fig. 11.31 represents typical bifurcational transitions taking place at strong coupling. In the lower part of this diagram, there are two regions of stable out-of-phase solutions O_L and O_R. The regions of their existence are nearly symmetric with respect to the line $J_2 = 0.075$. On the torus birth bifurcation line T, the stability of the O_L changes ($O_L \longrightarrow O_L^{++}$), and quasiperiodic oscillations appear in its vicinity. The region of this quasiperiodic state is denoted as T_L. On the curve H, the quasiperiodic solution undergoes crisis touching the stable manifold of the saddle cycle I^+. The detailed mechanism of this bifurcation will be discussed in the next section. The unstable cycle O_L^{++} merges with I^+ at SSN curves and disappears. Similarly, on the left H curve, the crisis of the T_R occurs. In the regions bounded by the two H curves, both states T_L and T_R coexist. To emphasize this, we highlighted a small part of existence areas of T_L and T_R by dark and light grey, respectively.

In the upper part of Fig. 11.31, the in-phase solution I is the only stable solution. It originates from a cusp C at $J_2 = 0.075$ and $\gamma = 0.5$. This cusp point corresponds to a pitchfork bifurcation, at which the saddle in-phase solution I^+ bifurcates into the stable in-phase solution I and the two unstable out-of-phase solutions.

The region of the stable in-phase solution is bounded by two SN curves on which the stable I and the unstable out-of-phase solution collide to disappear via

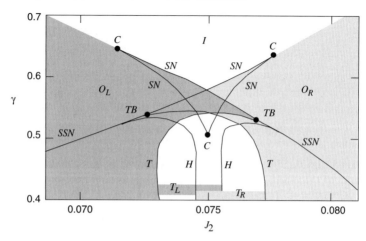

Fig. 11.31. (Color online) Bifurcation diagram illustrating the transitions between different synchronous states in two coupled Morris–Lecar systems (11.30) in the case of strong coupling γ. Notations are the same as in the Figs. 11.23 and 11.29

the saddle-node bifurcation. One can notice a star-like region formed by five co-dimension-two bifurcation points: the three cusp points C and the two Takens–Bogdanov points TB. At the upper cusps, the two SN curves merge. At these points, two stable cycles and one saddle cycle are involved in the bifurcation. Since I is the only stable solution for the sufficiently strong coupling, we can conclude that all transitions between different forms of synchronization states end up at the in-phase regime solution.

11.7.6 Mechanism of Torus Folding in the Vicinity of Unstable Orbit

In the previous Subsection we briefly mentioned the crises which occur when one of the quasiperiodic solutions T_L or T_R touches the stable manifold of the saddle cycle I^+. Below we show that such crisis is accompanied by the effects similar to dephasing, but is realized in the phase space of higher dimension.

In the region where both T_L and T_R coexist, they are separated by the stable and unstable manifolds of the saddle cycle I^+. Figure 11.32(a) displays the phase portraits of the T_L in the Poincaré section defined by the condition $w_2 = 0.29415$. The invariant closed curve for T_L is separated from T_R (not shown) by the stable and the unstable manifolds of the saddle point S being the section of the saddle cycle I^+. With increasing J_2, we approach the boundary crisis at the line H in Fig. 11.31. Nearby the boundary crisis, the invariant curve comes close to the saddle point S. Since the closed curve looks smooth, we expect a regular attractor. However, as shown in the insert, the trajectory is wiggling along the stable manifold W_s and the unstable manifold W_u of S while the *manifolds themselves are not deformed*. The folding of an invariant curve, corresponding to the quasiperiodic solution implies that the latter is close to the onset of chaos. Lyapunov exponents that are shown in

11.7 Synchronous Patterns in Coupled Morris–Lecar Models

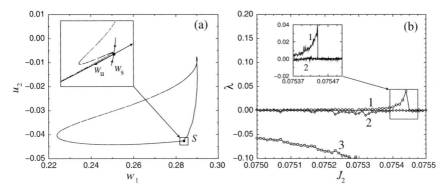

Fig. 11.32. a Poincaré section of a two-dimensional torus T_L in the vicinity of the saddle cycle I^+ (*point S*). A folded structure is well developed. b Two largest Lyapunov exponents as functions of control parameter J_2. Positive Lyapunov exponents indicate the onset of chaos from a quasiperiodic solution

Fig. 11.32(b) confirm this conclusion. The largest Lyapunov exponent is positive for J_2 close to 0.0754.

To understand how folding is formed in the vicinity of a saddle point S that has smooth manifolds, let us illustrate schematically the behavior of the Poincaré section of a phase trajectory in the vicinity of S. Figure 11.33 shows the phase space velocity projection onto the given plane, the darker areas corresponding to lower phase velocities. The trajectory is depicted as a sequence of Poincaré points. Let us compare two trajectories starting from two points X_n and X_m. After the first return into the secant surface they arrive at points X_{n+1} and X_{m+1}, respectively. With this, a point in the "slow" region X_n is mapped into a closely located X_{n+1}, whereas a point in the "fast" region X_m is mapped into a remote point X_{m+1}. Although the point X_m is behind X_n on the invariant curve, the trajectory starting from the former can overtake the trajectory launched from the latter. With this, the manifolds of the saddle point limit the area where the trajectory can arrive at. Altogether, these circumstances evoke the formation of a folding structure near the saddle point S. The proposed mechanism of torus folding is highly effective due to local inhomogeneity of the phase velocity in the neighborhood of a saddle cycle.

Contrary to the traditional route to chaos via the loss of torus smoothness [15], a significant part of the invariant curve in the Poincaré section seems to be smooth, and folding occurs only in a very small area nearby the saddle point. Thus, we can assume that a local singularity nearby the quasiperiodic motion can cause the appearance of chaos. Remarkably, similar arguments about the local singularity have allowed us to explain the dephasing effect discussed in Sect. 11.3, which is responsible for destabilization of in-phase regimes in weakly coupled anisochronous oscillators [104, 219].

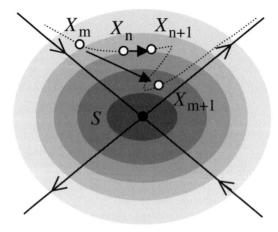

Fig. 11.33. Schematic presentation of a phase trajectory in the Poincaré section. Because of the inhomogeneity of the velocity filed in the vicinity of the saddle point S, the breakdown of the quasiperiodic solution occurs via folding on the torus surface

11.7.7 Remarks on Synchronization in Morris–Lecar Systems

It has been shown that a simple model of diffusively coupled neural oscillators is able to demonstrate a rich variety of synchronous states including the anti-phase, the out-of-phase, and the in-phase regimes. With this, for weak coupling, the anti-phase synchronization dominates. In the limit of the strong interaction, the in-phase regime is most probable in the cooperative dynamics. However, for moderate coupling, strong competition between inphasing and dephasing effects take place. This produces complex dynamical behavior with several stable synchronous states that can coexist. The typical transitions to chaos observed in such simple system are:

- A cascade of period-doubling bifurcations of a stable limit cycle on a torus surface
- Loss of torus smoothness that generally corresponds to Afraimovich–Shilnikov theorem, however in the case under study a heteroclinic surface formed by the manifolds of coexisting states provides a mechanism for re-injecting the trajectories into the vicinity of a chaotic set and thus gives rise to chaotic bursting
- Localized folding of a torus surface in the vicinity of a saddle cycle
- A sequence of torus doubling bifurcations that were not discussed here but was described in the original paper [228]

11.8 Summary

Throughout this chapter, we have been discussing how specific anisochronous features of an individual oscillator can affect its synchronization properties. Let us now summarize the main findings.

The dephasing effect plays an important role in synchronization of anisochronous oscillators.It occurs when some segment of a limit cycle in an individual subsystem approaches the vicinity of a saddle fixed point. At weak coupling, the dephasing effect destabilizes the in-phase solution. In the frames of our study we have considered an important case of synchronization near the homoclinic bifurcation that seems to be typical for a variety of oscillators, including models of biological cells.

We have shown that depending on the topology of the coupling, interacting anisochronous oscillators can demonstrate complex synchronous regimes when anti-phase and out-phase states coexist. In this context, we studied the structure of $1:1$ anti-phase locking region and revealed the main bifurcation scenarios. We have uncovered a specific bifurcation scenario for the resonant torus breakdown underlying the formation of chaotic bursting. Moreover, for sufficiently strong coupling, the transition from out-of-phase regimes to the in-phase locking can be associated with localized torus folding that occurs nearby a saddle cycle.

To summarize, anisochronicity can produce very prominent effects on synchronization of periodic oscillations. It is able to lead to a rich variety of new regimes that are not typical for isochronous oscillators. The knowledge of these effects can be very useful in studying the cooperative dynamics of real complex systems.

12 Phase Multistability

Many processes in nature are characterized by the coexistence of a number of limit states that can be reached from different initial conditions for a given set of parameters. Such phenomena known as multistability can be observed in almost all areas of science, including physics [55, 280], chemistry [124, 173], and physiology [60, 81]. In neuroscience, for instance, multistability is commonly considered as a mechanism for memory storage and temporal pattern recognition [113]. Multistability has also been reported for systems with time delays [142] and noise-induced patterns [141].

Multistability can also be related to synchronization phenomena. In Chap. 5 multistability involving asynchronous and synchronous states was shown to have a crucial effect on homoclinic transition to synchronization. Also, in Chap. 4 coexistence

318   12 Phase Multistability

of two synchronous states was discussed in connection with mutual synchronization of periodic oscillations.

In this chapter we study the phenomena called *phase multistability*. Generally, phase multistability assumes the coexistence two or more stable synchronous states, each corresponding to the same synchronization order $n : m$, and characterized by different phase shifts $\delta\Phi$ between oscillations in interacting systems. The fact that coexisting limit states correspond to different phase shifts gives the name for this phenomenon.

There are three main reasons leading to phase multistability [186]: (i) the complex wave shape of oscillations associated with subharmonics; (ii) specific geometry of coupling; and (iii) anisochronicity of oscillations. The case (ii) was illustrated in Chap. 4 where we considered evolution of the phase space structure in the system of two van der Pol oscillators with reactive coupling. Some examples of the case (iii) were discussed in Chap. 11 together with other phenomena induced by anisochronicity of oscillations.

In the sections below we focus on the major mechanism of phase multistability determined by complex wave forms of oscillations in interacting systems. The complex shape of oscillations can be associated with the presence of subharmonic components or with significant variations of the phase velocity along the orbit of the individual unit. Focusing on the mechanisms underlying the occurrence of phase multistability, we examine a variety of phase-locked patterns and universal transitions for different oscillatory regimes. In Sect. 12.1 we study phase multistability in systems demonstrating period-doubling route to chaos, in Sect. 12.2 self-modulated oscillations are considered, and in Sect. 12.3 bursting dynamics is investigated.

12.1 Period-Doubling Oscillations

Historically, the phenomenon of phase multistability is associated with synchronization in diffusively coupled oscillators that individually follow a period-doubling route to chaos [38, 242, 290, 291]. Spectrum of such oscillations contains subharmonics of the basic frequency ω_0, and for the same parameter values synchronization can be realized with several values of phase shift.

With this, the number of possible coexisting synchronous regimes is increased when more subharmonics of the basic frequency can be distinguished in the power spectrum. Remarkably that in such systems phase multistability can appear at negligibly small coupling between subsystems.

Let us try to understand how complex shape of the oscillations in interacting systems can lead to multistability. As an illustration, we consider Fig. 12.1 that schematically illustrates possible phase shifts for oscillations with different periods. Let us assume that we have two interacting identical subsystems without detuning coupled diffusively and demonstrating periodic oscillations with the same frequency. In this case, synchronization means that local maxima of the oscillations in both interacting systems occur at the same time moments. First, let us assume that both systems have oscillations with one local maximum per period (the left panel

12.1 Period-Doubling Oscillations 319

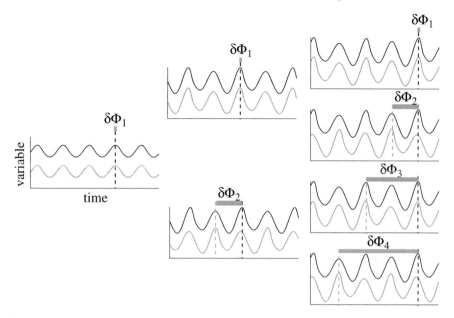

Fig. 12.1. (Color online) Schematic representation of different phase relations between oscillations corresponding to period-one (**left panel**), period-two (**middle panel**) and period-four limit cycles (**right panel**). The oscillations in the first and the second systems are denoted by *black* and *grey solid curves*, respectively. The *dashed lines* of the corresponding colors indicate the positions of the largest maxima of oscillations

of Fig. 12.1). We will call such oscillations "period-one oscillations." For such type of oscillations there exists only one possibility to be synchronized that corresponds to coinciding maxima in time and that is characterized by phase difference $\delta\Phi_1$. Next, we require both systems to undergo period-doubling bifurcation. Now, both interacting oscillators demonstrate period-two oscillations (the middle panel of the figure) and have two different local maxima per period. In this case, synchronization can be realized in two ways: (i) when maxima in both systems coincide, and when each larger maximum in one system corresponds to a smaller maximum in the other system. These two cases of synchronization will be characterized by phase shifts $\delta\Phi_1$ and $\delta\Phi_2$. Once both systems surmount another period-doubling bifurcation, their oscillations have four local maxima per period (the right panel of the figure), and therefore there are four possible versions of synchronization, each with its own phase shift.

For simplicity let us assume that changing the phase by 2π corresponds to time interval between the successive maxima of a time realization of oscillations. This will be valid, for example, if one introduces phases via Hilbert transform approach.[1] For the initial period-one oscillations with period T_0 a phase difference Φ_0 between

---

[1] This approach was described and corresponding, and the references were given, in Sect. 8.3.

320   12 Phase Multistability

the subsystems is equivalent to a phase difference $\Phi_0 + 2\pi k$, $k = 1, 2, \ldots$. For oscillations with doubled period $2T_0$ whose spectrum contains subharmonic $\omega_0/2$, two different limit cycles in the phase space of interacting systems correspond to the phase differences Φ_0 and $\Phi_0 + 2\pi$. Thus, for two synchronized oscillators whose spectrum includes subharmonics $\omega_0/2^k$ ($k = 1, 2, \ldots$) of the basic frequency, the phase difference between the interacting units can attain 2^k different values, i.e., $\delta\phi = \phi_0 + 2\pi m$, $m = 0, 1, 2, \ldots, 2^k - 1$. Obviously, when there is some non-zero detuning between the synchronized subsystems, the same principles apply, and the number of possible phase shifts can be predicted similarly.

Phase multistability can also take place for weak chaos, that demonstrates an N-band structure. The hierarchy of multistability in systems of identical interacting oscillators with weak diffusive coupling has been studied both numerically and experimentally in [38]. Evolution of the coexisting phase-shifted regimes with variation of control parameters is accompanied by different bifurcational transitions that depend on frequency mismatch and coupling strength [222, 242, 291].

12.1.1 Dynamics of Coupled Rössler Systems

Model

Since synchronization is a universal non-linear phenomenon, its key features are typically independent of a model. As an example, we consider the system of coupled Rössler oscillators in the form introduced in [247]:

$$
\begin{aligned}
\dot{x}_1 &= -\omega_1 y_1 - z_1 + \gamma (x_2 - x_1), \\
\dot{y}_1 &= \omega_1 x_1 + \alpha y_1, \\
\dot{z}_1 &= \beta + z_1 (x_1 - \mu), \\
\dot{x}_2 &= -\omega_2 y_2 - z_2 + \gamma (x_1 - x_2), \\
\dot{y}_2 &= \omega_2 x_2 + \alpha y_2, \\
\dot{z}_2 &= \beta + z_2 (x_2 - \mu),
\end{aligned}
\tag{12.1}
$$

where the parameters α, β and μ govern the dynamics of each subsystem. γ is the coupling parameter, $\omega_1 = \Omega + \Delta$ and $\omega_2 = \Omega - \Delta$ are the natural frequencies, and Δ determines the mismatch between these frequencies. Throughout this section we keep $\alpha = 0.15$, $\beta = 0.2$, $\Omega = 1.0$ and $\gamma = 0.02$. The equations (12.1) serve as a good model for real systems demonstrating period-doubling route to chaos, i.e., for electronic circuits [18, 112], chemical [95] and biological [182] systems.

To introduce an instantaneous amplitude and a phase of a chaotic oscillations of the system (12.1) one can use the following representation [212, 247]:

$$
\begin{aligned}
x_i(t) &= A_i(t) \cos \Phi_i(t), \\
\hat{x}_i(t) &= A_i(t) \sin \Phi_i(t).
\end{aligned}
\tag{12.2}
$$

Here, $A(t)$ and $\Phi(t)$ are the instantaneous amplitude and phase, respectively; $\hat{x}(t) = \hat{H}[x(t)]$ denotes Hilbert transform [204]. In the case when the dynamical variables

12.1 Period-Doubling Oscillations   321

$x(t)$ and $y(t)$ are in a linear relation (as for the Rössler system, for example) it is easy to introduce the following substitution:

$$x_i(t) = A_i(t) \cos \Phi_i(t),$$
$$y_i(t) = A_i(t) \sin \Phi_i(t). \tag{12.3}$$

Here, $A(t)$ and $\Phi(t)$ are the polar coordinates of the point $(x(t), y(t))$ in the (x, y) plane. When phase locking of chaotic oscillations occurs, the phase difference $\Phi_1 - \Phi_2$ is bounded, while outside the synchronization region it is an increasing or decreasing function of time [212, 247]. Note that phase locking is usually closely associated with the locking of basic frequencies in the power spectrum of chaotic oscillations (see Chap. 8).

Identical Systems

Let us study the dynamics of (12.1) as the parameter μ is varied in case of completely identical partial oscillators (i.e., with $\Delta = 0$) and with the fixed coupling strength $\gamma = 0.02$. As μ is increased, a sequence of bifurcations take place, leading from the initial cycle of period T_0 located in the invariant symmetric subspace U, defined as $x_1 = x_2$, $y_1 = y_2$, $z_1 = z_2$, to a set of coexisting attractors. Before arriving at chaos, a number of limit cycles coexisting in the phase space is increased. Let us denote the cycle with the period $2^n T_0$ and the phase shift $\Phi_0 = 2\pi m$ by the symbol $2^n C^m$ ($n = 1, 2, \ldots$; $m = 0, 1, 2, \ldots$). A chaotic attractor with 2^n bands arising from the cycle with the phase shift $\Phi_0 = 2\pi m$ is labeled as $2^n C A^m$, and a 2^n bands chaotic saddle we denote as $2^n C S^m$.

To illustrate different oscillatory regimes of the system and the transitions between them in Fig. 12.2 we show schematically the evolution of periodic and chaotic regimes as parameter μ is increased, while the coupling strength γ is fixed at 0.02. Note that branch **A** corresponds to the in-phase family of attractors (i.e., the phase shift between the oscillations is equal to zero and phase trajectories lie in U), while the branches **B**, **C** and **D** illustrate the out-of-phase regimes originated from $2C^1$, $4C^2$ and $8C^4$, respectively.

As μ increases, the in-phase limit cycle C^0 undergoes a period-doubling bifurcation. A cycle $2C^0$ of doubled period emerges smoothly. The cycle C^0 which becomes saddle continues to exist and undergoes another period-doubling bifurcation. As a result of this bifurcation a saddle cycle $2C^1$ of doubled period is born. This cycle does not lie in the symmetric subspace U any longer, but it is self-symmetric with the respect to the invariant manifold U (i.e., $x_1 = -x_2$, $y_1 = -y_2$, $z_1 = -z_2$). Cycle $2C^1$ becomes stable via the inverse subcritical pitchfork bifurcation as μ is increased further. In the same manner, each of the in-phase limit cycles $2^m C^0$ gives rise to the corresponding branch of out-of-phase regimes. For the above out-of-phase cycles the replacement of the next period-doubling bifurcation by torus birth bifurcation takes place. The torus birth bifurcation leads to quasiperiodicity, frequency locking and the emergence of new out-of-phase families of attractors which follow the period-doubling route to chaos. Above some critical value of μ, several

12 Phase Multistability

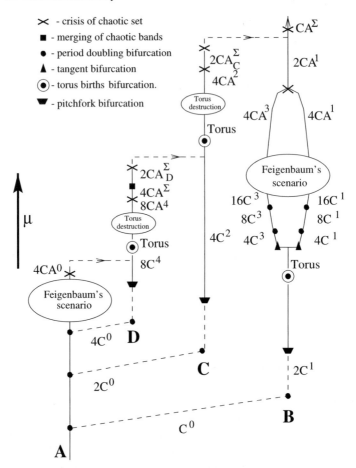

Fig. 12.2. Evolution of oscillatory regimes for the identical coupled Rössler systems. The *solid* and *dashed lines* correspond to bifurcational transitions of attractive and saddle solutions, respectively

chaotic attractors coexist. As μ increases further, there are the merging bifurcations where the number of bands in the chaotic attractor is halved. Besides this, a crises of chaotic limit sets leading to the merging of attractors of different branches take place. Finally, the only one-band global chaotic attractor CA^Σ, that includes the chaotic sets of all branches, emerges in phase space of the system.

A phase shift $2\pi m$ between the oscillations of subsystems that defines the corresponding branch of regimes can not be found using instantaneous phases $\Phi_{1,2}$ from (12.2) or (12.3). The instantaneous phases and their difference are determined with the accuracy of $\pm 2\pi k$, $k = 1, 2, \ldots$. Therefore, the phase differences, found using (12.2) or (12.3), for all branches is limited within the range $[-\pi, \pi]$ if their initial values are chosen inside this interval. To find the characteristic phase shift $2\pi m$ it is

12.1 Period-Doubling Oscillations   323

necessary to determine a time shift between oscillations in interacting systems for $\Delta = 0$, and then rewrite it in terms of phase difference. Taking into account that the out-of-phase regimes are located outside of the symmetric subspace U, we may introduce a time shift τ, such that states of the subsystems coincide but are lagged with respect to each other by τ. The value of τ can be calculated via the global minimum of a similarity function S as described in [248]:

$$S^2(\tau) = \frac{\langle (x_2(t+\tau) - x_1(t))^2 \rangle}{\sqrt{(\langle x_1^2(t) \rangle \langle x_2^2(t) \rangle)}}, \tag{12.4}$$

where the angular brackets denote averaging over time.

Let us consider in detail the evolution of chaotic attractors as the parameter μ is changed. Note that there are three types of crises which are labeled in Fig. 12.2 as crosses: transformation of a chaotic attractor into a chaotic saddle, merging of chaotic attractors of the same branch, and merging of a chaotic attractor of one branch with a chaotic attractor of another branch. Consider branch **A** that embraces the regimes whose trajectories belong to the symmetric subspace U. All stable regimes from this branch correspond to the case of complete synchronization. As μ is increased, on branch **A** a chaotic attractor which appears via a period-doubling cascade of the in-phase regimes bifurcates into a chaotic saddle $4CS^0$ at $\mu \approx 5.95$. As this happens, in the spectrum of Lyapunov exponents, in addition to the already existing positive exponent, another positive exponent appears. The latter corresponds to an additional unstable direction which is transversal to U. The transition $4CA^0 \rightarrow 4CS^0$ leads to the loss of complete synchronization. The mechanism of similar transitions was studied in [39, 209]. When an initial point on U is slightly perturbed, after a long transient time the phase trajectory tends to the stable cycle $8C^4$ of branch **D**. When μ is further increased, a sequence of bifurcations of this cycle leads to chaotic attractor $8CA^4$ (Fig. 12.3(a)) which at $\mu \approx 6.036$ undergoes a crisis by colliding with a chaotic saddle of branch **A**, as well as a band merging. As a result, branches **A** and **D** merge together which leads to the appearance of a chaotic attractor $4CA_D^{\Sigma}$ (Fig. 12.3(b)). This merging crisis is accompanied by "on-off" intermittency. Then $2CA_D^{\Sigma}$ appears from $4CA_D^{\Sigma}$. At $\mu \approx 6.06$ a chaotic attractor $2CA_D^{\Sigma}$ becomes a saddle. After this transition, phase trajectories switch to the stable cycle $4C^2$ which belongs to branch **C**. Chaotic attractor $4CA^2$ (Fig. 12.3(c)) appears from $4C^2$ via a sequence of bifurcations. At $\mu \approx 6.35$, the chaotic attractor $4CA^2$ merges with the saddle of branch **D**, and a new chaotic attractor $2CA_C^{\Sigma}$ emerges (Fig. 12.3(d)). At $\mu \approx 6.44$, this attractor becomes a saddle $2CS_C^{\Sigma}$ and a phase trajectory jumps to a chaotic attractor $2CA^1$ (Fig. 12.3(e)) of branch **B**. Then at $\mu \approx 6.70$, the chaotic attractor $2CA^1$ merges with a saddle of branch **C**. Thus, a sequence of crises ends as the only chaotic attractor CA^{Σ} (Fig. 12.3(f)) which involves chaotic trajectories from all branches.

It has been found that the behavior of the phase difference calculated from (12.3) is different for a variety of chaotic attractors inside the synchronization region. For chaotic attractor located in the symmetric subspace ($4CA^0$, for instance), it is constant in time ($\delta\Phi(t) = \Phi_1(t) - \Phi_2(t) = 0$). For out-of-phase attractors it is not

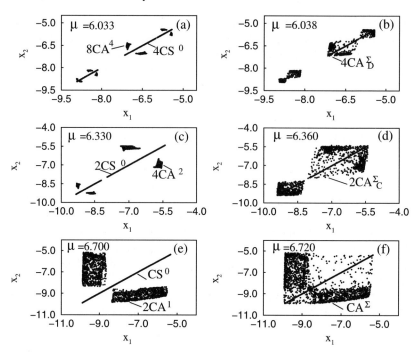

Fig. 12.3. Projections of Poincaré sections of chaotic sets in the identical coupled Rössler systems (12.1). Secant plane is specified as $y_1 = 0$. Label CS is used to identify a saddle set

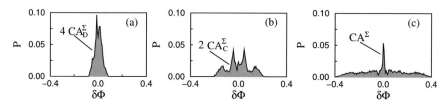

Fig. 12.4. (Color online) The distribution of phase difference $\delta\Phi$ for out-of-phase attractors: **a** $4CA_D^\Sigma$ at $\mu = 6.038$; **b** $2CA_C^\Sigma$ at $\mu = 6.36$; **c** CA^Σ at $\mu = 6.72$. Calculations were performed with the constant step $\delta\Phi = 2\pi/100$

equal to zero and varies chaotically in time. The width of the distribution of phase differences $P(\delta\Phi)$ characterizes how far the attractor is from the in-phase state. Figure 12.4 displays the distribution of phase differences for the chaotic attractors $4CA_D^\Sigma$, $2CA_C^\Sigma$, and CA^Σ. It is clearly seen that the merging of chaotic sets from different families (branches) leads to the expansion of the distribution function. However, note that $\delta\Phi$ remains bounded in the interval $[-\pi, \pi]$, since the described chaotic attractors are synchronous.

The chaotic attractor CA^Σ corresponds to the regime of hyperchaos. But the regime with two positive Lyapunov exponents appears before than CA^Σ is formed.

For example, the chaotic attractor $2CA^1$ of branch **B** which appears via merging of $4CA^1$ and $4CA^2$ has two positive Lyapunov exponents. For branches **C** and **D**, the transition to hyperchaos is observed when a torus is destroyed.

Effect of Frequency Mismatch

Now we introduce a mismatch between the basic frequencies in the system of coupled oscillators and study the evolution of multistability and of different forms of synchronization.

Figure 12.5(a), (b) represents the bifurcation diagrams of the synchronization region for attractors from two branches **A** and **B** (shown in Fig. 12.2), respectively. It has been found that a small frequency mismatch ($|\Delta| \leq 0.001$) almost does not affect the evolution of different oscillatory regimes which are observed in the case of vanishing mismatch Δ. Note that at $\Delta \neq 0$ the invariant subspace U does not exist any longer and the relations of symmetry for limit sets are not satisfied. Therefore, pitchfork bifurcations of limit cycles are replaced by tangent bifurcations[2] leading to the birth of saddle out-of-phase cycles [39].

When the frequency mismatch is further increased ($\Delta \geq 0.0015$), the period doubling bifurcations for cycles $2C^1$, $4C^2$, etc., are observed instead of torus birth

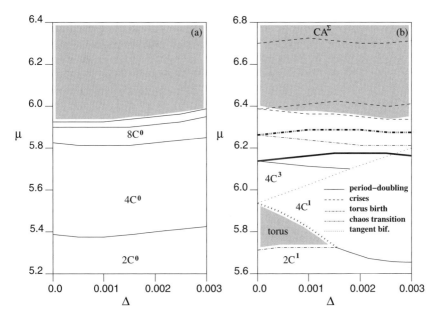

Fig. 12.5. Bifurcation diagram on ($\Delta - \mu$) parameter plane of the system (12.1) for the attractors of branches **A** (**a**) and **B** (**b**). Curves of different width correspond to different families of attractors

---

[2] Saddle-node bifurcation is sometime called *tangent* or fold bifurcation in literature.

326   12 Phase Multistability

bifurcations (Fig. 12.5(b)). Moreover, the types of merging crises of chaotic sets depend on the values of Δ. In presence of a frequency mismatch, there is no merging of chaotic attractors $4CA^1$ and $4CA^3$, but attractor $4CA^3$ becomes a saddle and then it merges with the attractor $4CA^1$. Transition to hyperchaos occurs before this crisis.

Let us consider the effect of frequency mismatch in terms of synchronization. Chaotic attractor $4CA^0$ of branch **A** that is located in the symmetric subspace U in the case of $\Delta = 0$, does not belong to U when $\Delta \neq 0$. Hence, complete synchronization is lost. However, for a weak frequency mismatch this regime remains topologically equivalent to the attractor in U. In this case referred to as lag synchronization [248] the oscillations of two systems coincide but shifted in time. For chaotic attractors of other families and attractors appearing via merging of chaotic sets from different branches neither complete nor lag synchronization can be achieved. They can demonstrate only phase synchronization.

Phase Transitions near the Boundary of Synchronization Region

Bifurcational mechanisms of the phenomena that take place at the boundary of chaotic phase synchronization are associated with the bifurcations of the saddle periodic orbits. Anishchenko et al. [18] have described this boundary as an accumulation of curves of tangent bifurcations of saddle cycles.[3] Pikovsky et al. [211] suggested (for model two-dimensional map) that attractor–repeller collisions take place at the transition to chaotic synchronization, thus drawing on the analogy with the tangent bifurcations of periodic orbits. Rosa et al. [246] consider the transition to phase synchronization as a boundary crisis mediated by bifurcations of non-stable periodic orbits on a branched manifold. We are interested in the transition between different coexisting regimes near the boundary of the phase synchronization region.

When the mismatch between the basic frequencies of interacting oscillators is introduced the regions of phase synchronization of chaos that are similar to Arnold tongues for periodic oscillations appear on the parameter plane. Hierarchy of multistability of synchronous regimes near the boundary differs from the case of $\Delta = 0$, see Fig. 12.6. Taking into account the different sequence of bifurcations for periodic solutions that has been described above, we focus on the peculiarities in the behavior of chaotic attractors. For a large mismatch, chaotic out-of-phase attractors of **B** and **C** branches become the saddles. When μ is increased, they merge with the in-phase attractor of branch **A**. Thus, attractor $4CA_A^\Sigma$ appears via merging of a chaotic attractor $4CA^0$ of branch **A** and a chaotic saddle of branch **C**. The band-merging crisis takes place and an attractor $2CA_A^\Sigma$ appears. At this moment the transition to hyperchaos occurs. Then the merging crisis of $2CA_A^\Sigma$ and a chaotic saddle of branch **B** originated from attractor $4CA^1$ leads to the single attractor $2CA^\Sigma$ in the phase space of the system. Figure 12.7 shows the projections of Poincaré sections of coexisting chaotic attractors $4CA_A^\Sigma$ and $4CA^1$ (Fig. 12.7(a)) and of attractor $2CA^\Sigma$ (Fig. 12.7(b)) born as a result of merging of chaotic sets from all branches.

---

[3] This is described in Sects. 8.4 and 8.5.

12.1 Period-Doubling Oscillations 327

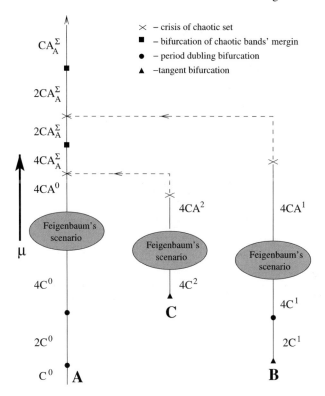

Fig. 12.6. (Color online) Evolution of oscillatory regimes in coupled Rössler systems (12.1) at frequency mismatch $\Delta = 0.0093$

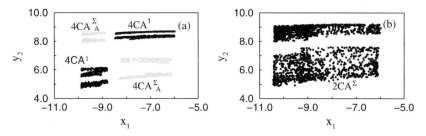

Fig. 12.7. Projections of Poincaré sections of chaotic sets from the system (12.1) when frequency mismatch $\Delta = 0.0093$ at **a** $\mu = 6.8$ and **b** $\mu = 7.2$

Figure 12.8 represents the bifurcation diagram of the synchronization region near its boundary. A *nested structure* of phase-synchronized regions for the attractors of branches **A** and **B** is observed. With this structure, the transition to non-synchronous behavior in the region of multistability (direction **a** in Fig. 12.8) is determined by the loss of stability for the most robust synchronous mode (branch **B**

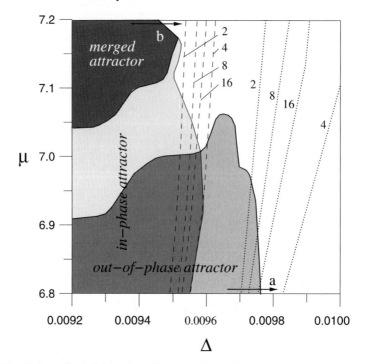

Fig. 12.8. (Color online) Bifurcation diagram near the boundary of synchronization region of coupled Rössler systems (12.1). *Dashed* and *dotted curves* denote bifurcations of periodic orbits on branches **A** and **B**. Numbers *2*, *4*, and *8* denote the periods of saddle cycles

in our case). Chaotic attractors of branch **A** remain structurally stable[4] when μ is increased. Hence, above the region of multistability the transition from complex chaotic regimes appeared after a series of merging crises, to non-synchronous dynamics (direction **b**) is observed.

The boundary of synchronization region is detected from the calculation of the spectrum of Lyapunov exponents and of the effective diffusion D of the phase difference is described as follows[5]:

$$D = \frac{\langle [\delta\Phi(t)]^2 \rangle - \langle \delta\Phi(t) \rangle^2}{t}. \qquad (12.5)$$

Figure 12.9 displays these characteristics of synchronization along direction **a** marked in Fig. 12.8 at $\mu = 6.8$. As it is clearly seen in Fig. 12.9(a), one of the negative Lyapunov exponents becomes equal to zero at the boundary of the synchronization region (a vertical dashed line). A similar behavior has been observed in mutually coupled Rössler systems with one-band chaos that were considered in Sect. 8.7 (compare with Fig. 8.32), and in a Rössler system with periodic forcing

---

[4] Do not change their topological properties with variation of a parameter.
[5] The concept of phase diffusion was discussed in Sect. 7.9.

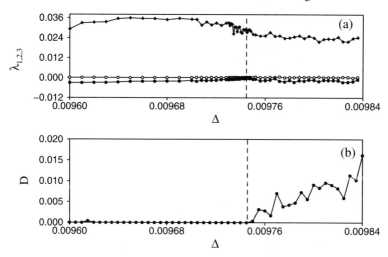

Fig. 12.9. a Three largest Lyapunov exponents $\lambda_{1,2,3}$ and **b** the coefficient of diffusion D of phase difference as functions of frequency mismatch Δ at $\mu = 6.8$ (direction **a** in Fig. 12.8)

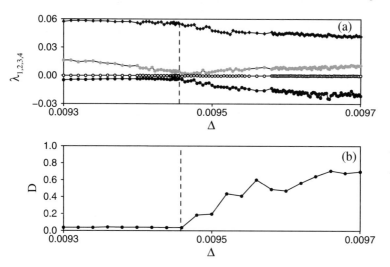

Fig. 12.10. (Color online) **a** The four largest Lyapunov exponents $\lambda_{1,2,3,4}$ and **b** the coefficient of diffusion D of phase difference as functions of frequency mismatch Δ at $\mu = 7.2$ (direction **b** in Fig. 12.8)

[289]. The coefficient of diffusion is vanishing inside the synchronization region but at the boundary it starts to grow (Fig. 12.9(b)). Similar calculations have been performed along direction **b** in Fig. 12.8 at $\mu = 7.2$. Figures 12.10(a), (b) show that the behavior of Lyapunov exponents is not changed, while the coefficient of diffusion is very sensitive to the transition to a non-synchronous regime.

Based on the results from Chap. 8 where the bifurcation mechanisms of phase synchronization are shown to be related to the bifurcations of saddle periodic orbits embedded in a chaotic attractor [18, 211, 246], we constructed the curves of tangent bifurcations of saddle cycles from branches **A** and **B** (dashed and dotted lines in Fig. 12.8, respectively). It is clearly seen that while multistability exists, the curves tend to be located near the synchronization boundary of each branch of attractors. However, as soon as merging crises occur, this is no longer valid. The question "what is the bifurcational transition from the merged synchronous regime which is characterized by two positive Lyapunov exponents, to non-synchronous behavior?" is still open.

12.1.2 Mapping Approach to Multistability

To construct a simplified model of the emergence of phase multistability let us introduce an analytical description of a high-periodic signal in the form [222]

$$x(t) = A(\phi(t)) \sin(\omega t). \tag{12.6}$$

Here, $\phi = \omega t$ is the phase of the oscillations, and $A(\phi) = \prod_{i=1}^{N}(1 - \sigma_i \sin(\omega t/2^i + i\pi/2))$ represents the instantaneous amplitude. ω is the natural (or fundamental) frequency of oscillations, N defines the period of the signal considered $T_N = 2^N(2\pi/\omega)$, and σ_i specify the amplitude of each of the subharmonic components. The term $i\pi/2$ is introduced to obtain a more obvious phase portrait of each period-doubling in our model. The function $x(t)$ described by (12.6) is illustrated in Fig. 12.11(a). As N increases, $x(t)$ provides a qualitative representation of a sequence of high-periodic cycles, leading in the limit to the birth of chaos via a cascade of period doublings.

For two synchronized oscillators coupled via the variables $x_1(t)$ and $x_2(t)$, each being described by an expression like (12.6), the phase difference can attain 2^N different values, i.e., $\Theta = \phi_1 - \phi_2 = 2\pi m$, $m = 0, 1, 2, \ldots, 2^N - 1$. Hence, coexistence of a large number of periodic attractors will occur. When approaching

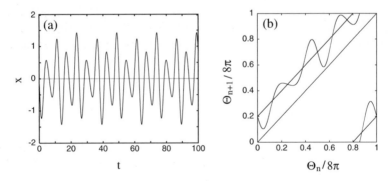

Fig. 12.11. a Realization $x(t)$ of the periodic orbit with period $4T_0$ simulated from the expression (12.6). **b** The model map (12.9) for the case of $N = 2$

12.1 Period-Doubling Oscillations   331

the boundary of the synchronization region, these attractors disappear one by one except for a single family whose bifurcations determine the transition to the non-synchronous regime. In order to understand the structure of this boundary in more detail we shall investigate a sequence of model maps.

For quasiperiodic oscillators, the phase difference Θ is known to develop according to an equation of the form [50]

$$\dot{\Theta} = \Delta - \gamma f(A_1, A_2) \sin \Theta, \tag{12.7}$$

where $f(\cdot)$ is a function of the amplitudes A_1 and A_2 that is defined by the type of interaction. Δ represents the mismatch between the basic frequencies and γ is the coupling strength.

In our case the oscillators have different instantaneous phases ϕ_1 and ϕ_2 while their amplitudes A_1 and A_2, as specified above, depend on the phases in the following way:

$$\begin{aligned}
A_1 &= A(\phi_1) = \prod_{i=1}^{N}\left(1 - \sigma_i \sin\left(\frac{\phi_1}{2^i} + i\frac{\pi}{2}\right)\right), \\
A_2 &= A(\phi_1 - \Theta) = \prod_{i=1}^{N}\left(1 - \sigma_i \sin\left(\frac{\phi_1}{2^i} - \frac{\Theta}{2^i} + i\frac{\pi}{2}\right)\right).
\end{aligned} \tag{12.8}$$

It is not possible to obtain an explicit relation for the phase difference between two chaotic oscillators. However, qualitatively we can consider the oscillators as high-periodic cycles of periods $T_N = 2^N 2\pi/\omega$, where ω is the natural frequency of the partial system (ω_1, for example). To obtain a discrete model, (12.7) is integrated over the characteristic time T of the system. This gives a model map in the form

$$\Theta_{n+1}^N = \Theta_n^N + \Omega - K F^N(\Theta_n^N) \bmod 2^N 2\pi, \tag{12.9}$$

where $\Theta_{n+1}^N = \Theta^N(t_0 + nT_N)$ and $\Theta^N \in [0, 2^N 2\pi]$. $\Omega = T_N \Delta$, and K is a measure of the strength of interaction. We may suppose, however, that the interaction strength depends on the phase differences in the same way as the amplitude of the individual subsystem depends on its phase. As a simple approach we shall therefore assume an expression of the form

$$F^N(\Theta_n^N) = \sin(\Theta_n^N) \prod_{i=1}^{N}\left(1 - \delta_i \sin\left(\frac{\Theta_n^N}{2^i} + i\frac{\pi}{2}\right)\right). \tag{12.10}$$

Equations (12.9) and (12.10) may be viewed as a generalized form of the well-known circle map[6] for simple oscillators [239]. Varying $N = 1, 2, 3, \ldots$, we obtain a family of maps, each being a model of synchronization for 2^N-periodic cycles. The case of $N = 2$ is illustrated in Fig. 12.11(b). The above equations are not normalized on the same scale because they are taken to the modulus $2^N 2\pi$, which is changed

---

[6] See Sects. 6.4 and 6.5.

12 Phase Multistability

with each period doubling. This allows us to preserve the values of Ω and K and to compare the results for different N. A similar approach to constructing a model map in the non-autonomous case was suggested by Pikovsky et al. [213].

With these preliminaries let us now investigate the structure of the boundary of the synchronization region for the main resonance $0:1$ (or $1:1$ for continuous-time systems). In terms of the map, the transition at that boundary corresponds to a tangent bifurcation. The condition for such a bifurcation to occur is

$$\Theta_*^N + \Omega - KF^N(\Theta_*^N) = \Theta_*^N,$$
$$\left.\frac{d(\Theta^N + \Omega - KF^N(\Theta^N))}{d\Theta^N}\right|_{\Theta^N = \Theta_*^N} = 1, \tag{12.11}$$

where Θ_*^N is the fixed point. Equation (12.11) immediately gives

$$KF^N(\Theta_*^N) = \Omega$$
$$\left.\frac{dF^N(\Theta^N)}{d\Theta^N}\right|_{\Theta^N = \Theta_*^N} = 0. \tag{12.12}$$

Hence, it is easy to see that for any value of Θ_*^N, the set of points corresponding to the tangent bifurcation forms a straight line in the (Ω, K) parameter plane. The number of roots of (12.12) defines the number of possible synchronous regimes. For the case of small N, (12.12) can be solved analytically. For larger N, the solution can be obtained numerically. Figure 12.12 shows the results for $N = 1$ (solid lines) and $N = 2$ (dotted lines). Each line corresponds to a tangent bifurcation for one of the fixed points of the map. Under variation of Ω, a pair of stable and unstable

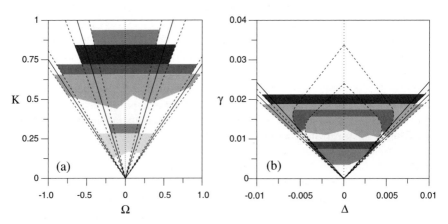

Fig. 12.12. (Color online) **a** Phase-locking regions for different families of attractors for the model map (12.9), (12.10) with $\delta_i = 0.45$, $i = 1, \ldots, N$. The *solid lines* correspond to $N = 1$ (two cycles of period-two coexist). The *dashed lines* correspond to $N = 2$ (four cycles of period-four coexist). **b** Nested structure of Arnold tongues for the coupled Rössler oscillators (12.1) with $\alpha = 0.15$, $\beta = 0.2$, and $\mu = 6.1$. The *solid* and the *dashed lines* correspond to the different coexisting families of regimes

fixed points are born on each line. A *nested structure* of Arnold tongues is clearly seen. For larger K, the stable fixed point can subsequently lose its stability through a period-doubling bifurcation. To find the corresponding parameter values, one only has to replace the zero in the right-hand side of (12.12) by $2/K$. However, in the present context we shall not consider the further bifurcations of the stable periodic solutions.

To verify the conclusions based on the model map dynamics, consider again the dynamics of coupled Rössler oscillators (12.1). In Fig. 12.12(b) the numerically obtained structure of four Arnold tongues is depicted. The control parameter μ was set $\mu = 6.1$ while the detuning Δ and the coupling strength γ were changed This figure clearly demonstrates good agreement with the results for our model map, at least for $\gamma < 0.01$ [222].

Thus, the maps (12.9), (12.10) for small enough K demonstrate 2^N stable (and a similar number of unstable) fixed points near the center of the synchronization region. In terms of continuous-time dynamical systems, a set of stable fixed points correspond to a set of possible synchronization regimes for the coupled oscillators. A two-dimensional torus exists both outside (ergodic) and inside (resonant) the synchronization region.[7] Entering the synchronization region corresponds to the birth of a pair of stable and saddle cycles, both lying on the torus surface. In these terms, the appearance and coexistence of other fixed points of the map represent the birth on the torus surface of additional pairs of stable and saddle cycles which do not intersect each other. Another interesting question arises: "Do the coexisting synchronous solutions actually lie on the same torus surface?" Note that this is not necessarily the case for continuous-time systems. In Fig. 12.13 the numerically obtained Poincaré section for the resonant torus surface is given. The parameters of two coupled Rössler systems correspond to the period-two limit cycle. Two stable coexisting solutions $C_{1,2}$ are observed in the plot, each paired with a corresponding saddle cycle $S_{1,2}$. Moreover, inspection of the figure clearly shows that all the solutions belong to the same closed curve, formed by the unstable manifold of the saddle cycles.

On this background we can draw the following conclusions concerning synchronization of large-period oscillations in coupled period-doubling systems: (i) There are 2^N coexisting synchronous solutions which differ from each other by phase shifts; and (ii) the synchronization region for these solutions consists of a set of tongues inserted into each other.

The question is now how the results described here manifest themselves in the case of two interacting chaotic oscillators.

It is well-known that for the period-doubling route to chaos the chaotic attractor has an N-band structure ($N = 1, 2, 4, \ldots$) within a range of control parameters. This structure is geometrically similar to the structure for the N-periodic cycles. Thus let us simulate an N-band chaotic attractor by means of the model map (12.9) with an added noise term. The logistic map may be used as the source of such random excitations:

---

[7] Phase-locking region.

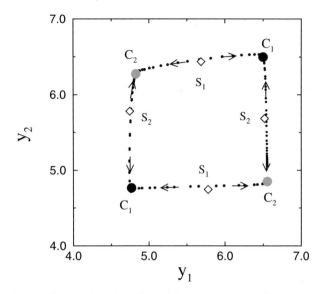

Fig. 12.13. (Color online) Poincaré section of the resonant torus for two coupled Rössler models (12.1). The secant was chosen as $x_1 = 0$. Control parameters are $\alpha = 0.15$, $\beta = 0.2$, $\mu = 5.0$, and $\gamma = 0.0273$. Points $C_{1,2}$ denote the stable limit cycles while $S_{1,2}$ are the saddle cycles. *Arrows* indicate the stable directions along the resonant torus surface

$$\Theta_{n+1}^N = \Theta_n^N + \Omega - KF^N(\Theta_n^N) + Bx_n \quad \mathrm{mod}\, 2^N 2\pi,$$
$$x_{n+1} = \lambda x_n (1 - x_n),$$
(12.13)

where the value of λ is fixed at 3.99. Note that we introduce the source of noise in the above way (not a Gaussian noise, for example) to maintain the multiband structure of the chaotic attractor. Within some range of the noise amplitude B, the attractors produced by this equation become irregular but they still coexist in the phase space of the system and their basins of attraction differ. When B is further increased, the attractors start to merge [67].

Figure 12.14 (left panel) shows a one-parameter bifurcation diagram for the case of an 8-band chaotic attractor. There are eight different synchronous chaotic regimes which coexist at small Ω. When the detuning parameter Ω increases, the coexisting chaotic attractors disappear one by one on the edges of their respective synchronization regions. At $\Omega \geq 0.535$ a single synchronous solution is still stable. Note how the "ghosts" of all eight synchronous solutions remain distinguishable inside the region of merged chaos at $\Omega > 0.6$. The number of possible synchronous regimes decreases in the same way as is observed in coupled Rössler systems (Fig. 12.14 (right panel)).

Hence, our conclusions with respect to synchronization of large-period orbits are also valid for weakly-chaotic solutions. Moreover, we may expect the nested structure of synchronization tongues to be preserved in the case of an N-band chaotic attractor and to remain similar to the structure for an N-periodic cycle.

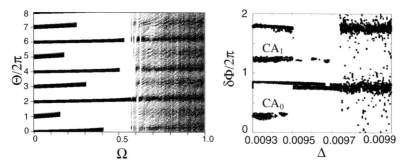

Fig. 12.14. *Left panel*: One parameter bifurcation diagram for the model map (12.9), (12.10) with $K = 0.5$, $\delta_i = 0.45$, $B = 1.2$, and $N = 3$. The figure shows how the coexisting noise inflicted periodic orbits one by one lose their synchronization. **Right panel**: One-parameter bifurcation diagram for two coupled Rössler systems (12.1) shows similar behavior for $\alpha = 0.15$, $\beta = 0.2$, $\gamma = 0.02$, and $\mu = 6.7$.

12.2 Self-Modulated Oscillations

Natural phenomena often involve dynamics with different time scales. Oscillations that are generated by a single self-oscillator and characterized by several different time scales are sometimes called *self-modulated oscillations*. They may be particularly significant in living systems. The thalamocortical relay neurons, for instance, can generate either spindle or delta oscillations [295]. The electroreceptors in paddlefish are found to demonstrate biperiodic dynamics [190]. The functional units of the kidney, the nephrons, demonstrate low-frequency oscillations arising from a delay in the tubuloglomerular feedback, together with somewhat faster oscillations associated with the inherent dynamics of the arteriole [227]. It has been shown that a system of two diffusively coupled oscillators operating in the $1:n$ regime of self-modulation (n is integer) reveals the same aspects of phase multistability [273] as the systems with period-doubling cascades [222]. For coupled identical oscillators one can expect n coexisting synchronous solutions that differ from each other by phase shifts. The corresponding synchronization region consists of a set of Arnold tongues embedded into each other or shifted with respect to each other. Let us consider these aspects of self-modulated systems in details.

12.2.1 Methods of Analysis

The description of synchronization phenomena observed in interacting oscillators may be divided into two stages. The first step is to consider the case when the coupling strength is sufficiently weak so that analytical methods can be applied. The second step is to examine the case of finite coupling strength and to show to what extent the results of the weak-coupling limit can be extrapolated. Since the definition of phase multistability involves the phase difference between the interacting oscillators, the phase variables will be the main quantities used to characterize the collective dynamics.

336   12 Phase Multistability

For the case of weak interaction, the effective coupling approach can be applied. This approach was considered in detail in Sect. 11.2 (in Chap. 11 devoted to synchronization of anisochronous oscillations). Here we use its simplified form.

The interaction of two identical oscillators with phases ϕ_1 and ϕ_2 can be quantified by the evolution of their phase difference $\Delta\phi = \phi_1 - \phi_2$. In the limit of weak interaction, the phase dynamics averaged over a period for one of the oscillators can be expressed by *effective coupling function* [151]

$$\frac{d(\Delta\phi)}{dt} = \Gamma(\Delta\phi) = \frac{1}{2\pi} \int_0^{2\pi} d\phi Z(\phi) P(\phi, \Delta\phi), \qquad (12.14)$$

where $P(\phi, \Delta\phi) = P(V_0(\phi), V_0(\phi + \Delta\phi))$ describes the rate of change of the state vector V of one oscillator due to the interaction with another oscillator with a phase difference $\Delta\phi$, and ZP is the phase shift along the limit cycle for the given perturbation. Note that the limit cycles in both systems are assumed to have similar shapes, i.e., to be topologically conjugate.

In mutually coupled oscillators, the entrainment manifests itself as a mutual phase shift. This can be analyzed purely in terms of the antisymmetric part $\Gamma_a(\Delta\phi)$ of the effective coupling function (12.14) [151]. The zeroes of $\Gamma_a(\Delta\phi)$ correspond to the phase-locked synchronous states ($\Delta\phi = \text{const}$) and their stability are determined from the slope of $\Gamma_a(\Delta\phi)$ at the respective states: a negative slope means a stable state, and vice versa. This method of effective coupling has been used in a number of applications [104, 203, 219].

When the coupling becomes strong enough to modify the geometry of the limit cycle, the phase reduction method can no longer be used. Direct numerical methods should then be applied. First of all, we calculate a set of points on the limit cycle modified by the interaction. Over a set of initial conditions covering the full length of the limit cycle, we follow the evolution of the initial phase shift $\Delta\phi(t)$ to some fixed value $\Delta\phi(t + \tau)$. Plotting these results together, i.e., $\Delta\phi(t + \tau)$ vs $\Delta\phi(t)$, we obtain a one-dimensional phase map with a discrete time step τ. The analysis of this map allows us to find the fixed points and estimate their stability.

Note that for the effective coupling method one can obtain the phase map in terms of Γ_a. Namely, for two coupled identical oscillators the phase difference behavior is given by [151]

$$\frac{d(\Delta\phi)}{dt} = 2\Gamma_a(\Delta\phi). \qquad (12.15)$$

Setting $dt \rightarrow \tau$ and $d(\Delta\phi) \rightarrow (\Delta\phi_{t+\tau} - \Delta\phi_t)$ for small enough τ one finds the expression

$$\Delta\phi_{t+\tau} \approx \Delta\phi_t + \tau 2\Gamma_a(\Delta\phi_t), \qquad (12.16)$$

to which our numerical calculations converge for vanishing coupling.

12.2 Self-Modulated Oscillations   337

12.2.2 Phase Dynamics of Coupled Oscillators

Model Equations

We apply the above approach to Anishchenko–Astakhov oscillator that can be implemented as an electronic circuit [15, 18] (see also Sect. 8.5.1) and is described by a simple set of dynamical equations

$$\dot{x} = mx - zx + y - bx^3,$$
$$\dot{y} = -x, \qquad\qquad (12.17)$$
$$\dot{z} = -gz + gx(x + |x|)/2.$$

Here, m, b, and g are control parameters. With different values of these parameters, a variety of regular and chaotic regimes can be observed [15]. Among these, the model (12.17) can operate in a regime of *self-modulation*. This autonomous regime is characterized by slow and fast oscillatory modes whose frequencies are in a $1:6$ ratio (Fig. 12.15).

In the models of coupled systems, the coupling terms are often taken to be proportional to the differences between the respective variables. For two coupled systems of the form (12.17), this implies the presence of terms of the form $(x_1 - x_2)$, $(y_1 - y_2)$, and $(z_1 - z_2)$ in the equations for the x, y, and z variables, respectively. The simplest case involves interaction through only one variable. Examples range from electronic circuits with a purely resistive coupling between the component circuits over mechanical oscillatory systems with inertial coupling to neuron models with electrical coupling. In more realistic circumstances, however, multivariable coupling seems to be more appropriate. For instance, the reactance in electronic circuits or the propagation time delay along neuronal axons may give rise to couplings through the velocity variable. Let us analyze the general case when the diffusive coupling is introduced in a vector form $\mathbf{K} = (K_x, K_y, K_z)$:

$$\frac{1}{\omega_{1,2}}\dot{x}_{1,2} = mx_{1,2} - z_{1,2}x_{1,2} + y_{1,2} - bx_{1,2}^3 + K_x(x_{2,1} - x_{1,2}),$$

$$\frac{1}{\omega_{1,2}}\dot{y}_{1,2} = -x_{1,2} + K_y(y_{2,1} - y_{1,2}), \qquad\qquad (12.18)$$

$$\frac{1}{\omega_{1,2}}\dot{z}_{1,2} = -gz_{1,2} + gx_{1,2}(x_{1,2} + |x_{1,2}|)/2 + K_z(z_{2,1} - z_{1,2}),$$

where $\omega_1 = 1$ and ω_2 defines the frequency mismatch. It may be advantageous to represent the vector coupling in terms of polar coordinates:

$$K_x = K \cos\theta \cos\beta,$$
$$K_y = K \sin\theta \cos\beta, \qquad\qquad (12.19)$$
$$K_z = K \sin\beta.$$

This is the approach that we shall use in the following analysis. Here, K denotes the coupling strength, and the angles $0 \le \theta \le \pi/2$ and $0 \le \beta \le \pi/2$ define the relative

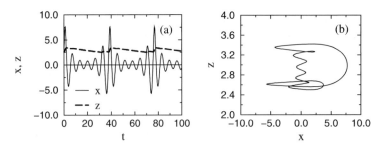

Fig. 12.15. Self-modulated regime 1 : 6 in a single Anishchenko–Astakhov oscillator (12.17): **a** realization and **b** phase portrait at $m = 2.90328$, $g = 0.012505$, and $b = 5 \times 10^{-5}$

weights of the three coupling terms. θ and β can be also viewed as the orientation angles of the coupling force in the three-dimensional subspace of each oscillator. Single-variable coupling is achieved when $(\theta = 0, \beta = 0)$, $(\theta = \pi/2, \beta = 0)$, or $(\beta = \pi/2)$.

Application of Effective Coupling Method

To reach the regime of self-modulated oscillations for the system (12.17), we fix $m = 2.90328$, $g = 0.012505$, and $b = 0.00005$. Figure 12.16 illustrates the effect of phase multistability through the effective coupling technique. Inspection of the Fig. 12.16(a) clearly shows that the calculated antisymmetric part of Γ for x and y allows one to detect six stable and six unstable solutions. Note that their number is equal to the number of local maxima over the period of oscillations (Fig. 12.15). Since the coupling is diffusive, the stable synchronous regimes in the coupled system imply the coincidence of local maxima of oscillations in the individual units. The system eventually settles down on one of the stable regimes depending on initial conditions. The coupling has little influence on the phase difference of the system when the oscillator is in the synchronized regime. If any phase shift from this state arises, the system will gradually be attracted back to synchronous state.

Coupling via the z variable demonstrates a completely different behavior. There is only one stable regime and this is an antiphase one. We suggest that this is related to the dephasing effect[8] [104, 219] caused by the vector field deformation in the vicinity of the saddle equilibrium point near the limit cycle. Variation of the z variable strongly affects the distance of the perturbed trajectory from this point and, hence, is responsible for its slowing down or acceleration. Moreover, $z(t)$ operates in a different regime as compared to $x(t)$ and $y(t)$, i.e., without any modulation (Fig. 12.15). When the vector of diffusive coupling is changed from x- or y-coupling towards z-coupling, a the transition between different sets of coexisting regimes is observed. Figure 12.16(b) shows how the multistable regimes successively disappear as a result of a smooth transition from x- to z-coupling provided by the variation of β (for $\theta = 0$).

---

[8] Dephasing effect was considered in Sect. 11.3.

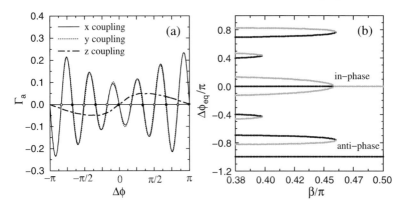

Fig. 12.16. (Color online) Phase analysis of the self-modulated regime of coupled Anishchenko–Astakhov oscillators (12.18). **a** Antisymmetric part Γ_a of effective coupling function; **b** Evolution of location and stability of coexisting regimes when the coupling vector is gradually changed from K_x to K_z. *Black circles* denote stable solutions

Mapping Approach

Let us consider the behavior of the coupled systems (12.18) for a strong interaction in order to compare the results with the case of vanishingly weak coupling.

As predicted by the phase reduction method, six phase-locked patterns at $K = 0.0005$ can be singled out (Fig. 12.17). Each state corresponds to one of six stable equilibrium points in the phase map described by (12.16), see Fig. 12.18(a). The time series of the multistable regimes are shifted with respect to each other while the phase portraits on the (x_1, x_2) plane indicate different out-of-phase regimes with respect to the symmetric phase space.

As the mismatch parameter ω_2 moves away from 1 the synchronous regimes sequentially lose their stability. The number of equilibrium points decreases via tangent bifurcations in terms of the map (Fig. 12.18(b) with insert). Figure 12.19 represents the bifurcation diagram of the possible synchronous regimes on the "frequency mismatch"–"coupling strength" parameter plane. For weak interaction, there are six stable (and the same number of unstable) solutions that differ from each other by a phase shift. There is a set of stability regions for different synchronous regimes whose structures are similar to those described in Sect. 12.1 for oscillators demonstrating the Feigenbaum period-doubling route to chaos [222]. In the present case, however, the tongues are not all inserted into each other, but some of them are shifted a little with respect to each other [273]. With increasing coupling, the solutions subsequently lose their stability through period-doubling bifurcations (dashed curves).

12.3 Bursting Dynamics

Bursting, i.e., complex behavior characterized by brief bursts of oscillatory activity interspersed in quiescent periods, is the primary mode of electrical activity for

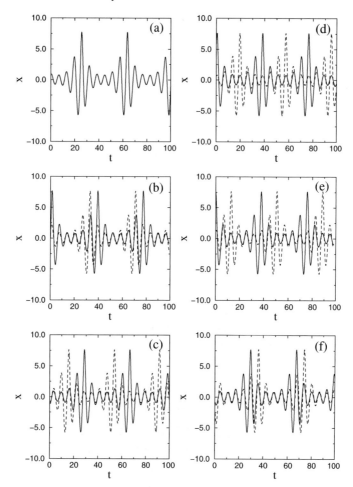

Fig. 12.17. Six phase-locked patterns in coupled Anishchenko–Astakhov oscillators (12.18) with different phase shifts **a** $\Delta\phi = 0.0$, **b** $\Delta\phi = 1.6553\pi$, **c** $\Delta\phi = 1.3134\pi$, **d** $\Delta\phi = 0.9928\pi$, **e** $\Delta\phi = 0.6710\pi$, and **f** $\Delta\phi = 0.3425\pi$, at $K = 5 \times 10^{-4}$ and $\omega_2 = 1.0$

a variety of nerve and endocrine cells [296]. Bursting patterns were found, e.g., in discharging cold fibers of cats [56] and in activity of shark sensory cells [57]. It is known that pancreatic β-cells under normal circumstances display a bursting behavior with alternations between an active (spiking) state and a silent state [69]. It is also established [200] that the secretion of insulin depends on the fraction of time that the cells spend in the active state, and that this fraction increases with the concentration of glucose in the extracellular environment. de Vries et al. [72] found asymmetrically phase-locked solutions to be typical in coupled heterogeneous β-cells while a set of coexisting out-of-phase regimes was observed for coupled Hindmarsh–Rose models [202, 203]. When at fixed parameters the initial conditions were changed

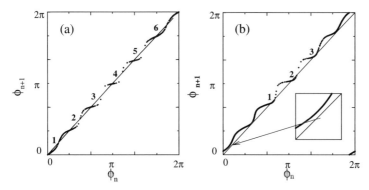

Fig. 12.18. The phase map (12.16) of the system (12.18) at $K = 5 \times 10^{-4}$. Note that as compared to (12.16) here subscript "t" correspond to "n" and "$t + \tau$" to "$n + 1$." **a** In case of identical systems ($\omega_2 = 1.0$) six stable equilibrium points correspond to six synchronous regimes. **b** When a frequency mismatch ($\omega_2 = 1.001$) is introduced, only three equilibrium points remain. K is fixed at 5×10^{-4}

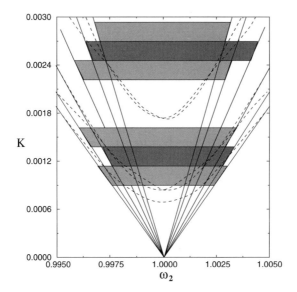

Fig. 12.19. (Color online) Synchronization regions for coexisting families of attractors ($m = 2.90328$, $g = 0.012505$, and $b = 5 \times 10^{-5}$). *Dashed lines* denote period-doubling bifurcations

the system switched from one burst-locked mode to another one. The mechanisms of phase multistability in coupled bursters are related to the complex wave forms of the oscillations, as well as to a version of the above-mentioned dephasing effect [230] (see Sect. 11.3).

12.3.1 Simple Qualitative Approach to Phase Multistability

The top traces $x_1(t)$ in Fig. 12.20(a) and (b) show typical examples of spike trains representing, for instance, the locations of local maxima for oscillations with complex wave forms or with bursting dynamics. While in Fig. 12.20(a) the spikes are equidistant in time, the spikes in Fig. 12.20(b) occur with different intervals.

Consider two identical oscillator with bursting dynamics that are coupled diffusively. One can easily count the number of possible synchronous regimes with different mutual phase shifts that are determined by different spikes in realizations $x_1(t)$ and $x_2(t)$ which occur simultaneously The results of a more formal analysis are summarized below. Note that this approach is also applicable in case when the bursting systems are not identical.

Equidistant Spike Train

- We consider a signal that is characterized by the firing interval $T_f = i \Delta t$ and a silence interval $T_s = j \Delta t$ (i, j are integers) with $\Delta t = $ const. The whole period is defined as $T = (i + j) \Delta t$.
- For two interacting signals $x_1(t)$ and $x_2(t)$, it is assumed that $i_1 + j_1 = i_2 + j_2 = N$. To be specific, let $i_2 < i_1$ and, thus, $j_2 > j_1$.
- If $j_1 < i_2 - 1$, the silent region overlaps with the spike train. Hence, the number of possible combinations is equal to $N = i_1 + j_1 = L$. The case with $j = 0$ and different spike amplitudes corresponds to the cases involving subharmonic components [38] and to those with self-modulated oscillations [273].
- If $j_1 \geq i_2 - 1$, the number of possible synchronous regimes is equal to $N = i_1 + i_2 - 1$ and increases with increasing i_1 and i_2.

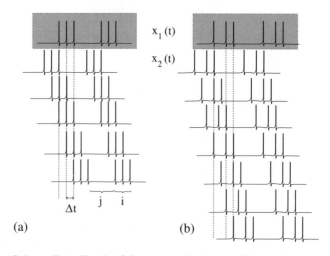

Fig. 12.20. (Color online) Sketch of the expected variants of the synchronous regimes for interacting bursting oscillations with three-spike trains. Note the difference between the cases when the interspike distances are **a** equal and **b** different, respectively

12.3 Bursting Dynamics 343

- If $i_1 + j_1 \neq i_2 + j_2$, while Δt is still the same for both spike trains, then a minimal period $T_{ij} = \Delta t (i_1 + j_1)(i_2 + j_2)$ exists, and the problem translates into the previous case. However, the particular configuration of silent regions and spike trains depends on the values of i_1, j_1, i_2, and j_2. The set of synchronous regimes can be estimated as $i_1 + i_2 - 1 \leq N \leq (i_1 + j_1)(i_2 + j_2)$.

We conclude that interacting equidistant bursting oscillators can provide even more synchronous states than self-modulated (period-doubled) oscillations of the same period T.

Non-equidistant Spike Train

This case is perhaps more realistic because a typical bursting scenario involves a gradual variation of the spiking frequency during a single burst. In such situation one can expect a different number of coexisting regimes in the interacting bursters as compared to their number in case of equidistant bursting.

Let one of the spikes in the train be located with a different time interval from the other spikes (Fig. 12.20(b)). This does not affect the fully in-phase regime. However, the stability of the phase-shifted regimes is likely to become weaker since the coincidence of spikes is not as good as in Fig. 12.20(a). At the same time, additional cases of coincidence for the "separated" peak appear. However, even although the tendency to synchronize may not be strong enough to provide additional stable synchronous states, at least they can produce the so-called "ghosts" where phase differences change slowly.

Limitations to the Above Approach

In Chap. 4 it was shown for mutually coupled van der Pol oscillators that only the in-phase synchronous regime is stable for weak dissipative coupling [34, 240]. But there is an interesting mechanism that can produce stable out-of-phase synchronous state in weakly coupled oscillatory units. In Chap. 11 dephasing was demonstrated to be responsible for antiphase synchronization in coupled Morris–Lecar neuron models and in coupled modified van der Pol systems [104, 219]. Models exhibiting this effect might have different details but they have a common structure of their phase space. The presence of a saddle equilibrium located nearby but outside the limit cycle is crucial. The latter create substantial inhomogeneity of phase velocity on, and in the vicinity of the limit cycle. When perturbed by coupling, the phase trajectories of the interacting units can be shifted towards, or away from, the saddle point and hence the dynamics can be slowed down or accelerated. Moreover, it has been found [230] that the mutual location of the equilibrium point and a limit cycle in the generalized FitzHugh–Nagumo model can be responsible for similar dephasing effects. In a certain region of the phase space the phase trajectory in the single cell model approaches the unstable equilibrium point quite closely. Thus, a weak perturbation can influence the motion of the phase point considerably.

344   12 Phase Multistability

It is not obvious how the above approach to synchronization of spike trains can be extended to the antiphase regime and to out-of-phase states. Finally, for some regimes the time intervals can be different between all spikes in a train. It is the purpose of the present section is to discuss this problem in detail.

12.3.2 Dynamics of Coupled Bursters

Model

As a basis for the present analysis we use the simplified model of a pancreatic β-cell suggested by Sherman et al. [265]:

$$\tau \frac{dV}{dt} = -I_{Ca}(V) - I_K(V, n) - I_S(V, S),$$
$$\tau \frac{dn}{dt} = \lambda \big(n_\infty(V) - n\big), \tag{12.20}$$
$$\tau_S \frac{dS}{dt} = S_\infty(V) - S,$$

where

$$I_{Ca}(V) = g_{Ca} m_\infty(V)(V - V_{Ca}),$$
$$I_K(V) = g_K n(V - V_K),$$
$$I_S(V) = g_S S(V - V_K),$$
$$\omega_\infty(V) = \frac{1}{1 + \exp[(V_\omega - V)/\Theta_\omega]}, \quad \text{with } \omega = m, n, \text{ and } S.$$

Here, V represents the membrane potential while, n may be interpreted as the opening probability of the potassium channels, and S accounts for the presence of a slow dynamics in the system. S is likely to be related to the intracellular Ca^{2+}-concentration, although the precise biophysical interpretation of this variable remains unclear. I_{Ca} and I_K are the calcium and potassium currents, $g_{Ca} = 3.6$ and $g_K = 10.0$ are the associated conductances, and $V_{Ca} = 25\,\text{mV}$ and $V_K = -75\,\text{mV}$ are the respective Nernst (or reversal) potentials. τ/τ_S defines the ratio of the fast (V and n) and the slow (S) time scales. The time constant τ for the membrane potential is determined by the capacitance and the typical total conductance of the cell membrane. With $\tau = 0.02$ s and $\tau_S = 35$ s, the ratio $k_S \equiv \tau/\tau_S$ is quite small, and the cell model is numerically stiff.

The calcium current I_{Ca} is assumed to adjust instantaneously to variations in V. For the fixed values of the membrane potential, the gating variables n and S relax exponentially towards their voltage dependent steady state values $n_\infty(V)$ and $S_\infty(V)$. Together with the ratio k_S of the fast to the slow time constants, V_S will be used as the main bifurcation parameter. This parameter determines the membrane potential at which the steady state value for the gating variable S attains half its maximum value. The other parameters are $g_S = 4.0$, $V_m = -20\,\text{mV}$, $V_n = -16\,\text{mV}$,

$\theta_m = 12\,\text{mV}$, $\theta_n = 5.6\,\text{mV}$, $\theta_S = 10\,\text{mV}$, and $\sigma = 0.85$. These values are all adjusted so that the model can reproduce the experimentally observed time series with reasonable accuracy. In accordance with the formulation used by Sherman et al. [265], all the conductances have been scaled relative to some typical conductance. Hence, we may also consider (12.20) as a model of a cluster of closely coupled β-cells that share the capacity and the conductance of the total membrane area.

Figure 12.21 provides an example of the evolution of V, n, and S obtained by simulating the cell model at the parameter values where it exhibits bursting behavior. Bifurcation analysis of the single Sherman model shows a variety of different spiking regimes [157]. An example of a two-dimensional bifurcation diagram is presented in Fig. 12.22. Near the bottom of this figure we observe Andronov–Hopf bifurcation curve. Below this curve, the model has one or more stable equilibrium points. Above the curve we find a region of complex behavior delineated by the period-doubling curve PD^{1-2}. Along this curve, the first period-doubling of the continuous spiking behavior takes place. In the heart of the region surrounded by PD^{1-2} we find an interesting squid-shape structure with arms of chaotic behavior (indicated in black) stretching down towards the Andronov–Hopf bifurcation curve. Each of the arms of the squid-shape structure separates a region of periodic bursting behavior with i spikes per burst from a region with regular behavior with $(i+1)$-spikes per burst. Each arm has a period-doubling cascade leading to chaos on one side and a saddle-node bifurcation on the other. It is easy to see that the number of spikes per burst becomes large as k_S approaches zero.

Simulation Results

Bursting dynamics that represents another example of fast-and-slow motion, differs from the described above oscillations in the period-doubled regimes since it contains a silent state. This implies that local maxima are distributed non-uniformly over the whole period, and the set of possible synchronous states is expected to have specific features. Let us develop a simplified qualitative analysis to understand how coexisting regimes arise. The basic assumption for such analysis is a tendency of coupled units to be synchronized with the coincidence of their local maxima. The more local maxima (spikes) coincide, the stronger the stability of the respective regime is.

To calculate the effective coupling function, it is necessary to define (i) the equations for the model to be coupled and (ii) the form of coupling. We assume that the coupling is of diffusive type and is expressed by the difference terms of the form $c(X_1 - X_2)$ where $X_1 = (V_1, n_1, S_1)^T$ and $X_2 = (V_2, n_2, S_2)^T$ are the state vectors of the individual cell models. c is the coupling matrix for which we assume the form $c = \text{diag}(1, 0, 1)$, indicating that coupling takes place via the first and the third variables. The membrane potentials are coupled resistively via electric currents that flow between the cells, and the third variables are coupled via the diffusive exchange of calcium between the cells [301]. We do not consider coupling via the gating variables n, since such a coupling appears less realistic from the biological

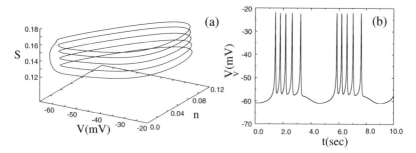

Fig. 12.21. Example of bursting oscillations in a single Sherman model (12.20) with five spikes per burst at $V_S = -39.0$ mV and $k_S = 0.00057$. (**a**) 3D phase plot; (**b**) realization of the membrane potential

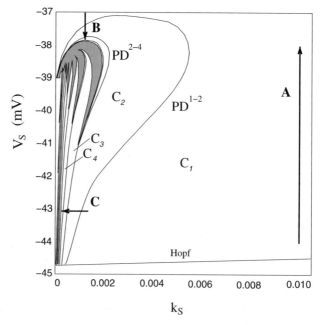

Fig. 12.22. (Color online) Two-dimensional bifurcation diagram outlining the main bifurcation structure in the (V_S, k_S) parameter plane for the single cell Sherman model (12.20). Note the squid-shape black region with chaotic dynamics. *Arrows* **A**, **B**, and **C** indicate different routes of parameter variation discussed in the text

point of view. Note that the coupling strength parameter is absent in the expression for c because the analysis assumes the coupling to be vanishingly weak.

Figure 12.23 illustrates how the number of detected stable synchronous regimes changes when varying the control parameter k_S along route **C** as indicated in Fig. 12.22. Along this route, the number of spikes in a train increases stepwise when crossing the bifurcation curves. The bifurcation mechanism in this direction was de-

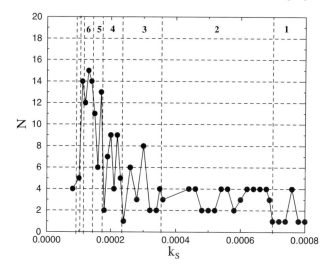

Fig. 12.23. The number N of phase-locked regimes vs the control parameter k_S under the period-adding scenario **C** (Fig. 12.22). Numbers in the upper part of the figure denote the number of spikes per burst

scribed by Mosekilde et al. [184]. One typically observes that the "i-spike per burst" solution destabilizes in a subcritical period-doubling bifurcation, and the "$(i+1)$-spike" solution arises in a saddle-node bifurcation. It is clearly seen from Fig. 12.23 that the maximal number of coexisting states N tends to grow with the increasing number of spikes in the train. However, the fluctuation of N is significant, and the whole plot looks quite random.

To understand how the number of synchronous regimes varies with k_S, let us consider the behavior of the effective coupling function as calculated for the seven-, eight-, and nine-spike trains (Fig. 12.24). We first note that the shape of the effective coupling function for V-coupling is much more complicated than for S-coupling. This is associated with the dynamics of the individual Sherman model where V and S are fast and slow variables, respectively. The spiking dynamics causes well-pronounced short-range oscillations of Γ_a around zero. Another interesting observation is that a smooth deformation of a long-range component of Γ_a with varying k_S (rather than changes in short-range oscillations of Γ_a) leads to the changes of the number of intersections with zero. An inspection of Fig. 12.24(c) shows that the region of short-range oscillations of Γ_a still exists, but the long-range structure dominates. As a result, the number of stable synchronous states for the nine-spikes per train bursting dynamics is only four.

The behavior described here supports the hypothesis that the dephasing effect can play a significant role for the long-range variation of Γ_a and, hence, cause the abrupt changes in the number of coexisting regimes. The strength of the dephasing effect can be indirectly measured by calculation of the minimal distance $D_{\min}$ between the limit cycle and the nearby equilibrium point (Fig. 12.25). Dephasing can

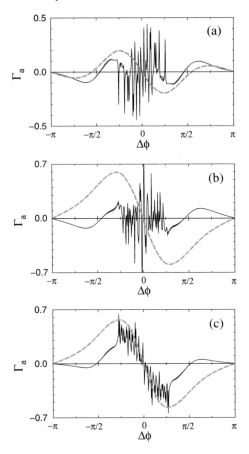

Fig. 12.24. (Color online) Asymmetric part Γ_a of the effective coupling function for the multi-spike bursting regimes. The *solid line* is for V-coupling while the *dashed line* is for S-coupling. **a** Seven spikes per train at $k_S = 0.00011$; **b** eight spikes per train at $k_S = 0.00009$; **c** nine spikes per train at $k_S = 0.00008$. Note how the slow variation of Γ_a in **c** causes the number of stable synchronization regimes to be quite small, even for V-coupling

explain the irregular changes of the set of coexisting regimes. To find some correlation, we introduce the quantity N/M characterizing how effectively the number of spikes in a train is transformed into the set of synchronous regimes. We compare the changes of this quantity with the change of the minimal distance $D_{\min}$ under variation of k_S. According to the simple quantitative analysis at the beginning of this Section, one can expect that $N/M \approx 2.0 - 1/M$ for the case of "perfect" bursting. In practice, the N/M curve jumps within the range $[0.666; 4.25]$. Moreover, one can observe a certain correlation between curves for N/M and for $D_{\min}$. This suggests that the phase multistability for the bursting regimes is govern by variation of distance between the limit cycle and equilibrium point rather than by the number of spikes per train.

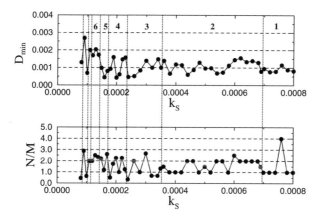

Fig. 12.25. There is a certain correlation between plots for the minimal distance D_{min} from the equilibrium point to the limit cycle (**upper panel**) and for the number of coexisting stable regimes, normalized to the number M of spikes per train (**lower panel**)

12.3.3 Multistability Induced by Dephasing

Let us return to the diagram in Fig. 12.22. There is no bursting to the right of the curve PD^{1-2}. Here, continuous spiking is the only stable mode. This regime is in many ways similar to the behavior of two dimensional models, like the van der Pol oscillator. Thus, a relatively simple pattern for the mutual synchronization of the cells is expected. However, inspection of the diagram for two coupled Sherman models (12.20) reveals different patterns of synchronous states even for weak diffusive coupling. For example, both in-phase and antiphase regimes can be stable, and an additional pair of out-of-phase solutions can occur. The reason for this variety of stable synchronous states is the dephasing effect that occurs due to the presence of a saddle equilibrium located nearby but outside, the limit cycle. In contrast to the two-dimensional oscillators, the Sherman model has a single equilibrium point inside the limit cycle. How can dephasing arise in this case?

To illustrate clearly the dephasing effect in Sherman oscillators, we reduce the model equations (12.20) to a two-dimensional model with only one fast (V) and one slow (S) variable (i.e., we assume the relaxation of the gating variable n to be very fast). This produces a model similar to the FitzHugh–Nagumo model in the general form

$$\tau \frac{dV}{dt} = -I_{Ca}(V) - I_K(V, n_\infty) - I_S(V, S) = f(V, S),$$
$$\tau_S \frac{dS}{dt} = S_\infty(v) - S = g(V, S). \tag{12.21}$$

Here the terms are the same as in (12.20), but n_∞ is used instead of n in the expression for I_K.

In Fig. 12.26 the mutual location of the limit cycle (white curve) and the unstable equilibrium point (*EP*) is illustrated together with contour plots of the phase

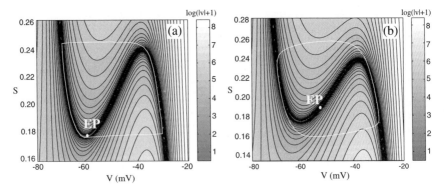

Fig. 12.26. Phase velocity contour plot for the reduced Sherman model (12.21) at **a** $V_S = -44.0$ mV, $k_S = 0.001$; **b** $V_S = -38.19$ mV, $k_S = 0.0175$

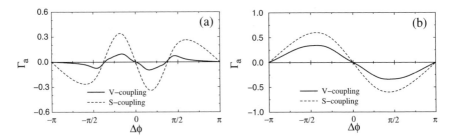

Fig. 12.27. Antisymmetric part for the effective coupling function, calculated for the reduced Sherman model at **a** $V_S = -44.0$ mV, $k_S = 0.001$; **b** $V_S = -38.19$ mV, $k_S = 0.0175$

velocity (solid lines with grey shading). There is an area of slow motion, determined by the location of the cubic-shape nullcline $f(V, S) = 0$. At the intersection of $f(V, S) = 0$ with the other nullcline $g(V, S) = 0$ there is a single point (*EP*) of zero phase velocity. It is clearly seen how the position of *EP* changes with varying control parameter k_S, and the sensitivity to a weak perturbation of the limit cycle changes as well. In Fig. 12.26(b), a deviation from the unperturbed cycle (white curve) should not produce a significant effect, while the motion along the limit cycle in Fig. 12.26(a) becomes inhomogeneous.

These qualitative observations are confirmed by the calculations of the effective coupling function (Fig. 12.27). At $V_S = -38.19$ mV, $k_S = 0.0175$, the equilibrium point is located away far from the limit cycle. In this case, the in-phase synchronous regime is stable, but the antiphase solution is unstable (Fig. 12.27(b)) for weak diffusive coupling via the V and S variables. This behavior is similar to synchronization of dissipatively coupled van der Pol oscillators (see Chap. 4), and the dephasing effect is not pronounced. As soon as the equilibrium point approaches the limit cycle (Fig. 12.27(a)), the antiphase regime becomes stable but the in-phase solution maintains its stability in contrast to the dephasing effect [104] described in Sect. 11.3.

Two new out-of-phase unstable regimes appear. Simultaneous coupling via both the V and S variables produces a qualitatively similar effect.

Thus, the coupled reduced models (12.21) exhibit the dephasing effect in a form different from the form described in [104, 219] and in Sect. 11.3. We expect that the dephasing effect will be preserved when we return to the full Sherman model (12.20). However, in coupled 3D systems it is difficult to make precise statements about the mutual configuration of a limit cycle and an equilibrium point based on a Poincaré section only. Useful information can be obtained by calculating the distance between the two objects in phase space. In Fig. 12.28 the variation of the minimal distance $D_{\min}$ between the limit cycle and the equilibrium point is plotted. It is clearly seen that this distance decreases with decreasing values of V_S. The inserts show examples of the Γ_a shape for selected values of V_S. For $V_S = -38.39$ mV the effective coupling function indicates "good" behavior, similar to the behavior observed in dissipatively coupled van der Pol oscillators: the in-phase state is the only stable solution for coupling via the V (solid line) or S (dashed line) variables. For $V_S = -43.25$ mV, Γ_a indicates both in-phase and antiphase regimes that are stable both for S-coupling and for V-coupling.

Note that the phase space structure of the Sherman model provides phase multistability even outside of the bursting region. The mechanism for this can be identified as a specific form of dephasing effect, related with a slowing down or acceleration of the trajectory in each coupled unit. Note that the described effect takes place for arbitrarily weak coupling and is the result of the phase space properties of the

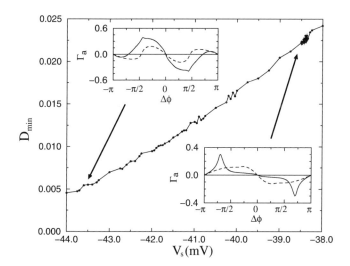

Fig. 12.28. Minimal distance $D_{\min}$ from the equilibrium point EP to the limit cycle plotted against the value of V_S (along the A route in Fig. 12.22). *Inserts* display the qualitatively different responses of the 3D Sherman model (12.20) to weak coupling via the V variable (*solid line*) or the S variable (*dashed line*). Note that the *solid line* in the *upper insert* is reduced by 20 in the vertical scale to fit the same plot as the *dashed line*

352   12 Phase Multistability

Sherman model rather than of specific features of the coupling. In the bursting area we expect the considered mechanism to interact with the effect of multicrest wave forms, producing additional complexity in the phase patterns.

12.4 Summary

Phase multistability provides a new insight on the variety and complexity of bifurcation transitions inside the synchronization region and near its boundary. The results presented in this chapter allow us to make a few general conclusions given below

- To estimate the number of stable synchronous states for a system of two weakly diffusively coupled models one has to take into account (i) the wave forms of the oscillations in different regions of parameter space (essential for period-doubling and self-modulated oscillations), and (ii) particular structure of the phase space of the system, involving regions of fast and slow motion, passing of trajectories close to singular points, etc. As a result, the dephasing effect can play an important role in the formation and evolution of coexisting regimes (essential for bursters).
- Mapping approach, as well as the method of effective coupling, serve a quantitative measure of phase dynamics that provides information on the phase properties of the interacting solutions and on the number of synchronous regimes.
- Synchronization region should be considered as a set of embedded Arnold tongues formed by coexisting phase-shifted regimes. Boundary of the synchronization region is related to bifurcations of the most stable synchronous regime.

13 Synchronization in Systems with Complex Multimode Dynamics

In the previous chapters we considered synchronization phenomena in coupled dynamical systems whose individual oscillations were characterized mostly by a single basic time scale. However, often the natural dynamics of interacting systems can be more complex, involving several *independent* time scales of either deterministic, or stochastic (statistical) origin. This feature is called multimode dynamics. Some effects of the multimode dynamics on synchronization phenomena were discussed in Chap. 12, where we showed that interaction of the self-oscillators, each being characterized by several time scales, can lead to phase multistability. In this chapter we get a deeper insight into how synchronization is developed in the systems whose natural dynamics is multimode from the viewpoint of evolution of different time scales in the interacting systems.

Illustrations of several types of multimode dynamics are given in Fig. 13.1. The simplest example is oscillations with two independent components (*modes*) corresponding to fast and slow motion (see Fig. 13.1(a)). In this case, the periods of the slow oscillations T_1 and of the fast oscillations T_2 serve as two characteristic time scales of the dynamical system. Another bright example of a system with multimode oscillations is the famous Lorenz system [172], whose phase dynamics is a combi-

13 Synchronization in Systems with Complex Multimode Dynamics

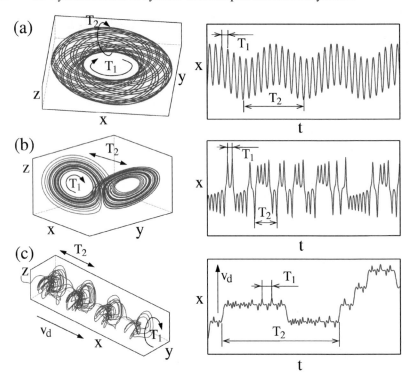

Fig. 13.1. Examples of multimode dynamics including: **a** a combination of fast and slow rotation with periods T_1 and T_2; **b** rotation with basic period T_1 and switching with mean switching time T_2; **c** rotation with basic period T_1, switching with mean switching time T_2 and drift with drift velocity v_d. The **left panels** represent phase portraits. The **right panels** show the corresponding realizations

nation of rotations around two symmetrically located fixed points and occasional jumps from the vicinity of one fixed point to the vicinity of the other (Fig. 13.1(b)). In such motion one can separate two independent time scales, the first being associated with rotation around, and the second with switching between, the vicinities of the fixed points: e.g. the basic period of rotation T_1, and mean time interval between jumps T_2. Generally, the system can have an infinite number of fixed points and can demonstrate the behavior that involves rotation around them as well as jumps between their vicinities. Moreover, jumps in one direction can be more probable than in another. In this case, as one can see from Fig. 13.1(c), besides the basic period of rotation T_1 and the mean time interval between jumps T_2, the motion is characterized by the drift velocity v_d that reflects a trend of the system state towards the direction in which jumps are more probable.

Multimode oscillations are widely spread in nature and in engineering. They are typical dynamical regimes, e.g., in lasers [140], in a phase-locked loop [176, 216], in electrochemical oscillators [145], and in semiconductor nanodevices [10]. Liv-

13.1 Synchronization of Chaotic Systems with Fast and Slow Time Scales   355

ing systems often exhibit dynamics with different time scales. The thalamocortical relay neurons, for instance, can generate either spindle or delta oscillations [295]. Recently, it has been found that the dynamics of electroreceptors in paddlefish can be biperiodic [190]. In [272] an individual nephron was described as a two-mode oscillator demonstrating relatively fast oscillations associated with the myogenic regulation of the arteriolar diameter, and slower oscillations related to a delay in the tubuloglomerular feedback.

Many models of bursting neurons [136], for example, can be split into slow and fast subsystems. Such an approach works very well when these subsystems can operate separately and the coupling is weak. Otherwise, the paradigm of coupled units seems to be less fruitful. Hence, the description of double-oscillatory nature of the original system by means of a single two-mode oscillator is useful when coupling is strong enough and the essential dynamical effects arise due to interaction between the subsystems.

Often the cooperative dynamics of coupled multimode systems can be considered from the viewpoint of synchronization of different components of motion. In [27] the authors considered synchronization of the systems with quasiperiodic oscillations, when each of the interacting systems demonstrates two independent time scales. In [20] synchronization of switching processes in coupled Lorenz systems has been studied. In [185, 216] the cooperative dynamics of coupled oscillators with drifts was explored.

In this chapter we consider the main principles of how different types of multimode behavior can be induced in chaotic and stochastic systems. We also focus our study on synchronization of oscillations characterized by several time scales of different origins.

13.1 Synchronization of Chaotic Systems with Fast and Slow Time Scales

13.1.1 Single System with Two Time Scales

First, consider the case when the phase dynamics of each subsystem is characterized by two time scales associated with rotation in the phase space. The model we are going to study consists of two oscillatory subunits, where a self-sustained oscillator drives a damped non-linear oscillator via both additive and multiplicative forcing. This model was proposed in [232] in order to describe bimodal oscillations, which are observed in nephron autoregulation [272]. From the viewpoint of physics, this process may be considered as a parametric perturbation of the fast oscillations. The model can be implemented with non-linear electronic circuits or coupled mechanical oscillators.

The equations read

$$\ddot{x} - \left(1 - x^2\right)\dot{x} + \omega^2 x = E + c\dot{v}, \tag{13.1}$$
$$\ddot{v} + d\dot{v} + v\Omega(v) = F(x, v), \tag{13.2}$$

356   13 Synchronization in Systems with Complex Multimode Dynamics

where the first equation represents a van der Pol-type oscillator with frequency ω. This oscillator is subjected to a constant force E and receives a feedback $c\dot{v}$ from the other subunit. The second equation describes a damped oscillator with a frequency $\Omega(v)$ represented by a non-linear function in the form $\Omega(v) = 1 + \beta e^v$ with $\beta \ll 1$. This form originated from observation of real nephron dynamics, but actually describes a fairly generic case: for small v, $\Omega(v) \approx 1$, but larger values of v produce a considerable upshift of the resonance frequency. The term $F(x, v)$ represents the forcing from the first oscillator. The specific form to be used includes both an additive and a multiplicative forcing

$$F(x, v) = a\tanh(x)(1 + \gamma v). \tag{13.3}$$

The function $\tanh(x)$ is used to describe saturation phenomena at both very positive and very negative values of x. Together with the non-linear frequency term $\Omega(v)$, $F(x, v)$ provides stabilization of the oscillation amplitude in the parametrically forced oscillator (13.2). ω^2 and E are used as control parameters while the other parameters are fixed at $c = 2.0$, $d = 0.1$, $\beta = 0.001$, $a = 0.474$, and $\gamma = 12.85$.

We can rewrite (13.1)–(13.2) as a set of four first-order ordinary differential equations in the following form:

$$\dot{x} = y, \tag{13.4}$$
$$\dot{y} = (1 - x^2)y - \omega^2 x + E + cu, \tag{13.5}$$
$$\dot{v} = u, \tag{13.6}$$
$$\dot{u} = -du - v\Omega(v) + F(x, v). \tag{13.7}$$

In the limit of vanishingly small values of c, the self-sustained dynamics of the system is bounded by the lines of an Andronov–Hopf bifurcation for subsystem (13.4)–(13.5), whose equation on the plane of parameters (ω^2, E) is

$$E = \pm\omega^2. \tag{13.8}$$

However, for finite values of c, self-sustained regimes occupy a wider area on the (ω^2, E) plane because of the positive feedback provided by the term cu. For larger values of c ($c = 2.0$ in this study), this region contains various periodic, quasi-periodic, and chaotic regimes. Among them, let us focus on the regime of chaotic dynamics that appears through a period-doubling cascade and whose main feature is the presence of two time scales originating from the slow dynamics of the subunit (13.4)–(13.5) and the fast dynamics of the subunit (13.6)–(13.7). An example of a realization of such oscillations is given in Fig. 13.2.

Since both characteristic time scales of the system dynamics are associated with rotation, the obvious way to estimate those time scales is to calculate the mean period of rotations for each subunit. Technically, it can be done by averaging the time intervals between the successive returns to some Poincaré sections formally introduced for each of subunits. For example, one can collect the time intervals between

13.1 Synchronization of Chaotic Systems with Fast and Slow Time Scales

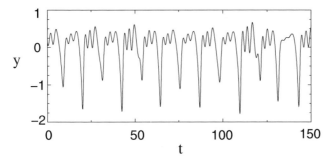

Fig. 13.2. A typical realization of two-mode chaos in (13.4)–(13.7) with $E = -0.4898$ and $\omega^2 = 0.5202$

the maxima of realizations $x(t)$ and $u(t)$, which would correspond to return times to Poincaré sections defined by $y = 0$ and $v = 0$, respectively. However, if for fast motion these return times bring a correct information about the velocity of rotations, for slow motion not every maximum reflects the corresponding rotation. As shown in Fig. 13.2, the feedback from the fast subunit modulates the slow dynamics making if difficult to choose the Poincaré section appropriately. In order to filter out the contribution from the fast component, two auxiliary equations are introduced as follows:

$$\dot{\xi} = \omega(x - \xi) \quad \text{and} \quad \dot{\eta} = \omega(\xi - \eta). \tag{13.9}$$

This way, we can correctly extract information about each of the two oscillatory modes by calculating return times τ_x and τ_v of the phase trajectories to the Poincaré secant surfaces defined by $\eta = 0$ and $u = 0$, respectively, in the (ξ, η) and (v, u) phase subspaces. By introducing the winding (rotation) number

$$r = \langle \tau_v \rangle / \langle \tau_x \rangle, \tag{13.10}$$

we can determine the ratio between the slow and fast frequencies that are associated with the first and the second subunits, respectively. By using sequences of τ_x and τ_v we can also introduce phase for each mode applying a method discussed in Chap. 8.

Figure 13.3 shows how dynamical characteristics of the systems (13.4)–(13.7) change with variation of E for $\omega^2 = 0.5202$. In Fig. 13.3(a) the dependence of three largest Lyapunov exponents on E is depicted. With increasing E, the system undergoes a cascade of period-doubling bifurcations, and at $E \approx -0.48989$ (Fig. 13.3(a)) one of the Lyapunov exponents becomes positive, i.e., the dynamics of the system becomes chaotic. Figure 13.3(b) presents a plot of winding number r versus E. It is clearly seen that for a significant range of E, r has a rational value $1/4$. This means that frequency locking occurs between two modes of the same system. Note that since $E \approx -0.48989$, the mode-locked regimes correspond to chaotic attractors. At $E \approx -0.4898$, the mode locking is destroyed and values of r start to "float" with variation of E in some range slightly above $1/4$. The destruction of mode locking is also confirmed by the calculation of the effective phase diffusion D_{eff}, which was

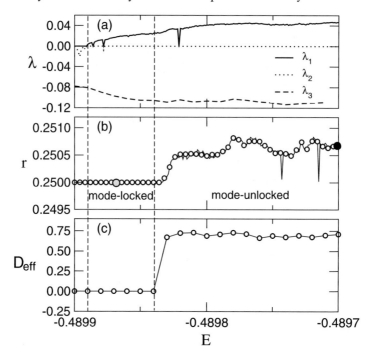

Fig. 13.3. **a** Three largest Lyapunov exponents, **b** rotation number r, and **c** phase diffusion coefficient D_{eff} as a function of control parameter E for a single system (13.4)–(13.7) with $\omega^2 = 0.5202$. While the largest Lyapunov exponent grows monotonically, r and D_{eff} indicate transitions inside the chaotic dynamics. *Grey circle* is put in the middle of parameter range with mode-locked chaos. *Black circle* corresponds to mode-unlocked chaos. The properties of the two types of chaos are compared in Fig. 13.4

introduced in Sect. 7.9. Although Lyapunov exponents do not reveal the qualitative changes between the chaotic attractors with rational and with non-rational rotation numbers, further inspection of the system dynamics shows that there is a clear difference between them.

In Fig. 13.4 the two columns compare the attractor characteristics before and after the mode unlocking transition at the points marked by grey circle ($E = -0.48987$) and by black circle ($E = -0.48970$) in Fig. 13.3. It is clearly seen that a 3D phase projection in Fig. 13.4(a) changes in a specific way with loops being added inside and around the main body of the attractor. Figure 13.4(b) shows a zoom on a part of the attractor in the (x, y) phase projection. The main difference between the two panels is the appearance in the right hand panel of additional small-sized structures in the bundle of trajectories that are indicative of the existence of a time scale faster than the time scale that defines the main shape of the attractor. Figure 13.4(c) shows a return time map with τ_v^n being τ_v calculated for nth return to the given Poincaré section. For a simple period-doubling chaos (left panel) this map has a clearly visible structure with segments, each being visited in a certain order. After the mode-

13.1 Synchronization of Chaotic Systems with Fast and Slow Time Scales 359

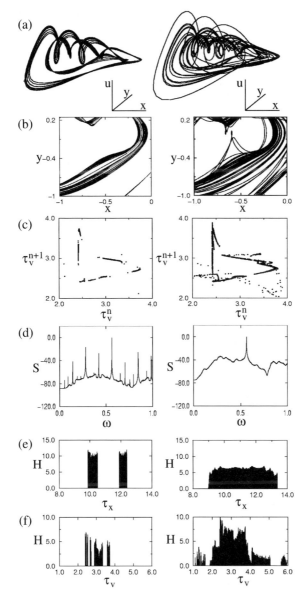

Fig. 13.4. Comparison of chaos characteristics in a single system (13.4)–(13.7) at $E = -0.48987$ (**left panels**) and at $E = -0.48970$ (**right panels**) corresponding to the *grey* and *black points* in Fig. 13.3, respectively. **a** 3D phase projection (x, y, u); **b** zoomed part of (x, y) projection; **c** return time maps; **d** power spectral densities in dB; **e** distribution H of return times τ_x; **f** distribution H of return times τ_v

360 13 Synchronization in Systems with Complex Multimode Dynamics

unlocking transition (right panel), however, the map becomes more disordered, with some segments merged and with many points outside the main part of the map. With this, the power spectra in Fig. 13.4(d) reveal the band structure of the chaotic attractor. Note that the transition being discussed occurs for a period-doubling chaos, thus one could expect the well-known band-merging bifurcations [15] to occur. However, in our case, the sequence of band-merging bifurcations is interrupted by a mode-unlocking transition described above. Figure 13.4(e), (f) indicate the changes in the distribution of return times τ_x and τ_v. Before the mode-unlocking transition, the Poincaré section has a well-pronounced band structure for both time scales. After the transition, the histogram for the slow time scale becomes uniform but clearly bounded. For the fast time scale, the histogram remains split into a few segments and spread over a wider interval. Summarizing the description above, the transition from a rational value of r to its floating behavior is accompanied by considerable changes in the attractor characteristics as indicated in Fig. 13.4.

Note that a similar phenomenon was observed in a case of generalized synchronization of two chaotic systems with frequency ratio $1:2$ [255]. However, in our case system cannot be split into two independent chaotic oscillators, and chaos appears due to the non-linear interaction between the functional units.

We have also compared the observed transition with the known evolution of a chaotic attractor to the so-called "Shilnikov chaos" [15]. Again a clear difference exists. In our case the trajectory does not visit the close vicinity of an unstable equilibrium point embedded in the attractor. At least there are no visible changes before and after the mode-unlocking transition. Accordingly, the statistics of mean return times (given in Figs. 13.4(e), (f)) is rather different from what we know for the Shilnikov attractor, for which the return time histogram extends to (infinitely) large times. In our case the return time histogram is smoothed, but bounded for both modes.

13.1.2 Coupled Systems with Two Mode Dynamics

Let us now consider how such systems, individually operating in the two-mode chaotic regime, can interact. We introduce a simple difference coupling term with a strength k. The equation for the x variable in (13.4) then becomes

$$\dot{x}_1 = y_1 + k(x_2 - x_1), \qquad \dot{x}_2 = y_2 + k(x_1 - x_2),$$

where subscripts indicate the first and the second interacting units. By calculating two rotation numbers r_x and r_v, each being the ratio between the similar time scales in the coupled units,

$$r_x = \frac{\langle \tau_{x1} \rangle}{\langle \tau_{x2} \rangle}, \qquad r_v = \frac{\langle \tau_{v1} \rangle}{\langle \tau_{v2} \rangle}, \tag{13.11}$$

we can separately describe the adjustment of the slow and fast modes. The simplest way to introduce a mismatch between the units would be to choose different values of ω_1^2 and ω_2^2. However, for the individual system (13.4)–(13.7) the curves of period-doubling bifurcation are generally parallel to the Hopf bifurcation curve

13.1 Synchronization of Chaotic Systems with Fast and Slow Time Scales

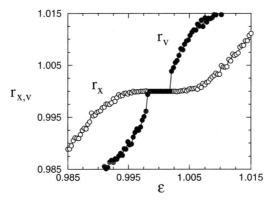

Fig. 13.5. Adjustment of two pairs of oscillatory modes is indicated by changes in the r_x and r_v rotation numbers with respect to the frequency mismatch ε. $\omega_2^2 = 0.5202$, $E_1 = E_2 = -0.48987$, and $k = 0.0035$

given by (13.8), and any variation of ω^2 will change not only the main frequency, but also the operating regime of the unit. Hence, it would be difficult to come to a reasonable conclusion about the interaction between the attractors of a particular type. In order to avoid this problem we have introduced a mismatch through an additional scale factor ε in the left-hand side of the equations for one of the interacting units, as suggested in [18], namely, $\varepsilon = 1.0$ corresponds to the case of identical units, while variations below and above 1.0 give rise to a detuning that does not influence the operating regime.

Let us consider the mutual adjustment of the oscillatory modes for the selected value of the coupling strength $k = 0.0035$. Figure 13.5 presents the variation of the rotation numbers for the slow r_x and the fast r_v time scales versus the frequency mismatch ε. There exists an interval of $\varepsilon \in [0.9984, 1.00176]$ where $r_x = r_v = 1.0$. This implies synchronous behavior with respect to both time scales. Both for larger and for smaller values of ε, the rotation number r_v diverges from 1.0 while r_x remains equal to 1.0 within a wider interval of $\varepsilon \in [0.9957, 1.00382]$. This demonstrates desynchronization between the fast oscillatory modes in the coupled units while the slow modes remain locked. This way, both partial synchronization (one of the two time scales is synchronized) and all-mode frequency locking of chaos can be observed. Note that the behavior of r_x and r_v near the edges of the locked region are different. By comparing the shape of the curves with the known synchronization pictures, one can draw analogies with the synchronization of periodic oscillations for r_v[1] and something similar to synchronization of noisy oscillations for r_x (see Chap. 7). We assume that this reflects different synchronization mechanisms for the fast and slow modes. Figure 13.6 represents the synchronization regions on the (ε, k) parameter plane for the slow and fast oscillatory modes separately. We can now clearly see that the two time scales have different widths of the Arnold

---
[1] Sharp tongue edges imply saddle-node bifurcations.

362   13 Synchronization in Systems with Complex Multimode Dynamics

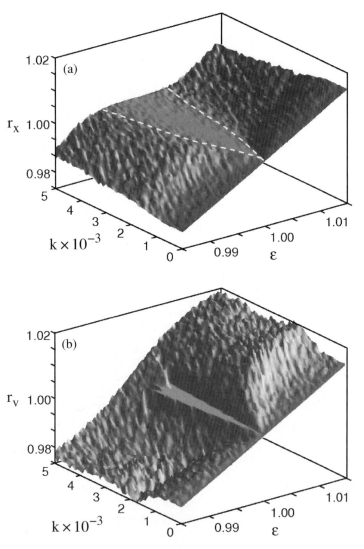

Fig. 13.6. 3D plots on the "frequency mismatch ε"–"coupling strength k" parameter plane for rotation numbers r_x and r_v separately. $\omega_1^2 = \omega_2^2 = 0.5202$ and $E_1 = E_2 = -0.48987$

tongues down to vanishingly small coupling strengths. An interesting observation can also be made for stronger coupling. For $k > 0.004$, the fast oscillatory mode is completely desynchronized and displays a gradual increase of r_v with increasing ε. This seems to be due to coupling through the slow x-variable. A stronger coupling increases the coupling-induced shift of the operating point for the two interacting units, and hence provokes the complete unlocking of the fast modes from the slow

13.2 Generation and Synchronization of Oscillations with Several Noise-Induced Modes   363

ones. Since the fast modes can interact only via the slow variable such a situation leads to desynchronization.

13.1.3 Conclusions

In conclusion to this section, we would like to note that the entrainment of two-mode chaotic regimes is realized in a more complicated way than synchronization of one-mode chaotic systems studied in Chap. 8. In particular, a mode-unlocking transition for chaos significantly influences the cooperative dynamics of the coupled units. Although obviously connected, the mutual behaviors of the slow and fast modes manifest many signs typical of independent time scales. As shown above, they are synchronized independently from each other at different parameter values. With this, slow components have wider region of synchronization in the parameter plane "time scale detuning–coupling strength."

Thus, when studying the cooperative dynamics in systems with multimode oscillations, one should keep in mind that synchronization criteria can depend on the particular mode of motion. With this, the concept of separation of time scales used in this section can be very helpful, since it allows one to apply the well-established techniques developed for simpler cases to the analysis of more intricate behavior.

13.2 Generation and Synchronization of Oscillations with Several Noise-Induced Modes

In the previous section we demonstrated that entrainment phenomena can appear between the modes of oscillations generated by a single deterministic system, which cannot be decomposed into two independent self-oscillators. Here, we consider similar case when multimode dynamics arises in *stochastic oscillators*, namely in excitable systems, whose oscillatory behavior is induced merely by noise.

Noise can have quite different effects when acting on self-oscillators or on excitable systems. General aspects of noise-induced transitions have been discussed in Chap. 9. We remind the reader that deterministic self-oscillators already possess their own time scales that can be modified by the random forcing [66, 298]. With this, influence of noise on an excitable system is more sophisticated. Without any perturbation, there is no response of the system at all, while too large random fluctuations just result in a noisy output. For an appropriate noise intensity, however, the behavior of the excitable system becomes highly regular. If one introduces some order parameter, e.g., correlation time, to characterize the coherence of oscillations, it will change non-monotonously with the increase of noise strength. That is, there will be some optimal level of noise, at which the coherence of the noise-induced oscillations is maximal. Such phenomenon is known as coherence resonance [86, 87, 166, 210, 241]. In some cases coherence resonance can be understood as the response of a non-linear dynamical system to noise excitation near the bifurcation of periodic orbit [190]. The main feature of this effect is that the power spectrum of the system after a bifurcation may be visible even before the bifurcation if noise is applied

364   13 Synchronization in Systems with Complex Multimode Dynamics

[298]. Thus, noisy precursors of the bifurcation, i.e., *noise-activated* time scales, are observed. However, the effect of coherence resonance can be found even if the excitable system does not possess any kind of self-oscillatory behavior. The corresponding mechanism is explained by means of different noise sensitivities for the excitation and relaxation times [210]. The trajectory in this case may be considered as a motion on a stochastic limit cycle [287] with the corresponding *noise-induced* eigenfrequency.[2] These oscillations are controlled by noise and significantly depend on the noise intensity and its statistics. Notably, noise-induced dynamics can be multimodal, i.e., can be characterized by several time scales [187, 190, 196, 229, 231].

In this section we focus on *noise-induced* rather than noise-activated oscillatory modes, i.e., on the time scales that are delivered and controlled merely by noise and that did not exist in the deterministic case. We provide an experimental observation of such multimode behavior and investigate the conditions for the generation and the entrainment of the specified modes.

13.2.1 Description of Experiment

Our study involves experiments on coupled monovibrator circuits. This electronic model [221] captures well the essential aspects of excitable systems. A single monovibrator (Fig. 13.7(a)) generates a single electric impulse whenever the input voltage exceeds the threshold level V_{th}. The circuit employs an operational amplifier that supplies a non-linear response to the voltage between the two inputs. An RC-chain is involved in the positive feedback that for a certain time locks the output circuit in an excited state via a gradual voltage change at the '+' input. The recharging time constant is $\tau_0 = -RC \ln \frac{1}{2}(V_{th}/U + 1)$, where $V_{th} \leq U$ and U is the voltage of power supply. Being excited by white Gaussian noise $\xi(t)$ of an appropriate intensity D, the circuit can reach the regime of coherence resonance [221]. The noise-induced oscillations become quite regular and the whole system (excitable unit + noise) can be considered as a coherence resonance oscillator whose behavior is described by a peak frequency governed by the noise and a phase introduced as the position on a stochastic limit cycle.[3]

13.2.2 Characterizing Collective Response by Spectra

To characterize the collective response of the system (Fig. 13.7(b) and (c)) we use the summarized output from all functional units. Figure 13.8 compares the realization from the noise source $\xi(t)$ with the more regular response of the excitable system in Fig. 13.7(c). In order to characterize spectral properties of the latter signal we consider its power spectrum $S(f)$ calculated over a set of L sampled realizations

$$S(f) = \frac{1}{L} \sum_{i=1}^{L} |P_i(f)|^2, \tag{13.12}$$

---

[2] See also Chap. 9, where a concept of stochastic limit cycle was discussed.
[3] Stochastic limit cycle was introduced in Chap. 9.

13.2 Generation and Synchronization of Oscillations with Several Noise-Induced Modes 365

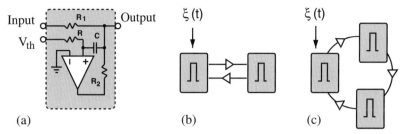

Fig. 13.7. Different implementations of excitable units. **a** Electronic circuit of a single monovibrator; **b** mutually coupled units; **c** units coupled in a circle

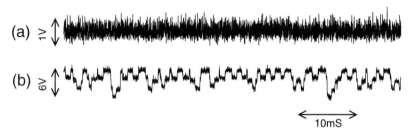

Fig. 13.8. a Realization of noise input $\xi(t)$ and **b** collective response from three coupled excitable units shown in Fig. 13.7(c). *Arrows* indicate voltage and time scales of the signals

where $P_i(f)$ is the fast Fourier transform calculated for ith realization from the system's output. With L large enough (we use about 200), the pronounced and smooth peaks can be detected for the excitable units in the regime of coherence resonance. When $S(f)$ is calculated from the summarized output signal of coupled units, all noise-induced time scales and their mutual entrainment can be observed.

13.2.3 Mutually Coupled Excitable Units

Figure 13.9 illustrates spectra corresponding to different patterns of collective behavior, at different values of the coupling strength g of two symmetrically coupled excitable units schematically shown in Fig 13.7(b). Without coupling ($g = 0$), the second (right-hand) unit can generate only randomly appearing impulses due to the presence of weak internal noise with intensity $D \cong 0.0005$ V^2. At the same time, the first unit generates a pronounced peak in the power spectrum. With increasing g, the second peak appears. Within a wide range of g, the *peak frequencies* are found to keep ratio of $1:2$ (Fig. 13.9(a)) and $1:1$ (Fig. 13.9(c)). This means that the frequency locking takes place. However, in a certain range of the parameter g, the resonance ratio between the noise-induced frequencies is broken down, and two peaks at incommensurate frequencies can be clearly distinguished in the power spectrum (Fig. 13.9(b)). The corresponding regions are clearly visible in the three-dimensional plot in Fig. 13.9(d). Hence, a two-mode behavior is observed with a resonant and a non-resonant ratios between the noise-induced frequencies. Such

366   13 Synchronization in Systems with Complex Multimode Dynamics

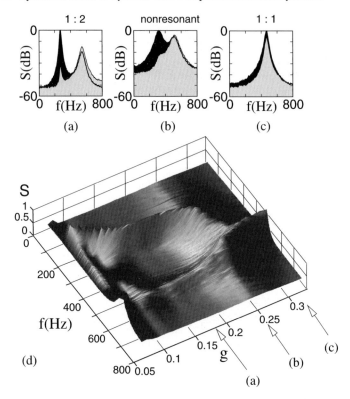

Fig. 13.9. Two-mode collective response in the system of two monovibrators mutually coupled as shown in Fig. 13.7(b), at $D = 0.475$ V^2 and different g. The evolution of the power spectrum $S(f)$ clearly shows the transitions from **a** 1:2 frequency-locking ($g = 0.18$) to **b** non-resonant two-mode behavior ($g = 0.25$), and finally to **c** 1:1 mode locking ($g = 0.325$); **d** three-dimensional plot illustrating frequency entrainment with varying coupling strength

behavior is similar to quasiperiodic motion in the deterministic case. Note that the multimode dynamics being considered here is induced merely by noise, since with vanishing random excitation none of the systems exhibit oscillations. Moreover, there is no a priori introduced detuning between the time scales of the systems. That is, coherence entrainment between interacting systems is also governed by noise.

Figure 13.10 illustrates how distinct phase patterns appear as coupling strength is fixed at $g = 0.1$. With varying noise intensity D, the frequencies of the noise-induced oscillations in the coupled systems move with respect to each other to give rise to oscillatory modes with two pronounced independent peaks in the power spectrum. At D ranging from 0.037 V^2 to 0.152 V^2, the 1:2 resonance behavior is observed (see Fig. 13.10(c)). At $D \in [0.788$ V^2, 1.07 V$^2]$, frequencies are locked in a 1:3 ratio (see Fig. 13.10(a)).

13.2 Generation and Synchronization of Oscillations with Several Noise-Induced Modes 367

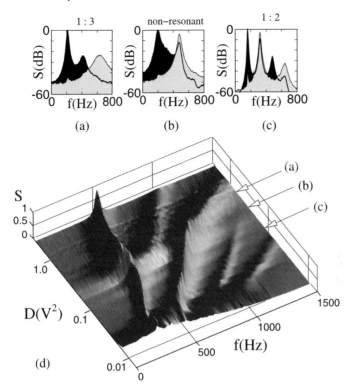

Fig. 13.10. Two-mode collective response in the system of two monovibrators mutually coupled as shown in Fig. 13.7(b), at $g = 0.1$. Spectrum is shown at different D: **a** 1 : 3 frequency-locking ($D = 0.77$ V^2); **b** non-resonant two-mode behavior ($D = 0.42$ V^2); **c** 1 : 2 mode locking ($D = 0.15$ V^2); and **d** three-dimensional plot illustrating frequency entrainment with varying coupling strength

In order to quantitatively characterize the effect of coherence resonance, different researchers described the inhomogeneity of the spectrum with different approaches, including calculation of the signal-to-noise ratio [86, 166, 287] and of the autocorrelation function [210]. We choose a method which in our case is more universal, namely the regularity of oscillations is characterized using their spectrum. First, each value of the spectrum $S(f_i)$ is divided by the integral of $S(f)$ to obtain a normalized spectrum $S_n(f_i)$

$$S_n(f_i) = \frac{S(f_i)}{\sum_{i=1}^{i=m} S(f_i)}, \tag{13.13}$$

where f_i are the frequencies at which the spectrum is estimated numerically. Next, Shannon entropy is calculated from the normalized spectrum S_n that contains m components

$$E = -\sum_{i=1}^{i=m} S_n(f_i) \ln(S_n(f_i)). \tag{13.14}$$

E takes zero value for a harmonic signal which is the most regular signal and has a spectrum in a form of a delta-peak. On the contrary, white noise is considered to be completely irregular with homogeneous spectrum, for which E reaches its maximal value

$$E_{\max} = -\sum_{i=1}^{i=m} \frac{1}{m} \ln\left(\frac{1}{m}\right) = \ln m. \qquad (13.15)$$

A measure of regularity β can be introduced as follows:

$$\beta = 1 - \frac{E}{E_{\max}}. \qquad (13.16)$$

Defined in this way, the β value reflects essentially the non-uniformity of the spectrum, varying from 1 for the purely harmonic oscillations to 0 for white noise.

For a single monovibrator the plot of β versus the noise intensity D has a single pronounced maximum, i.e., the system exhibits coherence resonance [221]. Since we deal with coherence resonance oscillators, we are particularly interested in establishing a relation between the regularity β of the noise-induced oscillations and the strength of interaction g. Figure 13.11 shows the behavior of β with increasing g both for the collective response of two mutually coupled monovibrators in Fig. 13.7 and for the individual units. It is clearly seen that the second (right-hand) unit produces the most regular output. It is remarkable that the local maxima of regularity β_2 correspond to the regions of $1:3$, $1:2$, and $1:1$ mode locking, where the relative widths of the peaks in the power spectrum are considerably smaller than at other values of g. The first unit is the subject of external random force $D\xi(t)$. Hence, its reaction to variations in g is insignificant, until the coupling becomes strong ($g > 0.3$). The regularity β_{12} of the collective response depends on g in a non-monotonic way. For very weak coupling, $\beta_{12} \approx \beta_1$ since the second system receives a weak input and produces almost no firing. At $g \in [0.05, 0.1]$, the β_{12} graph displays a considerable fall due to rather irregular spike generation in the second

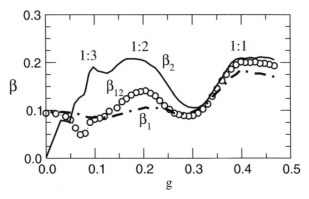

Fig. 13.11. Measures of regularity as a function of coupling strength ($D = 0.475 \text{ V}^2$) for the first (β_1) and second (β_2) units, and for their collective response (β_{12})

13.2 Generation and Synchronization of Oscillations with Several Noise-Induced Modes

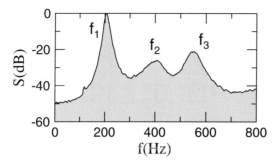

Fig. 13.12. Power spectrum illustrating three-mode collective behavior in a system of three interacting excitable units sketched in Fig. 13.7(c) at $D = 0.35$ V^2 and $g = 0.03$. Peak frequencies are estimated as $f_1 = 205.3$ Hz, $f_2 = 403.5$ Hz, and $f_3 = 549.1$ Hz

unit. When g is further increased, both units enter the regime of coherence resonance and β_{12} generally follows the behavior of β_1 and β_2, displaying maxima in the mode locking regions and being small in the non-resonant regimes. The main result of the above experiments is that symmetrically coupled identical excitable units can surprisingly produce multimode stochastic oscillations.

13.2.4 Three Coupled Excitable Units

To support the above proposition we consider a circle configuration that contains three functional excitable units (Fig. 13.7(c)). For a certain range of control parameters, a regime with three different frequencies is observed. It occurs as a mode locked state and as non-resonant behavior (Fig. 13.12). Thus, we can state that a three-unit system is able to generate a three-mode stochastic dynamics.

13.2.5 Two Mutually Coupled Excitable Units with Inhibitory Coupling

The coupling we considered above belongs to one of the simplest types. In neuronal excitable systems, a synaptic (i.e., delayed inhibitory or excitatory) interaction is more realistic. Let us now describe the two-mode stochastic behavior of system sketched in Fig. 13.13(a) that is actually an electronic model of the simplest breathing rhythm generator for a snail [251]. The circuit contains self- and mutually-inhibitory coupling chains that can increase the threshold voltages of the first (V_{th1}) and of the second (V_{th2}) units. Each coupling chain contains a rectifier and a low-pass filter with coupling strength g_{ij} and time constant τ_{ij}, where i, j are the unit numbers. Note that the self-inhibitory time constants were chosen to be equal and greater than the mutual-inhibitory time constants, i.e., $\tau_{11} = \tau_{22} > \tau_{12} = \tau_{21}$.

At small noise intensity D (which is the same for the two units), both excitable units keep silence most of the time, and their threshold voltages remain equal ($V_{th1} \approx V_{th2}$). At intermediate noise, the influence of coupling on threshold voltages becomes significant. With this, one of two units gets an "advantage" in suppressing

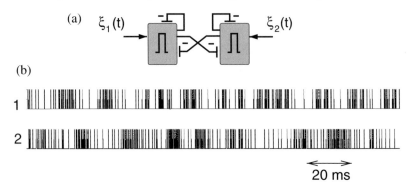

Fig. 13.13. a Two monovibrators with delayed inhibitory couplings imitate a simple neural circuit. **b** Stochastic spike trains generated by the first and the second excitable units. Antiphase behavior is registered in average

the firings in the other unit, since mutual inhibition makes the in-phase regime unstable. However, with intensive firing, the slow self-inhibitory chain with rate τ_{11} (or τ_{22}) comes into operation and suppresses the activity of the corresponding unit. This creates the best conditions for the excitation of the other unit. The process continues in a similar way, producing a behavior with time-varying firing rates for the two excitable units (Fig. 13.13(b)).

In this operating regime, two peaks in the power spectrum are clearly distinguished (Fig. 13.14(a)). The high frequency peak corresponds to noise-induced oscillations in the single system, while the low frequency peak reveals a new noise-induced oscillatory mode. Hence, the system of coupled excitable units generates a new oscillatory mode that is characterized by the values of τ_{ij} and by the relation between the noise intensity and the initial threshold voltages (V_{th1}, V_{th2}). Figure 13.14(b) shows how the frequency of these oscillations (empty circles) depends on the noise intensity. It is clearly seen that with increasing noise strength, both frequencies grow (i.e., they are noise-controlled), but the growth rates are different (i.e., they are essentially independent from each other). At strong noise, an excitable system can be immediately pushed away from the equilibrium state in spite of the threshold voltage. The low frequency peak in the power spectrum disappears, and the additional time scale no longer exists.

The regularity of the low-frequency stochastic oscillations is related to the process of pulse generation in each excitable unit. Hence it is determined by the effect of coherence resonance. Figure 13.14(b) illustrates that the output regularity β (filled circles) is suddenly increased when low frequency oscillations appear but the peak at the noise-induced eigenfrequency f_2 is washed out because of the threshold modulation.

Summarizing, in this section we have shown that a relatively simple system consisting of several *identical* excitable units, one of which is perturbed by random fluctuations, is able to demonstrate noise-induced multimode dynamics characterized by several independent time scales. With variation of noise intensity modes can

13.3 Synchronization of Chaotic Systems with Denumerable Set of Equilibrium States

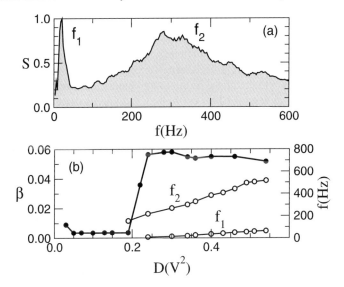

Fig. 13.14. Two-mode dynamics in the excitable system presented in Fig. 13.13(a). **a** Power spectrum of the sum of the outputs of two units; note well-pronounced peaks ($D = 0.34$ V^2). **b** Peak frequencies (*empty circles*) and regularity β (*filled circles*) vs noise intensity D

demonstrate mutual entrainment. Remarkably, the entrainment of modes is related to coherence resonance phenomenon. Actually, the system demonstrates a maximum of global coherence when entrainment of modes takes place.

The results presented in this section can also be useful for understanding and modeling the rhythmic biological phenomena, e.g., in systems of sensor neurons and pacemakers. In particular, possible advantages of multimode dynamics may include the following aspects: (i) increased sensitivity via coherence resonance and (ii) expanded flexibility—the presence and interaction of two distinct oscillatory modes enrich the dynamical patterns. This approach, involving excitable stochastic units with self- and mutually-inhibitory couplings, can be applied to simulate neural systems with distinct phase relations given a priori.

13.3 Synchronization of Chaotic Systems with Denumerable Set of Equilibrium States

Oscillatory behavior of dynamical systems can have very complicated character involving several dynamical modes of very different origins. Often each mode is characterized by its own time scale. An example of such complex behavior is shown in Fig. 13.1(c), which illustrates oscillatory motion involving rotation, jumps between the vicinities of different equilibrium states, and drift in the space. In this section we discuss synchronization phenomena that result from interaction of such systems.

An example of a dynamical system demonstrating this kind of multimode dynamics is a phase-locked loop. In electronics, a phase-locked loop is a feedback control circuit which generates a signal, whose characteristics depend on the frequency and phase of an input reference signal. A phase-locked loop circuit responds to both the frequency and the phase of the input signals by automatically adjusting the frequency of the generator being controlled, making the latter match the reference signal both in frequency and in phase. This type of systems is widely used in telecommunications, radiolocation, computers, and many other electronic applications where it is desirable to stabilize a generated signal, or to detect signals in the presence of noise.

A block diagram of one of the possible realizations of phase-locked loop is given in Fig. 13.15. The corresponding model equations read

$$\dot{x} = mx - zx + \sin(vy) + a,$$
$$\dot{y} = -x, \quad (13.17)$$
$$\dot{z} = -gz + gF(x).$$

The first two equations describe the action of the phase detector (block 3 in Fig. 13.15), the low-pass filter (block 4) and the amplifier (block 5) while the third equation accounts for the effect of the feedback loop (block 6) in the amplifier circuit. The function $F(x)$ is defined as

$$F(x) = \begin{cases} (\alpha + \epsilon)x^4, & x \geq 0, \\ (\alpha - \epsilon)x^4, & x < 0, \end{cases} \quad (13.18)$$

reflecting a non-ideal characteristic of a two-half-period detector, g defines the relaxation time of the feedback loop (block 6 in Fig. 13.15), $(m - z)$ determines a signal gain in the feedback loop. Remarkably, the model (13.17) can be considered

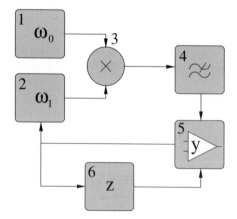

Fig. 13.15. Block diagram of a phase-locked loop: *1*—generator of a reference signal with frequency ω_0; *2*—generator with controlled frequency ω_1; *3*—phase detector; *4*—low-pass filter; *5*—amplifier; *6*—non-linear feedback loop of the amplifier

13.3 Synchronization of Chaotic Systems with Denumerable Set of Equilibrium States

as a modification of the Anishchenko–Astakhov oscillator[4] [15], which is a simple electronic circuit that demonstrates Feigenbaum type of chaos [15, 33]. The dynamics of (13.17) was studied in [216]. If $a \leq 1$, the system has a denumerable set of equilibrium states with coordinates $x = z = 0$, and y determined by the equation $\sin(\nu y) = -a$.

In our study we fix $a = 0.012$ and $\nu = -0.5$. With changing g and m, the system demonstrates a variety of dynamical regimes which are summarized in the bifurcation diagram in Fig. 13.16. For small positive values of m and g, the system possesses a countable set of stable limit cycles in the vicinity of each equilibrium state. Depending on the initial conditions, phase trajectories are attracted to different limit cycles. The range of the parameters where such cycles exist is denoted by C_1 in Fig. 13.16. As m grows, these cycles can undergo period-doubling bifurcations, and the system enters the region $C2$ where attractors are period-doubled cycles. A cascade of period-doubling bifurcations leads to the appearance of chaotic attractors which are associated with rotation of phase trajectory around each of equilibrium

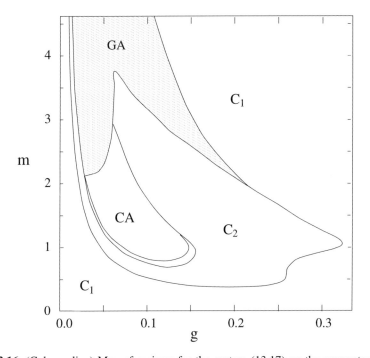

Fig. 13.16. (Color online) Map of regimes for the system (13.17) on the parameter plane (g, m). C_1 is an area of existence of a period-one limit cycle; C_2 is a domain of period-two limit cycle; CA is a region of chaotic attractor resulting from a cascade of period-doubling bifurcation; GA is an area where the system demonstrates complex oscillations accompanied by a drift in y-direction

---
[4] Coupled Anishchenko–Astakhov oscillators are considered in Chap. 8.

13 Synchronization in Systems with Complex Multimode Dynamics

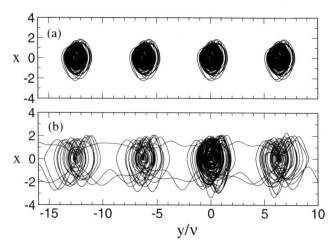

Fig. 13.17. Different types of chaos in the system (13.17) at $g = 0.05$: **a** coexisting chaotic attractors at $m = 1.3$ and **b** joint chaotic attractor for $m = 4.09$

points (region CA). Projections of such attractors on the phase plane (x, y) for the parameter values $m = 1.3$ and $g = 0.05$ are illustrated in Fig. 13.17(a). With variation of m, the sizes of the attractors grow until they touch the boundaries of their basins of attraction. As a result of the boundary crisis, a complex chaotic motion appears which includes fragments of the former chaotic sets (see Fig. 13.17(b)). This type of dynamics, which could generally be either chaotic or regular, exists in the area GA in Fig. 13.16.

The complex behavior of phase the trajectories shown in Fig. 13.17(b) involves several types of motion. First, this is a rotation around the fixed points whose characteristic time scale T can be introduced as a mean return time to the secant plane $x = 0$. Second, the phase trajectories jump from vicinity of one fixed point to the vicinity of another fixed point with the mean inter-jump time τ. Finally, due to the asymmetry of $F(x)$, jumps in one of the directions are more probable than in other direction, and therefore the phase state of the system slowly drifts towards the increasing values of y. This type of motion can be characterized by a time scale associated with the mean drift velocity v_d, which could be calculated as $(y(t + t_o) - y(t))/t_o$. Here, t_o is the observation time which is supposed to be quite long.

To study how different modes interact in two coupled systems, we consider the following model equations:

$$p_{1,2}\dot{x}_{1,2} = mx_{1,2} - z_{1,2}x_{1,2} + \sin(vy_{1,2}) + a + C(x_{2,1} - x_{1,2}),$$
$$p_{1,2}\dot{y}_{1,2} = -x_{1,2},$$
$$p_{1,2}\dot{z}_{1,2} = -gz_{1,2} + gF(x_{1,2}),$$

where the index of the variables means a number of an interacting subunit, $p_{1,2}$ define a time scale detuning between the interacting systems, terms $C(x_{2,1} - x_{1,2})$ pro-

13.3 Synchronization of Chaotic Systems with Denumerable Set of Equilibrium States

vide mutual diffusive coupling between the systems, and C governs the coupling strength. We fix $p_1 = 1$, $m = 4.0$, and $g = 0.05$.

The central questions of this study are (i) whether all dynamical modes of the interacting systems can be synchronized, and (ii) if yes, whether all of them are synchronized at the same values of parameter C. We fix $p_2 = 1.01$, and gradually increase the coupling C between the systems. In Fig. 13.18 the ratios of the time scales corresponding to the different dynamical modes are shown with variation of the parameter C. One can see that all ratios have a critical value of the parameter C, above which they are very close to 1.0. A small discrepancy from 1.0 is explained by numerical errors in estimation of these ratios. However, the critical value of C is different for different modes. With increasing C, the modes corresponding to the rotation are locked first at $C \approx 0.032$ (see Fig. 13.18(a)), and then the modes related to jumping and to the drift are both synchronized at the same value of $C \approx 0.6$ as shown in Fig. 13.18(b). The fact that jumps and drift are synchronized at the same values of the coupling strength can be explained as follows. The drift velocity in the system with jumps is proportional to the quantity $(N_r - N_l)/t_o$ [48, 217], where N_r and N_l are the numbers of jumps to the right and to the left, respectively, for a sufficiently long observation time t_o. Thus, once the jumps are synchronized, drift modes are synchronized as well.

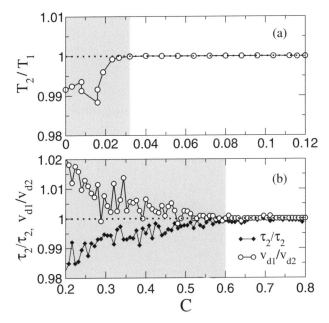

Fig. 13.18. a Ratios of the time scales characterizing different dynamical modes in interacting systems *vs* coupling strength C for $p_1 = 1$ and $p_2 = 1.01$. *White areas* correspond to the values of parameter C where the corresponding modes are synchronized

376   13 Synchronization in Systems with Complex Multimode Dynamics

Finally, we note that mechanisms of synchronization of different modes in deterministic systems similar to the ones considered above, could also be understood in term of unstable periodic orbits.[5] Namely, the synchronization of modes associated with rotations around equilibrium points is determined by bifurcations of the saddle periodic orbits winding around the separated fixed points, whereas synchronization of jumps results from the bifurcations of periodic orbits encompassing several fixed points [20].

13.4 Summary

In this chapter we considered a few representative examples of both deterministic and stochastic systems with irregular multimode dynamics of different origins. We have shown that different modes of the same oscillatory behavior can be entrained with each other in the way similar to the entrainment of the dynamics of several interacting self-oscillators.

In coupled multimode systems, the corresponding pairs of modes can be synchronized independently of each other. Thus, the synchronization criteria in the case of the multimode dynamics can be different for different modes involved. Decomposition of complex behavior into independent modes can therefore be a very useful technique for the analysis of cooperative dynamics in systems of interacting multimode oscillators.

---

[5] Synchronization of chaos in terms of periodic orbits is discussed in Sect. 8.6.

14 Synchronization of Systems with Resource Mediated Coupling

As one can learn from the previous chapters, the classical synchronization paradigm considers the interaction of two or more oscillators, each with its own source of energy and with the coupling being responsible for the frequency entrainment and mutual amplitude adjustments. If more than two oscillators are involved, the *geometry of coupling* also becomes important. The well-studied coupling geometries include *local coupling* [6], where oscillators in a lattice (or some other spatial arrangements) interact with their nearest neighbors, and *global coupling* [151] where each oscillator in an ensemble interacts with all other oscillators (or with a mean field produced by those oscillators).

378   14 Synchronization of Systems with Resource Mediated Coupling

Living systems cover all variety of different network connectivities. The first type of coupling may represent an interaction of heart muscle cells or pancreatic β-cells via gap junctions [178] when ions and small molecules can freely pass from one cell to its neighbors. This coupling typically produces waves or pulses that propagate along the interacting units. Examples of global coupling range from a system of coupled electrochemical oscillators [144] to metabolic oscillations in a suspension of yeast cells [68]. Typical phenomena associated with this coupling are global synchronization, oscillator death through mutual suppression of natural dynamics, and various forms of clustering. More recently, the studies of so-called *small-world networks* have attracted considerable interest [175, 297]. In this case, the interaction among the oscillators combine a local coupling with a few (more or less random) long range connections.

For the types of coupling discussed above, the mathematical description assumes that non-linear properties of the individual functional unit (i.e., its natural frequency and resistance to external perturbations) are governed by the unit's own parameters, while the interaction is specified through a separate set of parameters that characterize the coupling structure and the strength of interaction. Hence, one can distinguish the natural dynamics of the individual oscillators from the properties of the coupling network.

However, there are problems in physics, engineering, chemistry, and biology that cannot be considered within this paradigm. Namely, the coupling between oscillators takes place via the distribution of energy (or resources) that allows the individual oscillator to maintain its dynamics. In such a system, the energy (or primary resources) delivered to the individual subunit (and, hence, its behavior) depends on the energy consumed by all other oscillators in the system, both with respect to its mean value and to its temporal variations (amplitude and phase).

Let us consider a number of representative examples of systems with *resource mediated coupling* that are summarized in Table 14.1[1] to learn more about non-linear mechanisms underlying their main dynamical regimes.

The source of energy (or primary resources) can be local or distributed across the whole system. Depending on the consuming rate, the energy supply decreases or increases along a chain or branching structure. Hence, functional units operate at different regimes and, even if their parameters are identical, their amplitudes and frequencies may differ. From the viewpoint of synchronization one cannot separate the frequency mismatch and the coupling strength parameters any longer. The last column in the table indicates bifurcations associated with onset and termination of self-sustained oscillations in the system with increasing energy supply. In most of models, self-sustained dynamics is observed within a limited range of the control parameter. Only a certain group of oscillators in a network is under proper conditions to oscillate and/or to be synchronized. Depending on the total energy supply, this group can move along the network.

---

[1] Note that we use the term "Hopf bifurcation" instead of "Andronov–Hopf bifurcation" to save space in the table.

14.1 Neural Synchronization via Potassium Signaling   379

Table 14.1. Examples of systems with resource mediated coupling

System	Resource influx	Network type	Bifurcations
Neurons communicated via potassium signaling	Distributed	Global production and degradation	Subcritical Hopf and saddle-node/ Supercritical Hopf
Ensemble of electronic oscillators	Local	Negative gradient in branching or linear structure	Supercritical Hopf/ Supercritical Hopf
Cascaded microbiological oscillators	Local or distributed	Negative or positive gradient in linear structure	Supercritical Hopf/ Unbounded
Vascular-coupled nephron tree	Local	Negative gradient in branching structure	Supercritical Hopf/ Supercritical Hopf

To follow the main concept of our book, *From Simple to Complex*, we arrange our examples in a special order. We start with a system of two coupled neurons signalling each other via potassium in their common extracellular space. Then we proceed with a simple system of two (and more) coupled electronic oscillators sharing common power supply that is related to our daily life problems of electricity supply to a distributed network of consumers. Cascaded microbiological oscillators is our next example being an array of a large number of units with one-way nutrition supply: lateral and upstream. Finally, we arrive at the blood flow distribution in vascular-coupled units of kidney combining hemodynamic and electrochemical interactions.

14.1 Neural Synchronization via Potassium Signaling

The *resource mediated coupling* can play the role of energy source and can govern the cooperative dynamics of an ensemble of self-sustained units. The only necessary condition is that the individual oscillator should be sensitive to the total amount of the produced resource.

As an example of such interaction we consider interaction of closely located neurons affected by the temporally varying concentrations of extracellular ions produced by neighboring cells [234].

Not any type of ion represents a good candidate for this type of mechanism. For example, for calcium ions Ca^{2+}, the ratio of extracellular to intracellular concentration is about 10^3 and for sodium Na^+ and chloride Cl^- it is about 10 [136]. This implies that the transmembrane currents will cause very small changes of the extracellular concentrations of these ions. For potassium K^+, the situation is opposite. Because of its small concentration outside the cell (the extracellular to intracellular ratio is about 0.05), the transmembrane K^+ currents related to neuron firing can cause significant changes of the extracellular potassium concentrations. There is

380   14 Synchronization of Systems with Resource Mediated Coupling

now considerable evidence that extracellular potassium concentrations *in vivo* may fluctuate from a normal level of 3.5 mM and to 9 mM under conditions of high neuronal activity [70, 281]. Even more pronounced extracellular potassium variations can occur in various pathological cases when the glial cells fail to operate correctly [45]. It is known that moderate elevation of K^+ reduces neuronal excitability thresholds and may even induce spikes, while more severe elevations may reduce neuronal excitability [269]. The effects of elevated K^+ concentrations are not restricted to the immediate region of neuronal activity, however, as glia cells are connected by gap junctions to form a functional syncytium that allows spatial buffering of ions [159]. Moreover, tissue responses to ischemia include swelling of the intracellular space, shrinkage of the extracellular space, and an increase in the extracellular potassium concentration [108, 300]. Such increases in the extracellular potassium concentration may have important pathological consequences. For example, in cardiac ischemia, the increased K^+ may cause arrhythmia. Yi et al. [302] numerically studied the possible mechanisms underlying extracellular potassium accumulation during ischemia.

In the framework of this section we develop a simple model that describes the potassium signaling between two neighboring cells and to study the main features introduced by this coupling. Depending on the coupling parameters, both antiphase and in-phase synchronization can be observed. We explore the bifurcation transitions to and between different phase locked regimes with increasing mismatch.

14.1.1 Model

Our model is based on a four-dimensional set of equations for the leech P-neuron [40, 100]. These equations are similar to the well-known Hodgkin–Huxley equations, except for the precise formulation of the nonlinear functions and for the assumption of a much lower potassium conductance relative to the sodium conductance. This seems to be reasonable in our case, because interaction via the extracellular potassium concentration is expected to provide weak modulation of the neuron properties rather than more dramatic changes.

The environment we consider is schematically depicted in Fig. 14.1:

(i) We assume that there is a certain volume between the cells from which the ionic exchange with the outer bath is rate limited. For simplicity we assume that this volume is homogeneous and denote the potassium concentration here as [K].[2] With time, particularly during firing events in one or both neurons, the outward channel currents from the two cells deliver potassium to the extracellular space and [K] rises.

(ii) We neglect the associated intracellular changes of the potassium concentration and assume this concentration to remain constant. Na–K ATPase pumps K^+ back into the cells. This uptake is balanced by K^+ leakage when the potassium concentration is at equilibrium $[K]_0$.

---

[2] Notation [K] is related to chemical representation of ionic concentration and has no other mathematical meaning.

14.1 Neural Synchronization via Potassium Signaling

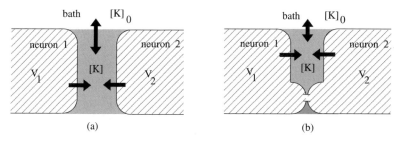

Fig. 14.1. (Color online) Schematic representation of the potassium signaling pathways between two closely located cells: **a** basic configuration; **b** with a gap junction between the cells

(iii) The exchange of K$^+$ ions with the surrounding bath is assumed to take place via a diffusion process, governed by the concentration difference.

Denoting the bath potassium concentration as [K]$_0$, we can simplify the model by incorporating all potassium uptake processes (including the influence of glial cells) in the effective diffusion rate γ.

Figure 14.1 shows two variants of such an environment to be compared. From a dynamical point of view, the main difference between the two variants is associated with the accumulating effect of the extracellular space as compared with direct exchange of ions in variant (b). The equation for the transmembrane voltage V reads

$$C_m \frac{dV_{1,2}}{dt} = -I_{1,2,K} - I_{1,2,Na} - I_{1,2,L} + I_{1,2,app} + I_{1,2,g}, \tag{14.1}$$

where subscripts 1, 2 denote the first and second cell, respectively. $C_m = 1.0\ \mu\text{F/cm}^2$ is the capacitance per unit area of the cell membrane.

The potassium, sodium, leakage, and gap junction currents are

$$I_{1,2,K} = g_K(n_{1,2})^2(V_{1,2} - V_{1,2,K}), \tag{14.2}$$

$$I_{1,2,Na} = g_{Na}(m_{1,2})^4 h_{1,2}(V_{1,2} - V_{Na}), \tag{14.3}$$

$$I_{1,2,L} = g_L(V_{1,2} - V_L), \tag{14.4}$$

$$I_{1,2,g} = g_g(V_{2,1} - V_{1,2}), \tag{14.5}$$

respectively. Here, the conductance $g_K = 6$ mS/cm^2 for both cells, $g_{Na} = 350$ mS/cm^2 (such a relatively high value is specific for leech P-neurons as shown in [40, 100]), and $g_L = 0.5$ mS/cm^2. The expression for the potassium current incorporates a dependence on the extracellular potassium concentration [K] via a time varying driving potential:

$$V_{1,2,K} = \frac{RT}{F} \ln \frac{[K]}{[K]_{1,2}}. \tag{14.6}$$

The intracellular potassium concentration is [K]$_1$ = [K]$_2$ = 147 mM. R, T, and F are the universal gas constant, the absolute temperature, and Faraday's constant,

382   14 Synchronization of Systems with Resource Mediated Coupling

Table 14.2. The expression for voltage dependent rate α and β in (14.7)

var.	α_{var}
$n_{1,2}$	$0.024(V_{1,2} - 17)/(1 - e^{-(V_{1,2}-17)/18})$
$m_{1,2}$	$0.03(V_{1,2} + 28)/(1 - e^{-(V_{1,2}+28)/15})$
$h_{1,2}$	$0.045e^{-(V_{1,2}+58)/18}$

var.	β_{var}
$n_{1,2}$	$0.2e^{-(V_{1,2}+48)/35}$
$m_{1,2}$	$2.7e^{-(V_{1,2}+53)/18}$
$h_{1,2}$	$0.72/(1 + e^{-(V_{1,2}+23)/14})$

respectively. The sodium and leak equilibrium potentials are assumed to be $V_{Na} = 60.5$ mV and $V_L = -49$ mV.

We assume that the activation variables obey standard Hodgkin–Huxley kinetics:

$$\frac{d\xi}{dt} = \alpha_\xi(1 - \xi) - \beta_\xi\xi, \tag{14.7}$$

where $\xi = n_1, m_1, h_1, n_2, m_2, h_2$. Expressions for the rates α and β associated with the individual variables are summarized as [40] (Table 14.2):

Balance of extracellular potassium concentration is expressed by the equation:

$$W\frac{d[K]}{dt} = \frac{(I_{1,K} + I_{2,K})}{F} + \gamma([K]_0 - [K]), \tag{14.8}$$

where W is the extracellular volume per unit area, measured in nl/cm^2. $I_{1,K}$ and $I_{2,K}$ are the electric potassium currents from (14.1). Divided by Faraday's constant they provide the ion flow, and $\gamma([K]_0 - [K])$ describes the diffusion of potassium to the bath. Throughout the study, W and γ are used as main control parameters.

Let us first investigate the dynamics of a single cell without coupling to the intercellular variations in [K], i.e., ($g_g = 0$, [K] = [K]$_0$, and subscripts 1, 2 numbering the two cells are omitted). The bifurcation diagram in Fig. 14.2(a) shows how an injected current I_{app} influences the dynamics of the individual model. With increasing I_{app}, a pair of stable and saddle limit cycles arise via the saddle-node bifurcation at $I_{app} = J_1$ while the resting state losses its stability via a subcritical Andronov–Hopf bifurcation at the slightly larger value $I_{app} = J_2$. Thus, the stable equilibrium and the self-sustained oscillations coexist in the interval $J_1 < I_{app} < J_2$. Further increase of I_{app} is accompanied by a gradual decrease of the spike amplitude. At $I_{app} = J_3$, there is an inverse supercritical Andronov–Hopf bifurcation and the system returns to the stable equilibrium state. At $I_{app} \leq J_1$, the neuron model exhibits excitable properties. The same sequence of bifurcations is observed under [K] variation (Fig. 14.2(b)). When increasing the external potassium concentration [K], the bifurcation points J_1, J_2, and J_3 are shifted to lower values. The black point in Fig. 14.2(b) indicates the choice of control parameters [K] = 4.0 mM, $I_{1,2,app} = 16.0$ µA/cm^2 used throughout the next section.

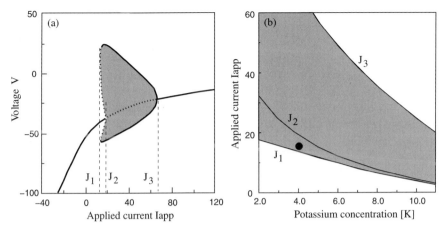

Fig. 14.2. (Color online) **a** Bifurcation diagram for the individual cell model at extracellular potassium concentration [K] = 4.0 mM. The *grey area* represents self-sustained oscillations. **b** Applied current I_{app} and bifurcational values J_1 (saddle-node bifurcation), J_2 (subcritical Andronov–Hopf bifurcation), and J_3 (inverse supercritical Andronov–Hopf bifurcation) vs the potassium concentration [K]. All currents and voltages are given in $\mu A/cm^2$ and in mV, respectively

14.1.2 Identical Cells: Competing In-phase and Antiphase Synchronization

As a first numerical experiment, we consider how two identical cells can adjust their dynamics via the intercellular potassium interaction (Fig. 14.1(a)). Identity of the cells is ensured by selecting $I_{1,app} = I_{2,app} = 16.0$ $\mu A/cm^2$. The extracellular volume W and the diffusion constant γ are used as control parameters. The obtained results are summarized in Fig. 14.3.

Weak Interaction (Large Values of W and γ)

Each cell has two stable regimes: self-sustained oscillations (limit cycle) and stable steady state. Thus, there are four stable regimes in the coupled systems (to the right of curve T in Fig. 14.3):

(i) Both cells can oscillate. For coupled systems with symmetry, two main synchronous states with in-phase and antiphase dynamics always exist but their stability depends on the specific choice of control parameters. In our case, the in-phase regime $V_1(t) \equiv V_2(t)$ is stable in a major part of the diagram (all values of $\gamma > 0.11$ nl/cm^2·ms). With decreasing diffusion parameter γ (or with increasing extracellular volume W), the in-phase regime losses its stability via a subcritical pitchfork bifurcation at line P_i. For high values of W, in contrast, the antiphase regime is stable. With decreasing W (or increasing γ), the antiphase regime undergoes a supercritical pitchfork bifurcation at the line P_a. A pair of out-of-phase limit cycles with reflection symmetry appears and evolves when W decreases further. Their region of stability is bounded by the saddle-node bifurcation curve SN.

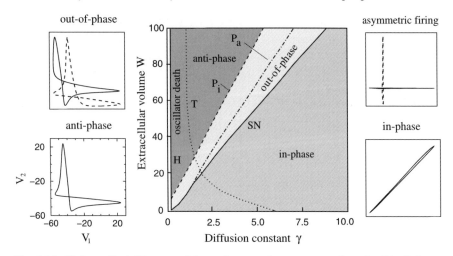

Fig. 14.3. (Color online) Diagram of the regimes on the parameter plane (γ, W). Cells are identical with $I_{1,app} = I_{2,app} = 16.0$ µA/cm^2. **Left** and **right panels** are representative phase projections on the plane (V_1, V_2). W is given in nl/cm^2. V_1 and V_2 are given in mV. γ is in nl/cm^2·ms. We use the following notations for bifurcation curves: H is Andronov–Hopf bifurcation, SN is saddle-node bifurcation, T is torus birth bifurcation, and $P_{i,a}$ is pitchfork bifurcation for in-phase and antiphase regimes, respectively

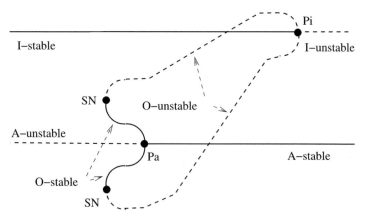

Fig. 14.4. Schematic bifurcational transition between in-phase and antiphase synchronous states along the vertical direction for $\gamma \in [2.5; 5.0]$ in Fig. 14.3

At this curve, each of the stable out-of-phase attractors collides with a saddle limit cycle and disappears. Note that the pair of saddle limit cycles is the same as that which merges with the stable in-phase limit cycles at the line P_i. The evolution of coexisting sets is illustrated in Fig. 14.4;

(ii) One of the cells oscillates but the other is in a steady state (asymmetric firing). There are two symmetrically located limit cycles (top right panel in Fig. 14.3).

14.1 Neural Synchronization via Potassium Signaling   385

For small values of W and γ the temporal variation of [K] becomes strong enough and oscillations in one of the cells make the steady state in the other cell unstable. Thus, the asymmetric firing regime losses its stability via a subcritical torus birth bifurcation (curve T in Fig. 14.3). Obviously, these curves coincide for the two asymmetric firing regimes.

(iii) Both cells can be in steady state corresponding to branch $J_1 < I_{app} < J_2$ in Fig. 14.2(a).

Strong Interaction (Small Values of W and γ)

When γ becomes smaller, both the in-phase and antiphase attractors undergo an inverse Andronov–Hopf bifurcation at $\gamma \approx 0.11$ nl/cm^2·ms. For lower γ values there are no oscillations in the coupled system. Inspection of the time variations of the intercellular potassium concentration [K] provides a reasonable explanation. At low diffusion, the mean value of [K] becomes so high, that the upper boundary of the oscillatory region in the individual cell is reached (J_3 line in Fig. 14.2(b)). For the selected parameters $I_{1,app} = I_{2,app} = 16.0$ μA/cm^2 and $\gamma = 0.11$ nl/cm^2·ms, the approached mean value of [K] $= 10.83$ mM falls outside of the oscillatory region for the individual model.

Note that increasing W and γ corresponds to weakening the interaction. Thus, there are *two weak coupling limits* with different dynamical patterns. Let us consider the limit cases:

If $W \to 0$ and γ remains finite, (14.8) gives

$$[K] = [K]_0 + \frac{I_{1,K} + I_{2,K}}{\gamma F}. \tag{14.9}$$

Suppose that the first subsystem starts to generate a spike while the second system lags behind. $I_{1,K}$ immediately increases [K] and, thus, depolarizes the second subsystem accelerating its firing. This gives a tendency for in-phase synchronization.

In the other limit case, where $W \to \infty$ and γ is finite, (14.8) gives d[K]/d$t \to 0$. Thus, for large enough W depolarization induced by the potassium release is less pronounced. It leads to the relation

$$I_{1,K} + I_{2,K} \approx \text{const.} \tag{14.10}$$

This means that the two currents are changed in opposite direction, $\Delta I_{1,K} \approx -\Delta I_{2,K}$. Increasing $I_{1,K}$ prevents the increase of $I_{2,K}$ and this tends to produce antiphase synchronization.

Hence, we conclude that coupling via the intercellular potassium concentration [K] can provide both "voltage interaction" (14.9) and "current interaction" (14.10) depending on the specific choice of the control parameters W and γ. In other words, potassium interaction demonstrates a dual nature from the synchronization viewpoint. The underlying biological mechanisms are the same. However, due to the threshold nature of ion-channel gating, the cellular response can be different, depending on the range of [K] variations.

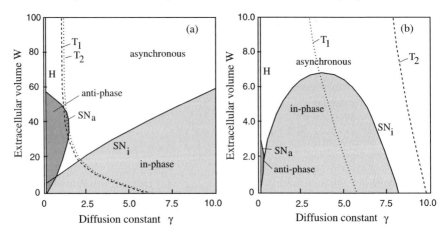

Fig. 14.5. (Color online) Diagrams of dynamical regimes on the parameter plane (γ, W). Cells are set to be heterogeneous with **a** $I_{1,\text{app}} = 16.0\ \mu\text{A/cm}^2$ and $I_{2,\text{app}} = 16.1\ \mu\text{A/cm}^2$, and **b** $I_{2,\text{app}} = 18.0\ \mu\text{A/cm}^2$, respectively. At finite mismatch both the antiphase and the in-phase synchronization areas become limited (compare with Fig. 14.3). W is given in nl/cm^2. γ is in nl/cm^2·ms. We use the following notations for bifurcation curves: H is Andronov–Hopf bifurcation, $SN_{i,a}$ is saddle-node bifurcation for in-phase and antiphase regimes, respectively, $T_{1,2}$ is torus birth bifurcation for two asymmetric solutions

14.1.3 Heterogeneous Cells: Dynamical Patterns

Let us now investigate how cell heterogeneity, introduced through a mismatch between the injected currents $I_{1,\text{app}}$ and $I_{2,\text{app}}$, will affect the synchronous regimes and their location in the parameter plane. Our results are summarized in Fig. 14.5. Note that as soon as a mismatch between the cells is introduced, pure in-phase or antiphase regimes no longer exist. The two solutions are deformed into close-to in-phase and close-to antiphase with phase shifts between the oscillations in the two cells near to 0 and π, respectively. To simplify the description in the following we will omit the words "close to."

Weak Mismatch

Let us fix $I_{1,\text{app}} = 16.0\ \mu\text{A/cm}^2$ and $I_{2,\text{app}} = 16.1\ \mu\text{A/cm}^2$. This produces small changes in the operating conditions of the individual cell as confirmed by the fact that the two torus bifurcation curves T_1 and T_2 in Fig. 14.5(a) still are located very close to each other (for identical cells they coincide). However, the picture of mutual adjustment of oscillations in two coupled cells changes dramatically. Now the main field of the diagram is occupied by an asynchronous regime. Mathematically, this corresponds to the existence of a two-dimensional torus (we have omitted the embedded weak resonances). The region of the in-phase operation still occupies a significant part of the diagram and is now bounded by the saddle-node bifurcation curve SN_i. The stable antiphase behavior can be observed in a small area within $W \in [0.0, 60.0]$ nl/cm^2.

Stronger Mismatch

Let $I_{1,\mathrm{app}} = 16.0$ μA/cm^2 and $I_{2,\mathrm{app}} = 18.0$ μA/cm^2. Note that two asymmetric firing regimes now have quite different areas of stability (bifurcational lines T_1 and T_2) reflecting the strong mismatch between the individual oscillators.

Figure 14.5(b) shows that the in-phase regime now occupies an area limited by $\gamma < 8.1$ nl/cm^2·ms and $W < 7.0$ nl/cm^2. Antiphase synchronization can only be observed in a narrow region close to the origin. Most part of the diagram is occupied by the asynchronous regime.

The reduction of the antiphase/in-phase synchronization regions with increasing mismatch can be explained by recalling that increasing W corresponds to a weakening of the interaction strength because of diminishing [K] variation. Thus, at a sufficiently large value of W, the introduced mismatch between the cell frequencies is able to desynchronize the oscillations.

To link our results with classical concept of synchronization we consider the parameter plane (mismatch $I_{2,\mathrm{app}}$–coupling γ) for $W = 30.0$ nl/cm^2 and $I_{1,\mathrm{app}} = 16.0$ μA/cm^2). Figure 14.6(a) shows how the antiphase synchronization region quickly reduces its width with increasing γ. At $\gamma = 2.5$ nl/cm^2·ms the transition between the antiphase and in-phase synchronous regimes occurs. The width

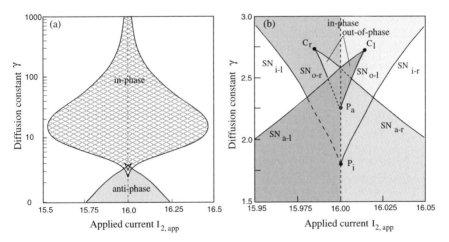

Fig. 14.6. (Color online) **a** Diagram of the regimes on the parameter plane ($I_{2,\mathrm{app}}, \gamma$) illustrates the transition between antiphase and in-phase synchronous patterns with varying diffusion rate. Note that we use logarithmic scale for γ. **b** Enlargement of bifurcation diagram. $I_{2,\mathrm{app}}$ is given in μA/cm^2. γ is in nl/cm^2·ms. We use the following notations: $SN_{r,l}$ is saddle-node bifurcation for different coexisting in-phase (i), antiphase (a), and out-of-phase (o) regimes, $C_{r,l}$ is cusp points for two out-of-phase solutions and $P_{i,a}$ is pitchfork bifurcation for in-phase and antiphase regimes, respectively. Multileaf structure is depicted by different colors

388   14 Synchronization of Systems with Resource Mediated Coupling

of the in-phase synchronous region changes with further increase of γ. This can be explained by competition between tendencies for in-phase and antiphase synchronization at intermediate values of γ and by asymptotic weakening of the coupling with further increase of γ.

Figure 14.6(b) represents the bifurcation scenario for the transition between antiphase and in-phase synchronization. A set of saddle-node bifurcation curves form the overlapping tongues. Note that for 15.98 μA/cm^2 $< I_{2,app} < 16.02$ μA/cm^2 there is no desynchronization with increasing γ. It is clearly seen that the in-phase solution is stable from the top of the diagram to the point P_i ($\gamma = 1.80$ nl/cm^2·ms) where it loses its stability in a pitchfork bifurcation. On the other hand, the antiphase solution remains stable from the bottom of the diagram to its pitchfork bifurcation point P_a at $\gamma = 2.313$ nl/cm^2·ms where two stable symmetric limit cycles are born. For $\gamma \in [2.313, 2.72]$ nl/cm^2·ms there are three stable and three unstable limit cycles involved in the considered transition: an in-phase and two out-of-phase regimes are stable while an antiphase and two other out-of-phase regimes are unstable. At $\gamma = 2.72$ nl/cm^2·ms both (left and right) horns formed by the out-of-phase solutions die in the cusp points C_r and C_l, respectively. Above these points only the stable in-phase limit cycle and a saddle antiphase cycle lie on the surface of the resonant torus in accordance with the classical structure of the Arnold tongue.

We can conclude that resource mediated coupling via potassium signaling gives the nearby cells opportunity to adjust their behavior and to interplay among different phase patterns. Different choices of the control parameters show that two pacemaker cells can control each other in different ways. At small extracellular volumes, the potassium release from one of interacting cells considerably depolarizes the other cell and thus accelerates its firing. This kind of interaction leads to in-phase synchronization. At large enough extracellular volume the depolarization induced by potassium release is less pronounced and the increase of the potassium current in one cell is more or less balanced by its decrease in another cell. As a result, the antiphase synchronization pattern is evoked. There is a wide range on control parameters where both tendencies to synchronized patterns are maintained.

14.2 Multimode Dynamics in Linear Array of Electronic Oscillators

The system of cascaded electronic oscillators is of interest both because the governing equations are well-established and because detailed experimental verification of simulation results can be easily obtained.

14.2.1 Model

To investigate the typical behavioral patterns in a resource consumption chain we look for a simple model that mimics the main properties of such a system. Let us consider an electronic circuit with a tunnel diode shown in Fig. 14.7. Here, the incoming I_{in} and outgoing I_{out} power supply currents are explicitly taken into account.

14.2 Multimode Dynamics in Linear Array of Electronic Oscillators 389

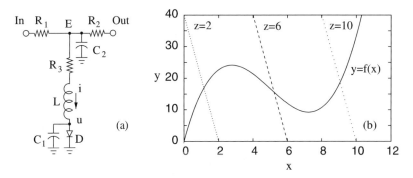

Fig. 14.7. **a** Circuit diagram and **b** the nullclines for the 2D limit case $R_{1,2} = 0$. In the dimensionless model equations (14.11), x represents the voltage u across the tunnel diode D, y is the current i in the inductor L, and z is the voltage E across the capacitor C_2

The voltage E plays the role of energy source for the oscillator system containing the elements R_3, L, C_1, and D. Self-sustained oscillations are maintained due to the N-shaped characteristics (negative differential resistance) of the diode D. The oscillation period is determined by the capacitance C_1 and the inductance L. Capacitor C_2 is introduced into account for possible accumulation of energy by an individual oscillator while the resistor R_1 is responsible for the finite replenishment rate and for losses because of transmission.

Using Kirchhoff's law for the circuit in Fig. 14.7(a) and introducing the new time variable $\tau = t/R^*C_1$ ($R^* = 1$ Ohm) we can write down the governing equations in dimensionless form:

$$\dot{x} = y - f(x),$$
$$\varepsilon\dot{y} = z - yR - x, \qquad (14.11)$$
$$\gamma\dot{z} = -y + \frac{1}{r}(e_{\text{in}} - 2z + e_{\text{out}}).$$

Here, $\varepsilon = L/R^{*2}C_1$ and $\gamma = C_2/C_1$. e_{in} and e_{out} are potentials in the "in" and "out" points, respectively. Parameters r and R are dimensionless representations of the resistors $R_1 = R_2$ and R_3. Variables x, y and z correspond to the voltage u across the diode, the current i through the inductor L, and the voltage E on the capacitor C_2, respectively. We assume that $C_2 \ll C_1$ so that z quickly follows variations of x and y. In the following calculations $\varepsilon = 0.1$ and $\gamma = 0.01$. $f(x)$ is assumed to be of cubic shape with the non-linearity chosen in the form

$$f(x) = 20x - 5x^2 + x^3/3. \qquad (14.12)$$

In the limit $r \to 0$, one can obtain $z = (e_{\text{in}} + e_{\text{out}})/2$. Hence, we obtain a 2D oscillator which is similar to the FitzHugh–Nagumo model [80] with z as a control parameter. Figure 14.7(b) illustrates the location of the nullclines in the system. Note that the x-nullcline coincides with the $f(x)$ function. It is clearly seen that for small

390   14 Synchronization of Systems with Resource Mediated Coupling

and large values of z, intersection of the nullclines occurs outside of the interval with negative slope of $f(x)$, i.e., the equilibrium point is stable. For intermediate values of z, there is a couple of Andronov–Hopf bifurcation points where a stable limit cycle is born, and extinguished.

In order to build a one-dimensional array of such functional units we take

$$e_{\text{in}} = z_{j-1}, \qquad e_{\text{out}} = z_{j+1}, \quad j = 1, \dots, N,$$

where j represents the number of the oscillator, and N is the total number of units. z_0 is constant bias voltage, hereafter denoted as Z_0 and corresponds to the primary energy supply. Free end of the chain is modeled by $z_{N+1} = z_N$.

14.2.2 Clustering

Organized in a chain, the units (14.11) become globally coupled via variation of the z_j variables. There is a gradual decrease of the mean value of z_j along the chain because of the voltage drop across each coupling resistor r. Note that the current along the chain splits into two currents at each unit. Thus it decreases along the chain, and the drop of z_j from unit to unit becomes smaller and smaller.

In the phase space of the whole system, the variation of the mean value of z_j affects the stability of the global equilibrium state that can be defined from

$$\dot{x}_j = 0, \qquad \dot{y}_j = 0, \qquad \dot{z}_j = 0, \quad j = 1, \dots, N.$$

The transition from damped behavior to self-sustained oscillations for a particular unit in the chain takes place through an Andronov–Hopf bifurcation in the (x_i, y_i) subspace. In Fig. 14.8 the real part for each pair of complex-conjugate eigenvalues is plotted against Z_0 for a chain of ten units (14.11). The third eigenvalue for each subsystem is not shown in the figure since it is strongly negative in the whole range of control parameters. With Z_0 increasing from 4.0, the eigenvalues one by one cross the imaginary axis and attain positive real values, and as Z_0 increases further, then again become negative. According to the number of eigenvalues with positive real part, the dimension of the unstable manifold D_u of the equilibrium point first raises and then decreases with increasing Z_0 (inset in Fig. 14.8).

From a physical point of view, these results imply the possibility of $D_u/2$-mode self-sustained dynamics in the whole system. In spite of the fact that we cannot formally assign a given pair of eigenvalues to a specific unit of the chain, it is clear that the first pair of eigenvalues crossing zero should be related to the first unit in a chain that receives the necessary energy input from Z_0. Similarly, the subsequent crossings of zero by different pairs of eigenvalues represent the subsequent transitions of oscillatory units from the damped state to self-sustained dynamics, or vice versa. This is the basis for the formation of a group of units along the chain, that we call an *oscillatory cluster*.

Let us now consider what happens in a longer chain of 50 units in terms of amplitudes of oscillations. In Fig. 14.9, the x_j variable of each unit is plotted against

14.2 Multimode Dynamics in Linear Array of Electronic Oscillators

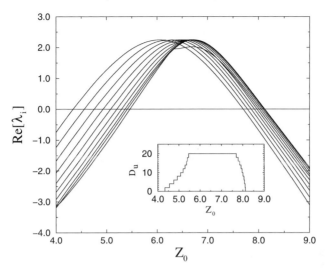

Fig. 14.8. The real part of the equilibrium state eigenvalues λ_i exceeds zero in a certain range of primary energy supply Z_0 (each curve represents one pair of complex-conjugate eigenvalues). The resulting dimension of the unstable manifold D_u versus Z_0 is shown in the *inset*

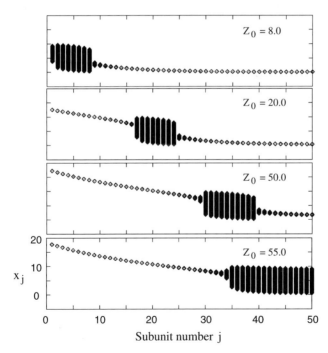

Fig. 14.9. With increasing primary energy supply Z_0, the oscillatory cluster changes its position along the chain and varies slightly in size. Parameters are fixed at $r = 0.001$, $R = 0.05$, $\varepsilon = 0.1$, and $\gamma = 0.001$

its position in the chain for different voltages Z_0 of the power supply. For relatively small voltages ($Z_0 = 8.0$), the first eight units display oscillations of considerable amplitude. For the next units, the mean value of z_j is not sufficient to support self-sustained dynamics.

For a larger value of the voltage ($Z_0 = 20.0$), inspection of the figure shows that the first 16 units no longer display significant temporal variations of x_j because the high mean value of z_j places the individual unit in a damped state according to Fig. 14.7(b). The next eight units demonstrate self-sustained dynamics while the rest are in a damped state. With increasing Z_0 ($Z_0 = 50$), the oscillating group shifts towards the end of chain and grows in size. Thus, we observe an oscillatory cluster shifting upstream or downstream the chain with variation of the energy source parameter Z_0. When approaching the low-voltage end of the chain, the cluster becomes fairly large before it completely disappears at $Z_0 \approx 59$.

While Z_0 defines the maximal level of z_j for units in the chain, the parameter r affects the voltage drop from unit to unit. Hence, r can also influence the position and size of the oscillatory cluster. Figure 14.10 reveals the relation between the variations of Z_0 and r.

In both cases, the size of the cluster remains relatively stable until the end of the chain is reached. Here the cluster becomes significantly longer before it disappears. This can partly be explained by the decreasing voltage drop per oscillator along the chain. When the cluster is located in the middle of a chain, the tailing oscillators consume electric power even though they are in a damped state. This produces an additional voltage drop and reduces the cluster size. Closer to the end of chain, the cluster has less "silent" energy consumption. The voltage drop between the subsequent units decreases, and the cluster length grows.

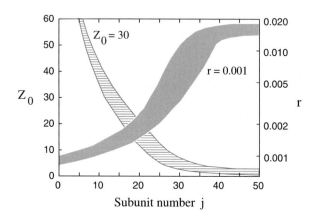

Fig. 14.10. (Color online) Position and size of the oscillatory cluster are given by the *grey region* for different Z_0 values (at $r = 0.001$), and by the *hatched region* for different r values (at $Z_0 = 30$). Other parameters are fixed at $R = 0.05$, $\varepsilon = 0.1$, and $\gamma = 0.001$

14.2.3 Intracluster Synchronization

With the emergence of a cluster of oscillatory, identical units, one might expect to observe synchronized behavior for the globally coupled identical oscillators. However, inspection of the cluster dynamics reveals a completely different picture.

Figure 14.11(a) shows how the mean period of oscillations is distributed along the cluster for different values of the voltage Z_0. In this figure $m = 1$ is assigned to the first oscillator with self-sustained dynamics in the chain. Thus, different positions along the chain correspond to different curves in Fig. 14.11(a). This allows us to compare the spatiotemporal structure of clusters for different values of the control parameter. It is remarkable that the period distribution maintains its form. For all the values of Z_0 shown, there are two oscillators with longer periods close to the beginning and the end of the cluster, and there is a clear minimum of the oscillation period near the center of the cluster. Both the first and the last oscillators in the cluster also display relatively low oscillation periods.

To explain the particular shape of the intracluster period distribution, we calculate how the period depends on the z value for an individual system (14.11) in the limit $r \to 0$. The results are given in Fig. 14.11(b). The period distribution along the cluster clearly follows the behavior of an individual unit with decreasing z. The observed structure is the result of the drop in z_j from unit to unit, combined with the variation of $f(x)$ in the region of differential negative resistance.

Hence, in spite of the presence of a coupling between *identical* cluster units, synchronization is not observed due to a *coupling-induced frequency mismatch* between the oscillators. For small values of r the frequency mismatch vanishes together with the drop of z_j between the subsequent oscillators. However, the coupling vanishes, too. The larger r is, the stronger the coupling will be, but at the same time the fre-

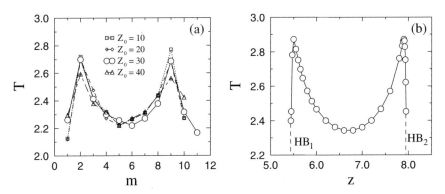

Fig. 14.11. a The distribution of oscillation periods T inside the cluster is preserved, while the whole cluster moves with the variation of Z_0. m represents the relative position within the cluster. Other parameters are set as follows: $r = 0.001$, $R = 0.05$, $\varepsilon = 0.1$, and $\gamma = 0.001$. **b** Dependence T versus z for the individual oscillatory unit reveals the origin of intra-cluster period distribution

quency mismatch becomes more pronounced between two neighboring units. This results in an asynchronous intracluster behavior in the considered parameter range.

As observed in Fig. 14.11(b), the maximal possible period drop between two units is not greater than $\Delta T \approx 0.53$ corresponding to the difference between the top and bottom points of the curve. Thus, the coupling-induced mismatch is limited to about 20%, while the coupling strength can be increased by an appropriate choice of control parameters. Let us select $R = 0.01$ and $r = 0.02$. This provides an abrupt drop of z_j, and the oscillatory cluster consists mostly of just two units. The cluster position and operating regimes are schematically given in Fig. 14.12 together with representative phase plots. The synchronized pairs of oscillator units are given in grey, while asynchronous behavior is denoted by hatched regions. For higher values of Z_0 the oscillatory cluster moves out of the chain to the right.

For some intervals of Z_0 (e.g., $Z_0 \in [10.4, 11.2], [12.2, 12.8]$), oscillations in clusters of two or three units are synchronized with a phase lag between two neighboring units as shown in the inserts. Note that the phase lag in the two-unit cluster increases with Z_0, and the cluster passes through the antiphase state somewhere in the middle of the Z_0 interval for synchronous behavior. For the two-unit cluster there is a clear explanation: when only two units in the chain display self-sustained dynamics, the influence of other units can be regarded as a shift of control parameters. Hence, we have a system of two oscillators coupled in a competitive way that typically leads to antiphase synchronization. Together with the coupling-induced frequency mismatch this provides antiphase rather than in-phase synchronization. At some value of Z_0 we find the most balanced regime (minimal frequency mismatch), and the synchronization regime becomes antiphase. The inset for $Z_0 = 16.8$ shows that there is also synchronization at various rational frequency ratios.

At some values of Z_0 the cluster changes its position. Such a translation is typically accompanied by an extension of the cluster to three units and a desynchronization between two or all three involved units (see the bottom inset in Fig. 14.12). When the first or last element of a cluster passes through the Andronov–Hopf bifurcation point (Fig. 14.11(b)), the coupling-induced frequency mismatch can be strong enough to desynchronize the intracluster behavior.

We can thus conclude that:

(i) Although the chain units are originally identical, the resource drop along the spatial coordinate introduces variations of the operating point and, hence, a frequency mismatch between neighboring units.

(ii) In the case of weak coupling, the cluster elements are generally out of synchrony, and the period distribution along the cluster follows the curve of period vs energy supply for the individual oscillatory unit. The resulting intracluster period/frequency distribution preserves its structure as the cluster moves along the chain with variation of the bias voltage. Due to the competitive nature of the coupling, there is a tendency for antiphase synchronization.

(iii) For strong enough coupling, intracluster synchronization is typical. However, shifts of the cluster position are accompanied by desynchronization of the cluster elements.

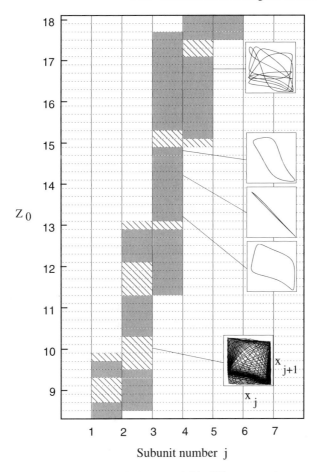

Fig. 14.12. (Color online) Small-sized clusters exhibit different synchronous or asynchronous patterns with varying primary energy supply Z_0 for strong interaction along the chain ($r = 0.02$, $R = 0.01$, $\varepsilon = 0.1$, and $\gamma = 0.001$). *Grey areas* denote synchronized behavior, and *hatched areas* correspond to asynchronous dynamics

14.3 Cascaded Microbiological Oscillators

Microbiological population dynamics plays an important role in biotechnological industry. The homogeneous, well-controlled bacterial cultures used in modern cheese production, for instance, are often quite sensitive to virus attacks, and significant efforts are made for searching more resistant cultures [275]. Based on the original work by Levin et al. [167], Baier et al. [41] formulated a multispecies model of interacting bacteria-virus populations and studied the development of a chaotic hierarchy with increasing number of bacterial variants. A resource distribution chain with several cascaded pools was first considered by Postnov et al. [218]. They showed how variations in substrate concentration in the overflow from the

396   14 Synchronization of Systems with Resource Mediated Coupling

upstream habitant can produce increasingly complicated dynamics in downstream units, and how regions of phase synchronization could arise along the chain. In the present section, these results will be placed in a more general perspective.

14.3.1 Model

We consider a one-dimensional array of population pools whose general model was considered in Chaps. 5 and 8 [218]. Each pool is the habitat for a three-variable predator-prey system, consisting of bacteria, infected bacteria, and viruses, represented in the equations by their concentrations B_i, I_i, and P_i, respectively (i denotes the pool number). Nutrition balance of inflow, outflow, and consumption provides a fourth equation for the substrate concentration S_i. Altogether this leads to the following set of coupled differential equations:

$$\frac{\mathrm{d}B_i}{\mathrm{d}t} = \frac{\nu B_i S_i}{S_i + K} - \rho B_i - \alpha \omega P_i B_i, \tag{14.13}$$

$$\frac{\mathrm{d}I_i}{\mathrm{d}t} = \alpha \omega B_i P_i - \rho I_i - I_i/\tau, \tag{14.14}$$

$$\frac{\mathrm{d}P_i}{\mathrm{d}t} = \phi - P_i \rho - \alpha B_i P_i - \alpha I_i P_i + \beta I_i/\tau, \tag{14.15}$$

$$\frac{\mathrm{d}S_i}{\mathrm{d}t} = \rho(S_{i-1} + \sigma_i) - \rho S_i - \frac{\gamma \nu B_i S_i}{S_i + K}, \tag{14.16}$$

where the term $\nu B_i S_i/(S_i + K)$ in the first and fourth equations describes standard Monod kinetics for bacterial growth. The Michaelis–Menten constant K represents the concentration of nutrients at which the growth rate is reduced to half its maximal value, and each cell division is assumed to be associated with a resource consumption γ. For all variables, negative terms proportional to ρ in the governing equations reflect the washing out from the habitat. According to our assumptions, however, only nutrients will be transmitted to the next pool. Infection of bacteria by viruses is described by the term $\alpha \omega B_i P_i$ in (14.13) and (14.14). Here, α is the kinetic rate constant, and ω is the probability that a virus particle successfully infects a cell, once it has affixed to its surface. The I_i/τ term in (14.14) and (14.15) describe a lytic response to the virus attack where, after a latent period τ of the order of 30 min, the infected cell bursts and releases an average of β new viruses. The term $-\alpha I_i P_i$ in (14.15) represents unsuccessful virus attacks on already infected cells. Coupling between the pools takes place only through the flow of nutrients with a total incoming rate of $\rho(S_{i-1} + \sigma_i)$, an outflow of $-\rho S_i$, and a consumption in the habitat of $\gamma \nu B_i S_i/(S_i + K)$. For the first habitat, $S_{i-1} \equiv S_0$ is assumed to be zero. σ_i represents a possible lateral nutrition source for the ith habitat.

The parameter values that we have used in the present analysis correspond to the values used in our previous studies [41]: $\nu = 0.024$ min^{-1}, $K = 10$ µg/ml, $\tau = 30$ min, $\omega = 0.8$, $\gamma = 0.01$ ng, $\beta = 100$. These values are also in general agreement with the experimental values obtained by Levin et al. [167] for particular strains of bacteria and viruses. The concentrations B_i, I_i, and P_i will be specified in units of

10^6/ml. Here we have used a value of $\alpha = 10^{-3}$ ml/min (as compared with the value $\alpha = 10^{-9}$ ml/min applied by Baier et al. [41]).

Like many other ecological models, our system involves positive feedback mechanisms related to the replication of bacteria and viruses. There are non-linear constraints associated both with the bacterial growth rate and with the infection rate, and there is a delay associated with replication of the viruses. The rate of dilution is a main determinant of dissipation in the system. In the absence of viruses, the single pool model displays an equilibrium point

$$B_0 = \frac{1}{\gamma}\left(\sigma - \frac{\rho K}{\nu - \rho}\right), \qquad S_0 = \frac{\rho K}{\nu - \rho}, \tag{14.17}$$

in which the rate of bacterial growth balances the wash out. For dilution rates $\rho > \rho_c = \sigma \nu/(K + \sigma)$, only the trivial equilibrium point $B_1 = 0$, $S_1 = \sigma$ exists.

As ρ is reduced below ρ_c, the equilibrium population of bacteria starts to increase. At the beginning, the cell concentration is still too small for an effective replication of viruses to take place, and the virus population remains nearly negligible. As the dilution rate continues to decrease, however, the virus population grows significantly. The model then undergoes an Andronov–Hopf bifurcation, and the system starts to perform self-sustained oscillations.

14.3.2 Spatial Dynamics

Depending on the population sizes attained in the upstream habitats, different degrees of depletion of the nutrient concentration will occur, and as the surplus resources continue to flow into the next habitat, this population pool will be modulated by a temporal nutrient supply that depends on the type of dynamics realized in the former pool.

Along the chain there will be a net consumption of resources. However, different choices of the surplus nutrient supply σ_i allow us to simulate different patterns of growth dynamics. Below we consider two important cases: lateral nutrition (Figs. 14.13(a) and (b)) and afferent nutrition (Figs. 14.13(c) and (d)). The dilution rate is assumed to be $\rho = 0.003$/min, and the nutrient concentration is specified in µg/ml.

In the first case, the choice of equal values of $\sigma_i = \sigma, i = 1, 2, 3 \ldots$, provides a separate influx of resources to each habitat. For low values of σ, the system attains a stable equilibrium state that extends along the entire chain. As the nutrient supply is increased, starting in the downstream end of the chain, habitat after habitat begins to perform self-sustained oscillations. There is an interval where all habitats starting from a given number exhibit a synchronous periodic behavior. As σ is further increased, the downstream habitats begin to show quasiperiodic, chaotic and hyperchaotic behaviors, and only the intermediate or first pools perform simple periodic oscillations. For $\sigma > 7$ µg/ml, only the motion of the first habitat remains periodic. Figure 14.13(b) shows the variation of the bacterial concentration B along the chain for $\sigma = 2.0$ µg/ml. It is clearly seen that the first five pools reach an

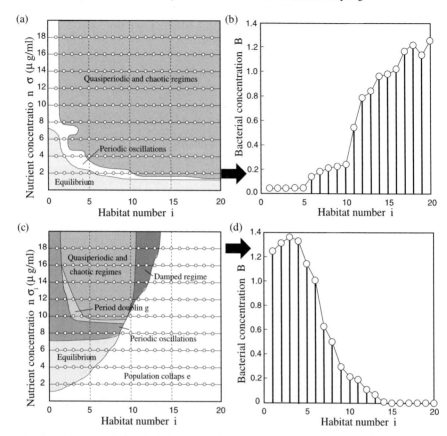

Fig. 14.13. (Color online) Overview of the behavior of a chain of 20 population pools: diagram of dynamical regimes (**left panels**) and variance of bacterial concentrations for selected σ values (**right panels**). **a** and **b** lateral nutrition: $\sigma = \sigma_i$ is assumed to be the same for all pools; **c** and **d** upstream nutrition: $\sigma_1 > 0$, $\sigma_i = 0$, $i = 2, 3, 4, \ldots, 20$

equilibrium state at very low B values. The bacterial populations survive here, but the concentrations are not high enough to generate self-sustained oscillations. Oscillatory behavior with increasing amplitude (grey lines indicate the variance of B_i) is observed starting from the 6th habitat. Pools from six to ten form a cluster with synchronized periodic oscillations. With gradually increasing resources, oscillations become chaotic in the 11th habitat and maintain this dynamics till the last, 20th, habitat.

The second interesting case is afferent nutrition where there is a single source of nutrition to the first habitat (Fig. 14.13(c) and (d)). Since each bacterial population consumes part of the incoming resources, the concentration decreases along the chain. Thus, for some habitat the available resources will not suffice to support self-sustained oscillations. However, the modulation of S may still be propagated with the flow providing an oscillatory forcing for habitats in the rest of the chain. In this

14.3 Cascaded Microbiological Oscillators   399

case, one can observe a limited region of self-sustained oscillations along the chain (Fig. 14.13(c)).

For low values of σ_1, a number of habitats at the beginning of the chain may be able to attain fairly high levels of bacterial concentration. However, these concentrations are not sufficient to produce self-sustained dynamics in interaction with the viruses. As more and more of the resources are consumed along the chain, the bacterial populations collapse ($B_i \approx 0$). At $\sigma_1 = 7.024$ µg/ml the first population reaches the point of Andronov–Hopf bifurcation and self-sustained dynamics arises in the chain. Since the variation of the natural frequency is weak and the coupling strength (proportional to modulation depth of resources flow) is quite high, one observes synchronization in the group of habitats that display self-sustained dynamics.

Further increase of σ_1 to 10.0–12.0 µg/ml reveals a different pattern where only the first two habitats maintain the regime of synchronous regular oscillations, while period-doubled regimes and the development of chaos can be observed further downstream.

Finally, for $\sigma_1 > 16.0$ µg/ml only the first habitat shows regular oscillations (because this is the only possible oscillating regime for an individual system), while the subsequent habitats are in a chaotic state. The case $\sigma_1 = 18.0$ µg/ml is illustrated in Fig. 14.13(d). By virtue of the resource consumption along the chain, the self-sustained chaotic behavior dies out after the 10th habitat. However, the next three habitats display some intermediate dynamical patterns, representing neither the self-sustained regime nor the population collapse. We can describe these states in terms of chaotic forcing across the Andronov–Hopf bifurcation point for the individual system. Thus, such habitats switch chaotically between a self-sustained regime when the nutrition amount is temporarily high enough, and a damped state. In Fig. 14.13(d) habitats 11, 12, and 13 display low, but finite amplitudes of variation in the bacterial concentrations. All downstream habitats are in the population collapse regime ($B_i \approx 0$).

Let us consider the development of frequency patterns for the case of lateral nutrition when the whole chain oscillates. To examine the formation of localized domains of chaotic synchronization we calculate the mean return time $\langle \tau_i \rangle$ to the Poincaré section in the individual unit subspace as described in Part I. Each oscillator now is characterized by the mean frequency $\langle f_i \rangle = 1/\langle \tau_i \rangle$.

Figure 14.14(a) shows the frequencies of all habitats as functions of the nutrition parameter σ. One can find that for $\sigma > 5.0$ most of frequencies run independently without a tendency to synchronization. In the enlargement for $\sigma \in [7.5; 9.6]$ µg/ml (Fig. 14.14(b)), synchronization is witnessed by the fact that several habitats show the same value of $\langle f \rangle$. As we follow the nearly straight line starting around $\sigma = 8.8$ µg/ml and extending to the lower right corner of the figure we can see how pool after pool falls out of synchrony. In the beginning all pools between number 2 and number 13 are synchronized. As σ is increased, however, starting with the pool number 13, one pool after the other loses synchronization. In the interval between $\sigma = 8.3$ and 8.6 µg/ml, increasing σ causes the pool number 20, the pool number 19, etc., to lose synchronization. At the same time, however, pools with lower

400   14 Synchronization of Systems with Resource Mediated Coupling

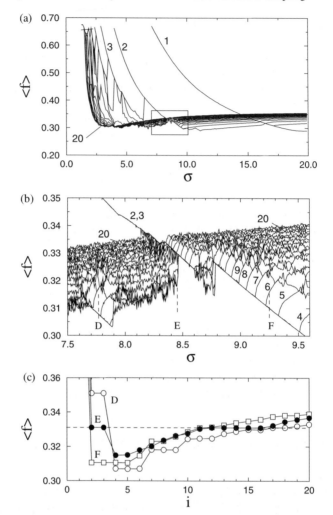

Fig. 14.14. Variation of mean frequency along the cascaded population system. Lateral nutrition parameter σ is assumed to be the same for all pools. **a** Overview of dynamics. There is no tendency to synchronization. **b** The enlargement illustrates the phenomenon of sliding synchronization regimes and partial synchronization between several pools. **c** Examples of partial synchronization. Note how synchronization with the second population pool can reappear at a considerable distance down the chain (curve E)

chain number gradually entrain with the behavior of the second and third population pools.

Three most representative regimes labeled as D, E, F are illustrated in Fig. 14.14(c). The variation of $\langle f \rangle$ along the chain is shown for three different values of the local resource supply: $\sigma = 7.785$ μg/ml (curve D), 8.452 μg/ml (curve E), and 9.235 μg/ml (curve F). Synchronization with the periodic motion of the first population pool does not occur with these values of σ. It is interesting to note, however, that synchronization with the more complicated dynamics of the second population pool is quite common. Curves D and E, for instance, demonstrate how the third pool synchronizes with the second pool. For curve D we can also observe synchronization between pools 4, 5, and 6, between pools 7, 8, and 9, and between pools 10, 11 and 12.

Intuitively, one would expect that once synchronization with a specific pool had failed, it could not be reestablished again further down the chain, i.e., information about the dynamics of a pool would be lost if the subsequent pool did not synchronize with it. By contrast, curve E shows how synchronization with the dynamics of pool number two can reappear far down the chain. In this figure the dynamics of pools 4–10 bear no obvious relation with the dynamics of the second pool. Nonetheless, pools 12–16 again synchronize with the pool number 2. Curve F also shows locking of several chaotically oscillating pools, only the synchronization domain has now moved all the way up along the chain.

Let us summarize some of the main findings:

(i) Changing resource supply along the chain of habitats can generate clusters of limited size with self-sustained dynamics while the rest of the chain is in equilibrium state.

(ii) Inside such clusters, units with different behavior (regular oscillations, quasiperiodic or chaotic behavior) can be detected.

(iii) Inspection of frequency patterns shows that generally there is no tendency to synchronization. There are regions of small size where locking patterns occur. Such locking patterns slide downstream or upstream with variation of nutrition parameter.

14.4 Synchronization Patterns in Kidney Autoregulation

While a chain of microbiological habitats provides a one-way coupling between the oscillators, the next example represents more complicated distribution network associated with an asymmetric but global coupling through the sharing blood flow in a nephronic system.

The blood filtration in kidney is processed with a large number of subunits (nephrons) connected to a complex branching structure of vessels called the *preglomerular vascular tree* with an inhomogeneous distribution of arteriolar lengths, nephron parameters, etc. [63]. Interaction between adjacent nephrons can occur due to vascularly propagated coupling mediated by electrochemical signals, and muscular contractions that travel along the arteriolar wall, and hemodynamic coupling by which an increased flow resistance in the afferent arteriole leading to one nephron forces a higher blood flow to the neighboring nephrons [118, 120].

Since the individual nephron is known to operate in a regime of self-sustained oscillations with the arterial pressure being a control parameter, the coupled nephrons can be considered in the framework of a resource distribution system.

14.4.1 Vascular-Nephron Model

In order to examine the typical mechanisms associated with the structure of preglomerular vascular tree we consider a simplified vascular network that allows us to made conclusions about the main operating regimes and the transitions between these regimes with varying parameters [233].

Our model of the vascular-nephron tree consists of a set of afferent arterioles branching off from a single interlobular artery as shown schematically in Fig. 14.15. The vascular-nephron tree is described in terms of the lengths of the arteriolar and arterial branches together with their hemodynamic resistances. It is assumed that the glomerulus of each nephron is connected to the corresponding branching point via an arteriole of length L_i^g, and of hemodynamic resistance R_i^g, $i = 1, \ldots, 12$. The arterial pressure P_a to be used in the model of individual nephron now becomes the driving blood pressure at the associated branching point P_j^b, $j = 2, \ldots, 7$. Connection between the branching points is described in terms of the branch lengths L_{jk}^b and their hemodynamic resistances R_{jk}^b $j, k = 1, \ldots, 7$. The same approach is used to describe the connection of branching point 1 to the terminal points with the constant pressure values P_{in} and P_{out}. This part of the vascular-nephron tree imitates the connection of the tree to higher-level arteries. The hemodynamic-coupled vascular tree is purely resistive. Transients associated with the distribution of the blood pressure along the branching points are negative, and we can calculate the static pressure distribution for any state of connected nephrons using linear algebraic equations written for each branching point. An example of such an equation for the 6th branching point reads

$$\frac{P_5^b - P_6^b}{R^b} - \frac{P_6^b - P_7^b}{R^b} + \frac{P_3^g - P_6^b}{R_3^g} + \frac{P_4^g - P_6^b}{R_4^g} = 0. \qquad (14.18)$$

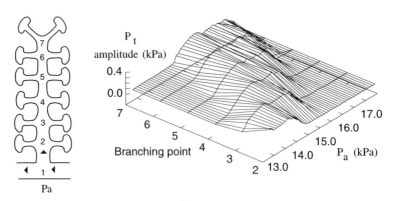

Fig. 14.15. *Left*: Sketch of vascular-coupled nephron tree including the interlobular artery, the afferent arterioles and the glomeruli. *Right*: Oscillation amplitude as a function of the arterial pressure and the position of the branching point along the preglomerular vascular tree

14.4 Synchronization Patterns in Kidney Autoregulation   403

Here, P_3^g and P_4^g represent the blood pressure in the glomerulus of the 3rd and 4th nephron. P_5^b, P_6^b, and P_7^b represent the blood pressure in the 5th, 6th, and 7th branching points, respectively. R^b denotes the hemodynamic resistance between the 6th and 7th branching points. This resistance is assumed to be the same for all branches. R_3^g and R_4^g are the hemodynamic resistances to the 3rd and 4th nephron, respectively. Note that R_3^g and R_4^g are not constant because they include the resistance of the active parts of the afferent arterioles.

Equations of this type for all branching points are obviously interdependent and, hence, produce a global *hemodynamic coupling* among nephrons in the vascular-nephron tree. The strength of this coupling is generally increasing with R_{jk}^b, but decreasing with R_i^g. Neighboring nephrons can also influence one another through a *vascular propagated electrical (electrochemical) signal*. To account for this mechanism, the total activation potential for kth nephron is assumed to be the sum of contributions from all other nephrons in the tree. Moreover, the electrical activation potentials are assumed to propagate along the vascular wall with an exponential decay. In this way, the vascular propagated interaction is delivered to each nephron as an additional part of its activation potential Ψ:

$$\Delta\Psi = \sum_{i=1,i\neq k}^{N} \Psi_i \exp\left(-\gamma\left(L_{ji} + L_k^g\right)\right), \tag{14.19}$$

where j is the number of the branching point to which the considered nephron with number k is connected. The matrix L_{ji} contains the lengths from a given branching point j to all nephrons $i = 1, \ldots, N$, and L_k^g is the length from the given nephron to the connected branching point.

Each individual nephron may be described by the following model [47, 182]:

$$\dot{P}_t = \frac{1}{C_{tub}}\{F_f(P_t, P_a, r) - F_{reab} - F_H\}, \tag{14.20}$$

$$\dot{r} = v_r, \tag{14.21}$$

$$\dot{v}_r = \frac{1}{\omega}\{P_{av}(P_t, P_a, r) - P_{eq}(r, \Psi(X_3, \alpha), T) - \omega d v_r\}, \tag{14.22}$$

$$\dot{X}_1 = F_H - \frac{3}{T}X_1, \tag{14.23}$$

$$\dot{X}_2 = \frac{3}{T}(X_1 - X_2), \tag{14.24}$$

$$\dot{X}_3 = \frac{3}{T}(X_2 - X_3). \tag{14.25}$$

The first equation determines the pressure variations in the proximal tubule in terms of the in- and outgoing fluid flows where F_f is the single-nephron glomerular filtration rate, reabsorption in the proximal tubule F_{reab} is assumed to be constant, F_H is the flow into the loop of Henle, and C_{tub} is the elastic compliance of the tubule.

The following two equations describe the dynamics associated with the flow control in the afferent arteriole. Here, r represents the radius of the active part of

404   14 Synchronization of Systems with Resource Mediated Coupling

the vessel and v_r is its rate of increase. d is a characteristic time constant describing the damping of the oscillations, ω is a measure of the mass density of the arteriolar wall, and P_{av} denotes the average pressure in the active part of the arteriole. P_{eq} is the value of this pressure for which the arteriole is in equilibrium with its present radius and muscular activation Ψ. The expressions for F_f, P_{av} and P_{eq} involve a number of algebraic equations that must be solved along with the integration of (14.20)–(14.25).

The remaining equations in the single-nephron model represent the delay T in the tubuloglomerular feedback (TGF) regulation. For a more detailed explanation of the model and its parameters, see, e.g., [11].

Thus the mathematical model of vascular-nephron tree that we investigate consists of (i) 12 sets of coupled ODEs describing individual nephrons, (ii) a set of linear algebraic equations that determine the blood pressure drop from one branching point to another, and (iii) algebraic relations for the vascular interaction.

Depending on the choice of the control parameters, the amplitudes of the pressure oscillations in the nephron tree are found to be different at different positions in the tree. Due to model symmetry, two nephrons connected to the same node have the same oscillation amplitudes. Thus, we can refer to the number of the branching point to describe the amplitude properties. Branching points 2, 3, and 4 may correspond to deep nephrons and branching points 6 and 7 to superficial nephrons. Experimentally, only the pressure oscillations in nephrons near the surface of the kidney have been investigated. However, we suppose that both deep (juxtamedullary) and superficial (cortical) nephrons can exhibit oscillations in their pressures and flows.

When varying the arterial pressure, different amplitude patterns can be observed in the multinephron model (14.18)–(14.25). For low values of the arterial pressure P_a, vanishing amplitude of the tubular pressure oscillations can be observed near the top of the tree (i.e., in branching points 5, 6, 7). The nephrons connected to these points operate in a damped regime like the population pools we discussed in the above example. In the individual nephron model, the self-sustained dynamics is bounded by two points of Andronov–Hopf bifurcation at $P_{a1} = 11.48$ kPa and $P_{a2} = 13.86$ kPa. The calculated value of the mean blood pressure in branching point 5 is lower than P_{a1}. Hence, neither nephrons connected to this point, nor nephrons downstream of it, oscillate. As blood pressure increases above P_{a2}, the upstream nephrons stop to oscillate. For intermediate values of P_a there is a cluster of nephrons with self-sustained dynamics in the middle section of the tree. Finally, for high enough values of $P_a (> 17$ kPa) only nephrons connected to the branching points 6 and 7 display oscillations because for all other nephrons the blood pressure in the corresponding branching point is too high.

Thus, the oscillatory amplitude patterns along a vascular-nephron tree have a reasonable explanation in terms of a drop in driving pressure from one branching point to the next one, causing a change of the operating regime of the individual nephrons.

14.4.2 Coupling-Induced Inhomogeneity

The vascular-nephron tree represents an extended network whose complex dynamics is controlled by a significant number of parameters. To focus our investigations we shall emphasize the generic aspects of cooperative behavior in the network rather than its physiologically relevant properties. From a structural point of view, the object we consider is a population of globally coupled two-mode oscillators [272]. Besides the relatively slow mode mediated by the tubuloglomerular feedback with its inherent delay of about 15 s (associated with the flow of fluid though the loop of Henle), there is a four to five times faster mode arising from the response of the smooth muscles in the arteriolar wall. A key question is to what extent the synchronization in the form of frequency and/or phase entrainment can be detected for such systems under variation of appropriate control parameters?

From the classical theory of synchronization it is known that there are two main parameters: the strength (or type) of the interaction and the degree of frequency mismatch. Our first question is, therefore, how the control parameters of the vascular network are related to the synchronization parameters. We have shown [227] that increasing vascular coupling leads to in-phase synchronization, while strong hemodynamic interaction can produce antiphase entrainment. We would expect similar results in the vascular-nephron tree. However, the influence of the arterial pressure P_a and of the hemodynamic resistances between the neighboring branching points is not trivial, since these parameters also affect the natural dynamics of the individual nephrons. Let us perform a few numerical experiments to clarify the situation.

Trial 1. Weak Hemodynamic Coupling

With the parameters used in Fig. 14.15, a choice of the arterial pressure of $P_a = 13.3$ kPa allows all nephrons to be in the oscillatory regime. Here, the hemodynamic resistance has been assumed to be $R^b = 0.002$ kPa·s/nl. The vascular coupling may then be varied by adjusting γ from 1.6 mm^{-1} (strong interaction) to 4.0 mm^{-1} (weak interaction). As defined above, R^b denotes the flow resistance between two successive branching points of the vascular tree (Fig. 14.15), and the parameter γ measures the length constant associated with the exponential decay of the vascular propagated coupling along the arterioles.

The frequency distribution among the nephrons is shown in Fig. 14.16 for the slow oscillatory mode. For $\gamma = 1.6$ mm^{-1}, the frequencies of the slow TGF mediated modes are locked at the same value $f_{slow} = 0.0275$ Hz for all nephrons. Hence, strong vascular coupling leads to perfect frequency locking along the tree. With decreasing coupling strength, the collective behavior becomes asynchronous. In Fig. 14.16(a) this is illustrated by the curves for $\gamma = 2.5$ and $\gamma = 3.0$ mm^{-1}. Surprisingly, we find that with further reduction of the coupling ($\gamma = 4.0$ mm^{-1}), all nephrons again demonstrate a synchronous state, now at $f_{slow} \approx 0.02874$ Hz. To explain this, let us consider the phase dynamics (Fig. 14.16(b) and (c)). In the first synchronous state (strong vascular coupling), in-phase relationships are clearly observed for the slow TGF oscillations (Fig. 14.16(b)), while the second state (weak

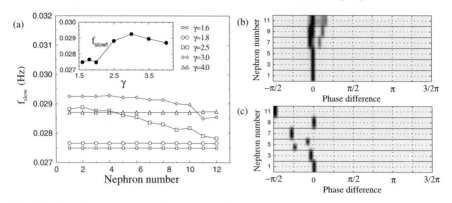

Fig. 14.16. a Slow frequency adjustment with varying vascular coupling. Inset shows the variation of f_{slow} for the first nephron versus γ. Phase entrainment of the slow mode at **b** $\gamma = 1.6$ mm^{-1} and **c** $\gamma = 4.0$ mm^{-1}. Phase differences are calculated with respect to the first nephron. ($R^b = 0.002$ kPa·s/nl and $P_a = 13.3$ kPa)

vascular interaction) corresponds to out-of-phase synchronization (Fig. 14.16(c)). In this case, the hemodynamic coupling probably plays the ordering role. The inset in Fig. 14.16(a) illustrates how the locking frequency within a certain range shifts non-monotonically with varying vascular coupling.

Trial 2. Stronger Hemodynamic Coupling

To examine the hypothesis that the hemodynamic interaction mechanism is responsible for the out-of-phase synchronous state at large γ, we increase the initial hemodynamic coupling to $R^b = 0.01$ kPa·s/nl and perform the same numerical experiment with increasing vascular interaction.

Let us focus on the changes in the slow dynamics of the nephrons. For strong vascular coupling ($\gamma = 1.6$ mm^{-1}) all nephrons are frequency-locked (Fig. 14.17). However, this synchronous state is out-of-phase (Fig. 14.17(b)) in contrast to trial 1. With decreasing vascular coupling one can observe asynchronous behavior detected both as frequency and phase divergence (Fig. 14.17(c)). Note that the pairs of nephrons connected to the same branching point remain synchronous, and the first six nephrons operate in synchrony, although the distribution looks a little washed out, but remains limited.

However, a globally synchronized state similar to the state shown in Fig. 14.16 is not achieved even for $\gamma = 10.0$ mm^{-1}. We conclude that a stronger hemodynamic coupling is unable to synchronize the slow mode oscillations in the whole vascular-nephron tree.

With increasing vascular coupling the nephrons are found to synchronize at a lower frequency than in the case of small vascular coupling. This effect cannot be explained solely in terms of the in-phase nature of the vascular coupling. To find an explanation of the observed behavior, let us consider how the vascular coupling influences the natural frequency of the individual nephron. Here, the natural frequency

14.4 Synchronization Patterns in Kidney Autoregulation

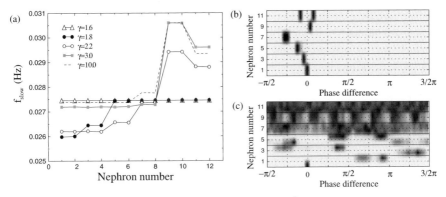

Fig. 14.17. (Color online) **a** Slow frequency adjustment for $R^b = 0.01$ kPa·s/nl with varying vascular coupling. Phase entrainment of slow dynamics at **b** $\gamma = 1.6$ mm^{-1} and **c** $\gamma = 3.0$ mm^{-1}. ($R^b = 0.01$ kPa·s/nl and $P_a = 13.3$ kPa)

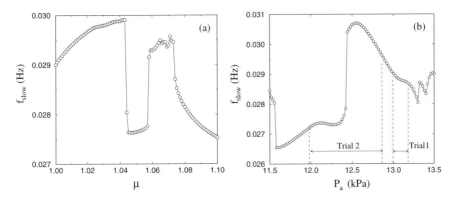

Fig. 14.18. a The strength of the muscular activation affects the oscillation frequency in a non-monotonic way. μ is an artificial scaling factor for the activation potential Ψ. **b** Frequency of individual nephron as a function of arterial pressure P_a

of the jth nephron is understood to be the frequency of the tubular pressure oscillations in the absence of interaction, but with the same driving pressure as the pressure at a branching point to which the nephron is connected, i.e., $P_a = P_j^b$. Since the vascular coupling acts via the activation potential Ψ of each nephron (14.19), it can influence the operating regime of the nephron. For in-phase synchronous state, one can estimate this influence via an artificial variation of Ψ in the individual nephron: $\Psi' = \mu\Psi$. Here, μ is a scaling factor: the stronger the vascular coupling is, the larger μ will be. Figure 14.18 illustrates how the nephron frequency depends on μ. It is clearly seen that the oscillation frequency changes in a non-monotonous way with increasing μ and becomes lower in average. Thus the vascular coupling in general reduces the nephron frequency (at least as long as it is introduced according to (14.19)).

408 14 Synchronization of Systems with Resource Mediated Coupling

The influence of the hemodynamic coupling is even more complicated. However, a similar approach can be used. Because the value of the hemodynamic resistance R^b is closely related to the drop of arterial pressure along the vascular tree, it can cause essential changes in the oscillatory properties of the individual nephrons.

Figure 14.18(b) shows how the nephron frequency depends on the driving pressure for the individual nephron. The curve is strongly non-monotonic and includes pieces of gradual increase, gradual fall, as well as values of abrupt change. Clearly, such behavior is related to the bifurcations of the nephron oscillatory regimes. Thus, a variation of the hemodynamic coupling can cause effects that are strongly non-monotonous and depend on P_a. In Fig. 14.18(b), trials 1 and 2 are presented in terms of P_j^b drop along the vascular tree. It is seen that the first trial corresponds to a relatively small frequency mismatch among the nephrons, while stronger coupling ($R^b = 0.01$ kPa·s/nl) leads to a faster drop of arterial pressure along the tree and, hence, to a larger frequency mismatch.

We conclude that:

(i) There is a clear evidence that clustering of oscillators in the vascular-nephron model occurs due to the limited blood pressure P_a interval for oscillatory behavior in the individual nephron model.

(ii) Dual nature of such interaction manifests itself (i) by bringing order in the form of synchronization and (ii) by inducing shift of operating regimes of each nephron along the tree that is equivalent to a mismatch between interacting systems.

Altogether, these mechanisms provide for a rather complex response of the system to variations in the coupling strengths. We refer the above described effect as *coupling induced inhomogeneity* that seems to be the essential feature of systems with resource mediated coupling.

14.5 Summary

We considered four examples of coupled oscillator systems where the coupling is mediated though the flow of resources that maintains the oscillatory state in the individual unit.

Since stimulation is one of the mechanisms that can cause changes of extracellular potassium associated with variation of extracellular space volume, and since the diffusion rate of potassium depends on the functional conditions of glial cells, these two quantities have been chosen as control parameters. In the framework of this approach, we demonstrated that changes of the extracellular potassium concentration can synchronize two nearby cells. With varying extracellular volume and diffusion rate, the dual nature of such resource mediated coupling is found to be responsible for competing in-phase and antiphase synchronization patterns.

The electronic example representing a relatively simple chain structure allowed us to demonstrate generic behavioral patterns under competing coupling. Localized (finite-sized) *clusters* of oscillating units that slide up and down the consumption

14.5 Summary   409

chain in response to the change of overall resource supply, and coupling-induced inhomogeneity, appear to be characteristic phenomena in such systems. For the general model of N identical oscillators, *coupling induced inhomogeneity* manifests itself either via an asynchronous intracluster behavior with a distribution of mean periods, resembling the variation of the period of the individual oscillator with the energy supply parameter, or (for stronger coupling) via small-sized clusters of out-of-phase synchronization that move along the chain. Note that the region of self-sustained dynamics, limited within a certain range of resource parameter, appears to be the necessary condition for such type of behavior.

The individual microbiological population pool displayed only a simple Andronov–Hopf bifurcation. However, different resource delivering environments provide different behavioral patterns. By increasing the lateral resource supply, the chain of population pools could be driven into a state of increasing complexity with clusters of chaotic, frequency-synchronized pools. The opposite case of afferent (downstream-only) nutrition provides a finite-sized cluster with self-sustained dynamics, outside which the oscillations die out.

The physiological example of a vascular tree involves a significantly more complicated coupling structure with the flow mediated hemodynamic coupling competing with a vascular propagated coupling of a very different nature. The nonlinearities of the physiological system considered allowed self-sustained oscillations only in a finite range of resource supplies. At high and low afferent blood pressures, the individual nephrons displayed stable equilibrium points. Hence, the cluster formation mechanism we observed for a chain of electronic circuits manifests itself also in physiological system. Inside the cluster we again encountered the coupling induced inhomogeneity that now activated rather complex patterns due to the complex response of the individual system to external driving.

Thus, there is a wide class of systems that are composed of a set of individual oscillators connected via a resource distributed network. In such systems, a number of spatially localized oscillatory modes is controlled by the amount of resources delivered to each individual oscillator, and its influx acts as a global parameter. It is remarkable that in this case the initially identical subunits show a tendency for desynchronization, so-called coupling-induced inhomogeneity, combined with spatial localization of subunits in oscillatory clusters.

The coupling that we have considered in this chapter is likely to be quite common in nature as well as in man-made systems. The generic nature of the resource dependent coupling suggests that it can serve as useful paradigm together with the well-known and widely used approaches assuming interaction via mechanisms unrelated to the resource supply.

15 Conclusions to Part II

In Part II of this book we endeavoured to demonstrate that synchronization is a seriously multifaceted phenomenon. Just a few main features include multistability, multimode behavior, the effect from coupling that is mediated by the resources available, and the non-trivial phenomena arising from anisochronous motion on the limit cycle.

412   15 Conclusions to Part II

And finally...

Across the whole book we have been trying to share with the reader everything (or almost everything) we know about synchronization: from the classical theoretical results on which we were trained ourselves, to the results of modern studies—which are perhaps not quite complete, but interesting nevertheless.

Before finishing this book, we would like to ask the reader:

Do you see how various manifestations and aspects of synchronization of oscillations of all kinds combine to form a single picture?

Do you agree that this approach provides a researcher with a powerful tool for the analysis of a huge variety of problems related to interactions between non-linear oscillators?

The authors will be happy with "I have to think about it";
and grateful to hear "Could be";
and totally delighted if the answer is "I do."

Nottingham and Loughborough (UK)
Lyngby (Denmark)
Saratov (Russia)
2005–2007

References

[1] H.D.I. Abarbanel, N.F. Rulkov, M.M. Sushchik, Phys. Rev. E **53**, 4528 (1996)

[2] R. Abraham, C. Shaw, *The Geometry of Behavior (Studies in Nonlinearity)* (Addison-Wesley, Reading, 1992)

[3] R. Abraham, Y. Ueda, *The Chaos Avant-Garde: Memoirs of the Early Days of Chaos Theory* (World Scientific, Singapore, 2001)

[4] M. Abramowitz, I.A. Stegun (eds.), *Handbook of Mathematical Functions with Formulas, Graphs, and Mathematical Tables* (Dover, New York, 1965)

[5] R. Adler, Proc. IEEE **61**, 1380 (1973)

[6] V.S. Afraimovich, V.I. Nekorkin, Int. J. Bifurc. Chaos **4**, 631 (1994)

[7] V.S. Afraimovich, L.P. Schilnikov, in *Methods of the Qualitative Theory of Differential Equations* (Gor'kov Gos. University, Gorky, 1983), p. 3; translated in: Am. Math. Soc. Transl. Ser. 2 **149**, 201 (1991)

[8] V.S. Afraimovich, V.S. Verichev, M.I. Rabinovich, Izv. Vyssh. Uchebn. Zaved. Radiofiz. **29**, 1050 (1986)

[9] B. Alving, J. Gen. Physiol. **51**, 29 (1968)

[10] A. Amann, A. Wacker, E. Schöll, Physica B **314**, 404 (2002)

[11] M.D. Andersen, N. Carlson, E. Mosekilde, N.-H. Holstein-Rathlou, in *Membrane Transport and Renal Physiology* (Springer, New York, 2001)

[12] A.A. Andronov, S.E. Khaikin, *Theory of Oscillations* (Gostekhizdat, Moscow, 1937) (in Russian)

[13] A.A. Andronov, A.A. Vitt, Zhurnal Prikladnoi Fiziki (J. Appl. Phys.) **7**(4), 3 (1930)

[14] A.A. Andronov, A.A. Vitt, S.E. Khaikin, *Theory of Oscillations* (Pergamon, Elmsford, 1966)

[15] V.S. Anishchenko, *Dynamical Chaos—Models and Experiments: Appearance Routes and Structure of Chaos in Simple Dynamical Systems* (World Scientific, Singapore, 1995)

[16] V.S. Anishchenko, D.E. Postnov, Pisma Zh. Tekh. Fiz. **14**, 569 (1988)

[17] V.S. Anishchenko, T.E. Vadivasova, D.E. Postnov, M.A. Safonova, Radiotekh. Elektron. **36**, 338 (1991)

[18] V.S. Anishchenko, T.E. Vadivasova, D.E. Postnov, M.A. Safonova, Int. J. Bifurc. Chaos **2**(3), 633 (1992)

[19] V.S. Anichshenko, M.A. Safonova, L.O. Chua, IEEE Trans. Circuits Syst. **40**, 792 (1993)

[20] V.S. Anishchenko, A.N. Silchenko, I.A. Khovanov, Phys. Rev. E **57**, 316 (1998)

[21] V.S. Anishchenko, A.B. Neiman, F. Moss, L. Schimansky-Geier, Phys. Uspekhi **42**, 7 (1999)

414   References

[22] V.S. Anishchenko, A.G. Balanov, N.B. Janson, N.B. Igosheva, G.V. Bordjugov, Int. J. Bifurc. Chaos **10**, 2339 (2000)

[23] V.S. Anishchenko, A.G. Balanov, N.B. Janson, N.B. Igosheva, G.V. Bordjugov, Discrete Dyn. Nat. Soc. **4**, 201 (2000)

[24] V.S. Anishchenko, V.V. Astakhov, A.B. Neiman, T.E. Vadivasova, L. Schimansky-Geier, *Nonlinear Dynamics of Chaotic and Stochastic Systems* (Springer, Berlin, 2002)

[25] V.S. Anishchenko, G.A. Okrokvertskhov, T.E. Vadivasova, G.I. Strelkova, New J. Phys. **7**, 76 (2005)

[26] V.S. Anishchenko, T.E. Vadivasova, G.A. Okrokvertskhov, G.I. Strelkova, Phys. Uspekhi **48**, 151 (2005)

[27] V. Anishchenko, S. Nikolaev, J. Kurths, Phys. Rev. E **73**, 056202 (2006)

[28] E.V. Appleton, Proc. Cambridge Philos. Soc. (Math. and Phys. Sci.) **21**, 231 (1922)

[29] A. Arneéodo, P.H. Coullet, E.A. Spiegel, Phys. Lett. A **94**, 1 (1983)

[30] V.A. Arnold, *Geometrical Methods in the Theory of Ordinary Differential Equations* (Springer, New York, 1996)

[31] V.I. Arnold, in *Nonlinear Waves*, ed. by A.V. Gaponov-Grechov (Nauka, Moscow, 1979)

[32] V.I. Arnold, *Catastrophe Theory* (Springer, New York, 1992)

[33] V.I. Arnol'd, V.S. Afraimovich, Yu.S. Iljashenko, L.P. Shil'nikov, *Dynamical Systems. Bifurcation Theory and Catastrophe Theory*, vol. 5 (Springer, New York, 1994)

[34] D.G. Aronson, E.J. Doedel, H.G. Othmer, Physica D **25**, 20 (1987)

[35] D.G. Aronson, G.B. Ermentrout, N. Kopell, Physica D **41**, 403 (1990)

[36] D.K. Arrowsmith, K.I. Taha, Meccanica **18**, 195 (1983)

[37] R. Artuso, E. Aurell, P. Cvitanović, Nonlinearity **3**, 325 (1990)

[38] V.V. Astakhov, B.P. Bezruchko, E.N. Erastova, E.P. Seleznev, Sov. Phys. Tech. Phys. **35**, 1122 (1990)

[39] V. Astakhov, M. Hasler, T. Kapitaniak, A. Shabunin, V. Anishchenko, Phys. Rev. E **58**(5), 5620 (1998)

[40] S.A. Baccus, Proc. Natl. Acad. Sci. USA **95**, 8345 (1998)

[41] G. Baier, J.S. Thomsen, E. Mosekilde, J. Theor. Biol. **165**, 593 (1993)

[42] A.G. Balanov, N.B. Janson, D.E. Postnov, P.V.E. McClintock, Phys. Rev. E **65**, 041105 (2002)

[43] A.G. Balanov, N.B. Janson, P.V.E. McClintock, Fluct. Noise Lett. **3**, L113 (2003)

[44] A.G. Balanov, N.B. Janson, V.V. Astakhov, P.V.E. McClintock, Phys. Rev. E **72**, 026214 (2005)

[45] B. Barbour, H. Brew, D. Attwell, Nature **335**, 433 (1988)

[46] K. Bar-Eli, Physica D **14**, 242 (1985)

[47] M. Barfred, E. Mosekilde, N.-H. Holstein-Rathlou, Chaos **6**, 280 (1996)

[48] R. Bartussek, P. Hänggi, J.C. Kissner, Europhys. Lett. **28**, 459 (1994)

[49] V. Blažek, Czechoslov. J. Phys. **18**, 1572 (1967)

[50] I. Blekhman, *Synchronization in Science and Technology* (ASME Press, New York, 1988)

[51] I.I. Blekhman, P.S. Landa, M.G. Rosenblum, Appl. Mech. Rev. **48**(11), 733 (1995)

[52] S. Boccaletti, J. Kurths, G. Osipov, D.L. Valladares, C.S. Zhou, Phys. Rep. **366**, 1 (2002)

[53] R.I. Bogdanov, Func. Anal. Appl. **9**, 144 (1975)

[54] K.F. Bonhoeffer, J. Gen. Physiol. **32**, 69 (1948)

[55] M. Brambilla, L.A. Lugiato, V. Penna, F. Prati, C. Tamm, C.O. Weiss, Phys. Rev. A **43**, 5114 (1991)

References   415

[56] H.A. Braun, H. Bade, H. Hensel, Pflügers Arch. **386**, 1 (1980)
[57] H.A. Braun, H. Wissing, K. Schäfer, M.C. Hirsch, Nature **367**, 270 (1994)
[58] British Patent 184282, 13 May 1921
[59] N.F. Britton, *Essential Mathematical Biology* (Springer, Berlin, 2003)
[60] C.C. Canavier, D.A. Baxter, J.W. Clark, J.H. Byrne, J. Neurophysiol. **69**, 2252 (1993)
[61] M.L. Cartwright, J. Inst. Elec. Eng. (London) **95**, 88 (1948)
[62] M.L. Cartwright, J.E. Littlewood, J. Lond. Math. Soc. **20**(77), 180 (1945)
[63] D. Casellas, F.J. Dupont, T. Kaskel, T. Inagami, L.C. Moore, Am. J. Physiol. **265**, F151 (1993)
[64] T. Chakraborty, R.H. Rand, Int. J. Non-Linear Mech. **23**, 369–396 (1988)
[65] J.D. Crawford, Rev. Mod. Phys. **63**, 991 (1991)
[66] J.P. Crutchfield, B.A. Huberman, Phys. Lett. A **77**, 407 (1980)
[67] J.P. Crutchfield, J.D. Farmer, B.A. Huberman, Phys. Rep. **92**, 45 (1982)
[68] S. Danø, F. Hynne, S. De Monte, F. d'Ovidio, P.G. Sorensen, H. Westerhoff, Faraday Discuss. **120**, 261 (2002)
[69] P.M. Dean, E.K. Matthews, J. Physiol. **210**, 255 (1970)
[70] J.W. Deitmer, C.R. Rose, T. Munch, J. Schmidt, W. Nett, N.-P. Schneider, C. Lohr, GLIA **28**, 175 (1999)
[71] R. Descartes, *Le Monde, ou Traite de la lumiere* (Treatise on the World) (Arabic Books, New York, 1979); translation and introduction by S. Mahoney
[72] G. de Vries, A. Sherman, H.-R. Zhu, Bull. Math. Biol. **60**, 1167 (1998)
[73] E.J. Doedel, H.B. Keller, J.P. Kernevez, Int. J. Bifurc. Chaos **1**(3), 493 (1991); **1**(4), 745 (1991)
[74] M.I. Dykman, R. Mannella, P.V.E. McClintock, S.M. Soskin, N.G. Stocks, Phys. Rev. A **42**, 7041 (1990)
[75] W.H. Eccles, Electrician **89**, 503 (1923)
[76] I.I. Fedchenia, R. Mannella, P.V.E. McClintock, N.D. Stein, N.G. Stocks, Phys. Rev. A **46**, 1769 (1992)
[77] M.J. Feigenbaum, J. Stat. Phys. **19**, 25 (1978)
[78] M.J. Feigenbaum, Phys. Lett. A **74**, 375 (1979)
[79] M.J. Feigenbaum, L.P. Kadanoff, S.J. Shenker, Physica D **5**, 370 (1982)
[80] R.A. FitzHugh, Biophys. J. **1**, 445 (1961)
[81] J. Foss, A. Longtin, B. Mensour, J. Milton, Phys. Rev. Lett. **76**, 708 (1996)
[82] J. Freund, A. Neiman, L. Schimansky-Geier, Europhys. Lett. **50**, 8 (2000)
[83] J.A. Freund, L. Schimansky-Geier, P. Hanggi, Chaos **13**, 225 (2003)
[84] H. Fujisaka, T. Yamada, Progr. Theoret. Phys. **69**, 32–47 (1983)
[85] D. Gabor, J. IEE (London) **93**, 429 (1946)
[86] P.C. Gailey, A. Neiman, J.J. Collins, F. Moss, Phys. Rev. Lett. **79**, 4701 (1997)
[87] H. Gang, T. Ditzinger, C.Z. Ning, H. Haken, Phys. Rev. Lett. **71**, 807 (1993)
[88] A.W. Gillies, Quart. J. Mech. Appl. Math. **7**(2), 152 (1954)
[89] R. Gilmore, Rev. Mod. Phys. **70**, 1455 (1998)
[90] P. Glandsdorff, I. Prigogine, *Thermodynamic Theory of Structure, Stability and Fluctuations* (Wiley, New York, 1971)
[91] L. Glass, Chaos **1**, 13–19 (1991)
[92] L. Glass, M.C. Mackey, *From Clocks to Chaos: The Rhythms of Life* (Princeton University Press, Princeton, 1988)
[93] L. Glass, R. Perez, Phys. Rev. Lett. **48**, 1772–1775 (1982)
[94] L. Glass, M.R. Guevara, J. Belair, A. Shrier, Phys. Rev. A **29**, 1348–1357 (1984)
[95] A. Goryachev, R. Kapral, Phys. Rev. Lett. **76**, 1619 (1996)

416 References

[96] I. Goychuk, J. Casado-Pascual, M. Morillo et al., Phys. Rev. Lett. **97**, 210601 (2006)
[97] I.S. Gradshteyn, I.M. Ryzhik, in *Tables of Integrals, Series and Products*, 6th edn., ed. by A. Jeffrey (Academic Press, San Diego, 2000)
[98] C. Grebogi, E. Ott, J.A. Jorke, Phys. Rev. Lett. **48**, 1507 (1982)
[99] C. Grebogi, E. Ott, J.A. Yorke, Phys. Rev. A **36**, 35223524 (1987)
[100] R. Guantes, G.G. de Polavieja, Phys. Rev. E **71**, 011911(1–4) (2005)
[101] J. Guckenheimer, P. Holmes, *Nonlinear Oscillations, Dynamical Systems, and Bifurcations of Vector Field* (Springer, New York, 1993)
[102] J. Guckenheimer, A.R. Willms, Physica D **139**, 195 (1999)
[103] S.K. Han, D.E. Postnov, Chaos **13**(3), 1105 (2003)
[104] S.K. Han, C. Kurrer, Y. Kuramoto, Phys. Rev. Lett. **75**, 3190 (1995)
[105] S.K. Han, T.G. Yim, D.E. Postnov, O.V. Sosnovtseva, Phys. Rev. Lett. **83**, 1771 (1999)
[106] P. Hänggi, R. Bartussek, in *Nonlinear Physics of Complex Systems*. Lecture Notes in Physics (Springer, Berlin, 1996)
[107] D. Hansel, H. Sompolinsky, Phys. Rev. Lett. **68**, 718 (1992)
[108] A.J. Hansen, Acta Physiol. Scand. **102**, 324 (1978)
[109] K. Hashimoto, T. Ikegami, J. Phys. Soc. Jpn. **70**, 349 (2001)
[110] C. Hayashi, *Nonlinear Oscillations in Physical Systems* (McGraw-Hill, New York, 1964)
[111] J.F. Heagy, S.M. Hammel, Physica D **70**, 140 (1994)
[112] J.F. Heagy, T.L. Carroll, L.M. Pecora, Phys. Rev. E **50**, 1874 (1994)
[113] J. Hertz, A. Krogh, R.G. Palmer, *Introduction to the Theory of Neural Computation* (Addision-Wesley, New York, 1991)
[114] H. Herzel, Z. Angew. Math. Mech. **68**, 582 (1988)
[115] J.L. Hindmarsh, R.M. Rose, Proc. Roy. Soc. Lond. B **221**, 87 (1984)
[116] A.L. Hodgkin, A.F. Huxley, J. Physiol. Lond. **117**, 500 (1952)
[117] P.J. Holmes, D.A. Rand, Q. Appl. Math. **35**, 495 (1978)
[118] N.-H. Holstein-Rathlou, Pflügers Arch. **408**, 438 (1987)
[119] N.-H. Holstein-Rathlou, P.P. Leyssac, Acta Physiol. Scand. **126**, 333 (1986)
[120] N.-H. Holstein-Rathlou, K.-P. Yip, O.V. Sosnovtseva, E. Mosekilde, Chaos **11**, 417 (2001)
[121] E. Hopf, Commun. Pure Appl. Math. **1**, 303 (1948)
[122] F.C. Hoppensteadt, J.P. Keener, J. Math. Biol. **15**, 339 (1982)
[123] G. Hu, J.Z. Yang, W.Q. Ma, J.H. Xiao, Phys. Rev. Lett. **81**, 5314 (1998)
[124] K.L.C. Hunt, J. Kottalam, M.D. Hatlee, J. Ross, J. Chem. Phys. **96**, 7019 (1992)
[125] C. Huygens, *Horologium oscillatorium sive de motu pendulorum ad horologia aptato demonstrationes geometricae* (Paris, 1673)
[126] E.M. Izhikevich, Int. J. Bifurc. Chaos **10**, 1171 (2000)
[127] E.M. Izhikevich, SIAM J. Appl. Math. **60**(5), 1789 (2000)
[128] D.S. Jones, B.D. Sleeman, *Differential Equations and Mathematical Biology* (CRC Press, Boca Raton, 2003)
[129] F. Julicher, A. Ajdari, J. Prost, Rev. Mod. Phys. **69**, 1269 (1997)
[130] P. Jung, Phys. Rep. **243**, 175 (1993)
[131] L. Junge, U. Parlitz, Phys. Rev. E **62**, 438 (2000)
[132] K. Kaneko, *Collapse of Tori and Genesis of Chaos in Dissipative Systems* (World Scientific, Singapore, 1986)
[133] A. Katok, B. Hasselblatt, *Introduction to the Modern Theory of Dynamical Systems* (Cambridge University Press, Cambridge, 1995)

References 417

[134] Y. Katznelson, *An Introduction to Harmonic Analysis*, 2nd corr. edn. (Dover, New York, 1976)

[135] M. Kawato, R. Suzuki, J. Theor. Biol. **86**, 547 (1980)

[136] J. Keener, J. Sneyd, *Mathematical Physiology* (Springer, New York, 1998)

[137] I.G. Kevrekidis, R. Aris, L.D. Schimdt, S. Pelikan, Physica D **16**, 243 (1985)

[138] I.G. Kevrekidis, R. Aris, L.D. Schmidt, Physica D **23**, 391 (1986)

[139] I.G. Kevrekidis, L.D. Schmidt, R. Aris, Chem. Eng. Sci. **41**, 1263 (1986)

[140] Ya.I. Khanin, *Fundamentals of Laser Dynamics* (Cambridge International Science Publishing, Cambridge, 2005)

[141] S. Kim, S.H. Park, C.S. Ryu, Phys. Rev. Lett. **78**, 1616 (1997)

[142] S. Kim, S.H. Park, C.S. Ryu, Phys. Rev. Lett. **79**, 2911 (1997)

[143] A.C. King, J. Billingham, S.R. Otto, *Differential Equations Linear, Nonlinear, Ordinary, Partial* (Cambridge University Press, Cambridge, 2003)

[144] I.Z. Kiss, Y. Zhai, J.L. Hudson, Science **296**, 1676 (2002)

[145] I.Z. Kiss, Q. Lv, L. Organ, J.L. Hudson, Phys. Chem. Chem. Phys. **8**, 2707 (2006)

[146] C. Knudsen, J. Sturis, J.S. Thomsen, Phys. Rev. A **44**, 3503 (1991)

[147] M. Komuro, in *Singular Phenomena of Dynamical Systems*. Surikaiseki Kenkyusho Kokyuroku, vol 1118 (1999), pp. 96–114 (in Japanese)

[148] N. Kopell, G.B. Ermentrout, Commun. Pure Appl. Math. **39**, 623 (1986)

[149] N. Kopell, D. Somers, J. Math. Biol. **33**, 261 (1995)

[150] N. Krylov, M.M. Bogoliubov, *Introduction to Non-linear Mechanics* (Izd. Akad. Nauk Ukr. SSR, Kyïv, 1936) (English translation: Princeton University Press, Princeton, 1947)

[151] Y. Kuramoto, *Chemical Oscillations, Waves and Turbulence* (Springer, Berlin, 1984)

[152] V.A. Kuzmenko, A.M. Badanova, I.M. Syrkina, Fiziol. Chelov. (Hum. Physiol.) **6**, 936 (1980) (in Russian)

[153] P.I. Kuznetsov, R.L. Stratonovich, V.I. Tikhonov, *Non-linear Transformation of Stocastic Process* (Pergamon, Oxford, 1965)

[154] S.P. Kuznetsov, Sov. Tech. Phys. Lett. **9**, 41 (1983)

[155] S.P. Kuznetsov, Radiophys. Quantum Electron. **28**, 681 (1985)

[156] Yu.A. Kuznetsov, *Elements of Applied Bifurcations Theory* (Springer, New York, 2004)

[157] B. Lading, E. Mosekilde, S. Yanchuk, Yu. Maistrenko, Prog. Theor. Phys. Suppl. **139**, 164 (2000)

[158] Y.-C. Lai, C. Grebogi, J.A. Jorke, in *Applied Chaos*, ed. by J.H. Kim, J. Stringer (Willey, New York, 1992)

[159] P.R. Laming, Neurochem. Int. **36**, 271 (2000)

[160] P.S. Landa, *Avtokolebanija v sistemakh s konechnym chislom stepenei svobody* (Self-Oscillations in the Systems with a Finite Number of Degees of Freedom) (Nauka, Moscow, 1980) (in Russian)

[161] P.S. Landa, *Nonlinear Oscillations and Waves in Dynamical Systems* (Kluwer Academic, Dordrecht, 1996)

[162] P.S. Landa, Ya.B. Duboshinsky, Sov. Phys. Usp. **32**, 723 (1989)

[163] P.S. Landa, M.G. Rosenblum, Appl. Mech. Rev. **46**, 414 (1993)

[164] L.D. Landau, Dokl. Acad. Nauk SSSR **44**, 339 (1944)

[165] K.J. Lee, Y. Kwak, T.K. Lim, Phys. Rev. Lett. **81**, 321 (1998)

[166] S.G. Lee, A. Neiman, S. Kim, Phys. Rev. E **57**, 3292 (1998)

[167] B.R. Levin, F.M. Stewart, L. Chao, Am. Nat. **111**, 3 (1977)

[168] J.S. Levine, N. Aiello, J. Benford, B. Harteneck, J. Appl. Phys. **70**, 2838 (1991)

418 References

[169] B. Lindner, M. Kostur, L. Schimansky-Geier, Fluct. Noise Lett. **1**, R25 (2001)
[170] B. Lindner, J. Garcia-Ojalvo, A. Neiman, L. Schimansky-Geier, Phys. Rep. **392**, 321 (2004)
[171] D.A. Linkens, Bull. Math. Biol. **39**, 359 (1977)
[172] E.N. Lorenz, J. Atmospheric Sci. **20**, 130 (1963)
[173] R.S. MacKay, J.A. Sepulchre, Physica D **82**, 243 (1995)
[174] A.N. Malakhov, *Fluctuations in Self-Oscillatory Systems* (Nauka, Moscow, 1956) (in Russian)
[175] N. Mathias, V. Gopal, Phys. Rev. E **63**, 021117 (2001)
[176] V.V. Matrosov, Radiophys. Quantum Elec. **49**, 239 (2006)
[177] G. Mayer-Kress, H. Haken, J. Stat. Phys. **26**, 149 (1981)
[178] P. Meda, I. Atwater, A. Goncalves, A. Bangham, L. Orci, E. Rojas, Q. J. Exp. Physiol. **69**, 719 (1984)
[179] N. Minorsky, *Nonlinear Oscillations* (Van Nostrand, Princeton, 1962)
[180] F.C. Moon, *Chaotic Vibration. An Introduction for Applied Scientists and Engineers* (Wiley, New York, 1987)
[181] C. Morris, H. Lecar, Biophys. J. **35**, 193 (1981)
[182] E. Mosekilde, *Topics in Nonlinear Dynamics: Applications to Physics, Biology and Economic Systems* (World Scientific, Singapore, 1996)
[183] E. Mosekilde, H. Standdorf, J.S. Thomsen, G. Baier, in *Cooperation and Conflict in General Evolutionary Process*, ed. by J.L. Casti, A. Karlqvist (Wiley-Interscience, New York, 1995)
[184] E. Mosekilde, B. Lading, S. Yanchuk, Yu. Maistrenko, BioSystems **63**, 3 (2001)
[185] E. Mosekilde, Yu. Maistrenko, D. Postnov, *Chaotic Synchronization. Applications to Living Systems*. World Scientific Series on Nonlinear Science, Series A, vol. 42 (World Scientific, Singapore, 2002)
[186] E. Mosekilde, D.E. Postnov, O.V. Sosnovtseva, Prog. Theor. Phys. Suppl. **150**, 147 (2003)
[187] E. Mosekilde, O.V. Sosnovtseva, D. Postnov, H.A. Braun, M.T. Huber, Nonlinear Sci. **11**, 449 (2004)
[188] J.D. Murray, *Mathematical Biology* (Springer, Heidelberg, 1989)
[189] A. Neiman, Phys. Rev. E **49**, 2484 (1994)
[190] A. Neiman, D.F. Russell, Phys. Rev. Lett. **86**, 3443 (2001)
[191] A. Neiman, P.I. Saparin, L. Stone, Phys. Rev. E **56**, 270 (1997)
[192] A. Neiman, A. Silchenko, V. Anishchenko, L. Schimansky-Geier, Phys. Rev. E **58**, 7118 (1998)
[193] A. Neiman, L. Schimansky-Geier, F. Moss, B. Shulgin, J. Collins, Phys. Rev. E **60**, 2845 (1999)
[194] J.C. Neu, SIAM J. Appl. Math. **37**, 307 (1979)
[195] S. Newhouse, D. Ruelle, F. Takens, Commun. Math. Phys. **64**, 35 (1978)
[196] A. Nikitin, Z. Néda, T. Vicsek, Phys. Rev. Lett. **87**, 024101 (2001)
[197] T. Nomura, S. Sato, S. Doi, J.P. Segundo, M.D. Stiber, Biol. Cybern. **69**, 429 (1993)
[198] G.V. Osipov, B. Hu, C. Zhou, M.V. Ivanchenko, J. Kurths, Phys. Rev. Lett. **91**, 024101 (2003)
[199] E. Ott, *Chaos in Dynamical Systems* (Cambridge University Press, Cambridge, 1992)
[200] S. Ozawa, O. Sand, Physiol. Rev. **66**, 887 (1986)
[201] E.-H. Park, M.A. Zaks, J. Kurths, Phys. Rev. E **60**, 6627 (1999)
[202] S.H. Park, S.K. Han, S. Kim, C.S. Kim, S. Kim, T.G. Kim, ETRI J. **18**, 161 (1996)
[203] S.H. Park, S. Kim, H.-B. Pyo, S. Lee, Phys. Rev. E **60**, 2177 (1999)

References 419

[204] P. Parter, *Modulation, Noise, and Spectral Analysis* (McGraw-Hill, New York, 1965)

[205] B.B. Peckham, Nonlinearity **3**, 261 (1990)

[206] L.M. Pecora, T.L. Carroll, Phys. Rev. Lett. **64**, 821 (1990)

[207] A.S. Pikovsky, Z. Phys. B **55**, 149 (1984)

[208] A.S. Pikovsky, Radiotekh. Elektron. **10**, 1970 (1985)

[209] A. Pikovsky, P. Grassberger, J. Phys. A **24**, 4587 (1991)

[210] A.S. Pikovsky, J. Kurths, Phys. Rev. Lett. **78**, 775 (1997)

[211] A. Pikovsky, G. Osipov, M. Rosenblum, M. Zaks, J. Kurths, Phys. Rev. Lett. **79**, 47 (1997)

[212] A.S. Pikovsky, M.G. Rosenblum, G.V. Osipov, J. Kurths, Physica D **104**, 219 (1997)

[213] A. Pikovsky, M. Zaks, M. Rosenblum, G. Osipov, J. Kurths, Chaos **7**, 680 (1997)

[214] A. Pikovsky, M. Rosenblum, J. Kurths, *Synchronization: A Universal Concept in Nonlinear Science* (Cambridge University Press, Cambridge, 2001)

[215] D.E. Postnov, Dissertation for the degree of a candidate of physics and mathematical sciences, Saratov State University, Russia, 1990 (in Russian)

[216] D.E. Postnov, A.G. Balanov, Izv. Vysch. Uchebn. Zaved. Appl. Nonlinear Dynamics **5**(1), 69 (1997)

[217] D.E. Postnov, A.P. Nikitin, V.S. Anishchenko, Pisma Zh. Tekh. Fiz. **22**, 24 (1996)

[218] D.E. Postnov, A.G. Balanov, E. Mosekilde, Adv. Complex Systems **1**, 181 (1998)

[219] D. Postnov, S.K. Han, H. Kook, Phys. Rev. E **60**, 2799 (1999)

[220] D.E. Postnov, A.G. Balanov, N.B. Janson, E. Mosekilde, Phys. Rev. Lett. **83**, 1942 (1999)

[221] D.E. Postnov, S.K. Han, T.G. Yim, O.V. Sosnovtseva, Phys. Rev. E **59**, R3791 (1999)

[222] D.E. Postnov, T.E. Vadivasova, O.V. Sosnovtseva, A.G. Balanov, V.S. Anishchenko, E. Mosekilde, Chaos **9**, 227 (1999)

[223] D.E. Postnov, A.G. Balanov, O.V. Sosnovtseva, E. Mosekilde, Int. J. Mod. Phys. B **14**, 2511 (2000)

[224] D.E. Postnov, O.V. Sosnovtseva, S.K. Han, T.G. Yim, Int. J. Bifurc. Chaos **10**, 2541 (2000)

[225] D.E. Postnov, A.G. Balanov, O.V. Sosnovtseva, E. Mosekilde, Phys. Lett. A **283**, 195 (2001)

[226] D.E. Postnov, D.V. Setsinsky, O.V. Sosnovtseva, Tech. Phys. Lett. **27**, 49 (2001)

[227] D.E. Postnov, O.V. Sosnovtseva, E. Mosekilde, N.-H. Holstein-Rathlou, Int. J. Mod. Phys. B **15**, 3079 (2001)

[228] D. Postnov, S.K. Han, O. Sosnovtseva, C.S. Kim, Differ. Equ. Dyn. Syst. **10**, 115 (2002)

[229] D.E. Postnov, O.V. Sosnovtseva, S.K. Han, W.S. Kim, Phys. Rev. E **66**, 016203 (2002)

[230] D.E. Postnov, O.V. Sosnovtseva, S.Y. Malova, E. Mosekilde, Phys. Rev. E **67**, 016215(10) (2003)

[231] D. Postnov, O. Sosnovtseva, D. Setsinsky, Fluct. Noise Lett. **3**, L275 (2003)

[232] D.E. Postnov, A.V. Shishkin, O.V. Sosnovtseva, E. Mosekilde, Phys. Rev. E **72**, 056208 (2005)

[233] D.E. Postnov, O.V. Sosnovtseva, E. Mosekilde, Chaos **15**, 1 (2005)

[234] D.E. Postnov, L.S. Ryazanova, E. Mosekilde, O.V. Sosnovtseva, Int. J. Neural Syst. **16**, 99 (2006)

[235] T. Poston, I. Steward, *Catastrophe Theory and Its Applications* (Pitman, London, 1978)

[236] I. Prigogine, R. Lefever, Chem. Phys. **48**, 1695 (1968)

420 References

[237] M.D. Prokhorov, V.I. Ponomarenko, V.I. Gridnev, M.B. Bodrov, A.B. Bespyatov, Phys. Rev. E **68**, 041913 (2003)
[238] S. Rajasekar, Phys. Rev. E **51**, 775 (1995)
[239] D. Rand, S. Ostlund, J. Sethna, E.D. Siggia, Phys. Rev. Lett. **49**, 132 (1982)
[240] R.H. Rand, P.J. Holmes, Int. J. Non-Linear Mech. **15**, 387 (1980)
[241] W.-J. Rappel, S.H. Strogatz, Phys. Rev. E **50**, 3249 (1994)
[242] J. Rasmussen, E. Mosekilde, Ch. Reick, Math. Comput. Simul. **40**, 247 (1996)
[243] R.J.S. Rayleigh, *The Theory of Sound*, vol. 2 (MacMillan, London, 1896)
[244] J. Rinzel, G.B. Ermentrout, in *Methods in Neuronal Modeling*, ed. by C. Koch, I. Segev (MIT Press, Cambridge, 1989)
[245] M.C. Romano, M. Thiel, J. Kurths, I.Z. Kiss, J.L. Hudson, Europhys. Lett. **71**, 466 (2005)
[246] E. Rosa, E. Ott, M.H. Hess, Phys. Rev. Lett. **80**(8), 1642 (1998)
[247] M. Rosenblum, A. Pikovsky, J. Kurths, Phys. Rev. Lett. **76**, 1804 (1996)
[248] M.G. Rosenblum, A.S. Pikovsky, J. Kurths, Phys. Rev. Lett. **78**, 4193 (1997)
[249] M.G. Rosenblum, J. Kurths, A. Pikovsky, C. Schäfer, P. Tass, H.-H. Abel, IEEE Eng. Med. Biol. **17**, 46 (1998)
[250] M.G. Rosenblum, A.S. Pikovsky, J. Kurths, I.Z. Kiss, J.L. Hudson, Phys. Rev. Lett. **81**, 264102 (2002)
[251] M.R. Rosenzweig, A.L. Leiman, S.M. Breedlove, *Biological Psychology* (Sinauer, Sunderland, 1996)
[252] O.E. Rössler, Phys. Lett. A **57**, 397 (1976)
[253] O.E. Rössler, Phys. Lett. A **71**, 155 (1979)
[254] D. Ruelle, F. Takens, Commun. Math. Phys. **20**, 167 (1971)
[255] N.F. Rulkov, M.M. Sushchik, L.S. Tsimring, H.D.I. Abarbanel, Phys. Rev. E **51**, 980 (1995)
[256] S.M. Rytov, Sov. Phys. Usp. **129**, 279 (1979)
[257] S. Rzeczinski, N.B. Janson, A.G. Balanov, P.V.E. McClintock, Phys. Rev. E **66**, 051909 (2002)
[258] R.Z. Sagdeev, D.A. Usikov, G.M. Zaslavsky, *Nonlinear Physics* (Harwood Academic, New York, 1988)
[259] C. Schäfer, M.G. Rosenblum, H.H. Abel, Nature **392**, 239 (1980)
[260] F. Schilder, B. Peckman, J. Comput. Phys. **220**, 932–951 (2007)
[261] E. Schöll, *Nonlinear Spatio-temporal Dynamics and Chaos in Semiconductors*. Nonlinear Science Series, vol. 10 (Cambridge University Press, Cambridge, 2001)
[262] H.G. Schuster, *Deterministic Chaos. An Introduction* (VCH, Weinheim, 1988)
[263] D. Setsinsky, Dissertation, Saratov State Univeristy, 2004
[264] A. Sherman, J. Rinzel, Proc. Natl. Acad. Sci. USA **75**, 2471 (1992)
[265] A. Sherman, J. Rinzel, J. Keizer, Biophys. J. **54**, 411 (1988)
[266] B. Shulgin, A. Neiman, V. Anishchenko, Phys. Rev. Lett. **75**, 4157 (1995)
[267] D. Sigeti, W. Horsthemke, J. Stat. Phys. **54**, 1217 (1989)
[268] A. Silchenko, T. Kapitaniak, V. Anishchenko, Phys. Rev. E **59**, 1593 (1999)
[269] G.G. Somjen, P.G. Aitken, J.L. Giacchino, J.O. McNamara, in *Advances in Neurology*, vol. 44, ed. by A.V. Delgado-Escueta et al. (Raven Press, New York, 1986)
[270] O.V. Sosnovtseva, A.G. Balanov, T.E. Vadivasova, V.V. Astakhov, E. Mosekilde, Phys. Rev. E **60**, 6560 (1999)
[271] O.V. Sosnovtseva, A.J. Formin, D.E. Postnov, V.S. Anishchenko, Phys. Rev. E **64**, 026204 (2001)

References   421

[272] O.V. Sosnovtseva, A.N. Pavlov, E. Mosekilde, N.-H. Holstein-Rathlou, Phys. Rev. E **66**, 061909 (2002)

[273] O.V. Sosnovtseva, D.E. Postnov, A.M. Nekrasov, E. Mosekilde, Phys. Rev. E **66**, 036224(9) (2002)

[274] A. Spirkin, *Dialectical Materialism* (Progress Publishers, Moscow, 1983); www.marxists.org/reference/archive/spirkin/works/dialectical-materialism/index.html

[275] J. Stadhouders, G.J.M. Leenders, Neth. Milk Dairy J. **38**, 157 (1984)

[276] R. Stratonovich, *Topics in the Theory of Random Noise*, vol. 1 (Gordon & Breach, New York, 1963)

[277] R. Stratonovich, *Topics in the Theory of Random Noise*, vol. 2 (Gordon & Breach, New York, 1963)

[278] R.L. Stratonovich, *Topics in the Theory of the Random Noise* (Gordon & Breach, New York, 1981)

[279] S.H. Strogatz, *Nonlinear Dynamics and Chaos: With Application to Physics, Biology, Chemistry, and Engineering* (Addison-Wesley, Reading, 1994)

[280] N.G. Sun, G.P. Tsironis, Phys. Rev. B **51**, 11221 (1995)

[281] E. Syková, Prog. Biophys. Mol. Biol. **42**, 135 (1983)

[282] F. Takens, Publ. Math. IHES **43**, 47 (1974)

[283] D.Y. Tang, N.R. Heckenberg, Phys. Rev. E **55**, 6618 (1997)

[284] D.Y. Tang, R. Dykstra, M.W. Hamilton, N.R. Heckenberg, Chaos **8**, 697 (1998)

[285] M.A. Taylor, I.G. Kevrekidis, Physica D **51**, 274 (1991)

[286] J. Testa, J. Pérez, C. Jeffries, Phys. Rev. Lett. **48**, 714 (1982)

[287] H. Treutlein, K. Schulten, Ber. Bunsen. Phys. Chem. **89**, 710 (1985)

[288] H. Treutlein, K. Schulten, Eur. Biophys. J. **13**, 355 (1986)

[289] T.E. Vadivasova, A.G. Balanov, O.V. Sosnovtseva, D.E. Postnov, E. Mosekilde, Phys. Lett. A **253**, 66 (1999)

[290] T.E. Vadivasova, O.N. Sosnovtseva, A.G. Balanov, Tech. Phys. Lett. **25**, 906 (1999)

[291] T.E. Vadivasova, O.V. Sosnovtseva, A.G. Balanov, V.V. Astakhov, Discrete Dyn. Nat. Soc. **4**, 231 (2000)

[292] B. van der Pol, Radio Rev. **1**, 701, 704, 754 (1920)

[293] B. van der Pol, Phil. Mag. **3**, 65 (1927)

[294] B. van der Pol, J. van der Mark, Phil. Mag. **6**, 763 (1928)

[295] X.-J. Wang, Neuroscience **59**, 21 (1994)

[296] X.-J. Wang, J. Rinzel, in *The Handbook of Brain Theory and Neural Networks*, ed. by M.A. Arbib (MIT Press, Cambridge, 1995), pp. 686–691

[297] D.J. Watts, *Small Worlds* (Princeton University Press, Princeton, 1999)

[298] K. Wiesenfeld, J. Stat. Phys. **38**, 1071 (1985)

[299] A.A. Witt, S.E. Khaikin, J. Tech. Phys. **1**, 428 (1931)

[300] G.X. Yan, J. Chen, K.A. Yamada, A.G. Kleber, P.G. Corr, J. Physiol. **490**, 215 (1996)

[301] S. Yanchuk, Yu. Maistrenko, B. Landing, E. Mosekilde, Int. J. Bifurc. Chaos **10**, 2629 (2000)

[302] C.-S. Yi, A.L. Fogelson, J.P. Keener, C.S. Peskin, J. Theor. Biol. **220**(1), 83 (2003)

[303] M.A. Zaks, E.-H. Park, M.G. Rosenblum, J. Kurths, Phys. Rev. Lett. **82**, 4228 (1999)

Index

amplitude 139
analytic signal 200
Andronov–Hopf bifurcation 22, 390
anharmonicity 266
Anishchenko–Astakhov oscillator 205, 337
anisochronicity 266
anti-phase state 272, 299
antiphase synchronization 301, 383
asymptotic phase 269
attractor 17, 95

bacteria–virus model 117, 229, 396
basic frequency 199
basin of attraction 95, 110
beat frequency 24, 67, 72
beating 24
Bessel functions 170
bidirectional coupling 75
Bogdanov–Takens point 301, 312
Bonhoeffer–van der Pol oscillator 274
boundary crisis 110, 288, 305, 308
boundary of synchronization region 326,
    328
bursting 305, 339, 345

cardiorespiratory synchronization 145
cardiovascular system 183
chaotic bursting 305
circle map 142
clustering 390
coherence resonance 188, 243, 363
coherence resonance oscillator 244, 364
complete synchronization 198, 226, 323
conservation of probability 166
control 17
controlled breathing 145
correlation between two processes 155
correlation function 153

correlation of ergodic process 53
correlation time 154
coupling network 378
coupling vector 266
coupling-induced frequency mismatch 393
coupling-induced inhomogeneity 405
covariance 153
crisis of attractor 323, 326
cross-correlation 155
cross-covariance 156

dephasing 270, 272, 349
deterministic chaos 191
detuning 23, 30
diffusive coupling 267
dimension of an oscillator 95
Dirac delta-function 55
direct coupling 76
dissipation 13
dissipative coupling 80

effective coupling function 268, 336, 338
effective synchronization 181, 198
electrocardiogram 145
electronic experiment 72, 99, 205, 249, 364
ergodic process 53
ergodic torus 57
excitable system 240, 363

family of attractors 321
fast-and-slow motion 290, 355
FitzHugh–Nagumo model 274, 349
fluctuational terms 158, 159
Fokker–Planck equation 159, 163, 166
fold bifurcation 325
frequency detuning 23, 30
frequency locking 214, 216
full-scale experiment 62, 99, 144, 183, 205,
    249

424   Index

generalized synchronization 198

hemodynamic coupling 401
Hindmarsh–Rose model 276
homoclinic bifurcation 109, 111, 227, 272, 278, 279, 300, 302
hyperchaos 227, 324, 326

imperfect phase synchronization 220
in-phase state 272, 299, 321
in-phase synchronization 301, 383
instantaneous amplitude 140
instantaneous frequency 140
instantaneous phase 251
interspike interval 122
intra-cluster synchronization 393
isochronous limit cycle 274

Krylov–Bogoliubov method of averaging 27

lag synchronization 226, 227, 326
LC-circuit 62

main frequency 58
Mandelstam's ideas 18
mean (average) of random process 150, 151
mean frequency 176
mean return time 231
mean square of random process 151
mechanisms of synchronization 56, 60
mode 353, 371, 405
mode locking 357
model map 331
monovibrator 244
Morris–Lecar system 257, 272, 276
multistability 17, 107, 144
mutual coupling 75
mutual synchronization 10, 75

nearly sinusoidal oscillations 22
Neimark–Sacker bifurcation 51, 202
nephron autoregulation 355
nested structure 327, 333
noise 149
noise-activated oscillations 240, 242, 364
noise-induced oscillations 240, 242, 363
noise-induced spectrum peak 188
non-hyperbolic attractor 196
non-linearity 14

normalized spectrum 367
nullcline 290

one-variable diffusive coupling 268
order of synchronization 123
oscillation death 10, 84
oscillator 12
oscillatory cluster 392
oscilloscope 64
out-of-phase state 300

period-doubling bifurcation 194
phase 122, 139, 199
phase difference 25, 27, 216
phase diffusion 180, 182, 241, 328
phase dynamics 177, 253
phase locking 56
phase multistability 10, 318
phase slip (jump) 177
phase velocity field 266
phase-locked loop 372
Poincaré return time 121
position-coupling 280, 282
potassium signaling 380
power 157
power spectral density 54, 156
probability density for phase difference 174
probability density of a random process 150, 152, 155

quasiharmonic oscillations 25, 26
quenching 84

random process 150
reactive coupling 89
regularity 188, 247, 368
relaxation limit 290
resonant torus 57
resource mediated coupling 377, 378, 409
respiration 145
robustness 17
robustness of synchronization 126
Rössler oscillator 192, 212, 222, 233, 320
rotation number 123, 231, 357
RR-interval 183

saddle cycle 195
saddle torus 311
saddle–node bifurcation 301
scalar coupling 282

Index   425

self-modulated oscillations 335
self-modulation 337
self-organization 17
self-oscillations 12
separatrix 95
Sherman model 344
signal-to-noise ratio 246
sine circle map 142
skeleton of a chaotic attractor 195
slow channel 293
spectrum 54, 156
spike train 342
stationarity, wide-sense 154
stationary process 154
stochastic limit cycle 241, 248
Stratonovich algorithm 159
stroboscopic section 56, 129
superimposed phase plane 267
suppression of natural dynamics 60, 101, 216, 255
symmetric subspace 321

synchronization via homoclinic bifurcation 120

Takens–Bogdanov point 107, 111, 301, 312
tangent bifurcation 325
time average 53
torus folding 312
truncated equations 31, 77, 158, 165

van der Pol oscillator 21, 76, 123, 149, 212, 274
van der Pol oscillator, modified 278
variance of random process 151
variance of stationary process 157
vascular-nephron tree 402
vector of coupling 282
velocity-coupling 280, 282

weakly non-linear oscillator 22
white noise 157, 364
Wiener–Khintchine theorem 54, 156
winding number 123, 231, 357

Printing: Krips bv, Meppel, The Netherlands
Binding: Stürtz, Würzburg, Germany